# Russia and the Arms Trade

**sipri**

**Stockholm International Peace Research Institute**
Frösunda, S-169 70 Solna  Sweden
Telephone:  46 8/655 97 00
Telefax:  46 8/655 97 33
Email:  sipri@sipri.se
Internet URL:  http://www.sipri.se

# Russia and the Arms Trade

*Edited by*
Ian Anthony

OXFORD UNIVERSITY PRESS

1998

Oxford University Press, Great Clarendon Street, Oxford OX2 6DP

Oxford New York

Athens Auckland Bangkok Bogotá Bombay Buenos Aires
Calcutta Cape Town Dar es Salaam Delhi Florence Hong Kong Istanbul
Karachi Kuala Lumpur Madras Madrid Melbourne Mexico City
Nairobi Paris Singapore Taipei Tokyo Toronto Warsaw
and associated companies in
Berlin Ibadan

Oxford is a registered trade mark of Oxford University Press

Published in the United States
by Oxford University Press Inc., New York

British Library Cataloguing in Publication Data
Data available

Library of Congress Cataloging-in-Publication Data
Data available

ISBN 0–19–829278–3

Typeset and originated by Stockholm International Peace Research Institute
Printed in Great Britain
on acid-free paper by
Biddles Ltd, Guildford and King's Lynn

# Contents

# Preface

The Soviet Union by the time it collapsed was the single largest supplier of conventional weapons to other countries. It also used arms exports to support armed groups in other countries that challenged the authority of their governments. In the Soviet period the only mechanisms for controlling these exports were decisions taken by a secretive and unaccountable executive. Discussion and debate were impossible. Since the dissolution of the Soviet Union, the Russian Government is very different from its predecessor. Now there is greater openness on the part of government officials, a multi-party political system and a free press while interest groups with specialist knowledge are allowed to express their views—including those on national and international security—without fear of the consequences.

This study, initiated and directed by Ian Anthony, leader of SIPRI's Arms Transfers Project, involved a group of leading experts who analysed different aspects of conventional arms transfers both during the Soviet period and since the collapse of the Soviet Union. They present Russian arms exports in the context of the new forces shaping international arms transfers.

From the material presented in this volume it is clear that Russia has rejected both the extreme positions regarding arms transfers that can in theory be adopted by states: to sell arms to everybody or sell arms to nobody. In common with most countries that manufacture arms the Russian Government will decide, on a case-by-case basis, whether or not to approve a given export.

One consequence of wider political participation has been that the executive can no longer ignore or override the views and interests of different interest groups. However, in making decisions about whether or not to export arms Russia has not yet developed a stable mechanism for balancing the sometimes conflicting interests of foreign policy, economic policy and military security. For a short period economic interests were perhaps paid too little attention by Russian decision makers very concerned with their political relations—in particular their relations with the United States. Now the pendulum may swing so far that economic factors become the main or even exclusive basis for arms export decisions. Russia may be in danger of giving too little attention to its international commitments and even its own military security.

It can also be seen from this volume that, in spite of the recent positive changes, the size and pattern of Russia's conventional arms exports are not transparent. The questions how much money Russia receives through arms exports and how that money is distributed between different agencies and enterprises are extremely controversial. Russia, in fact, as Ian Anthony concludes, has not yet found an effective mechanism for integrating the contributions of the various government agencies, committees and commissions in a manner which produces decision by consensus.

As in the past, SIPRI will continue to try to measure the size and flow of the arms trade and report its findings.

Adam Daniel Rotfeld
Director of SIPRI
October 1997

# Acknowledgements

I would like to acknowledge the efforts of several people in the preparation of this book. Plainly, it would not have been possible without the patient participation of the contributing authors in this effort to pin down a constantly moving target, and to them SIPRI extends sincere thanks.

A significant part of this book rests on the large bank of information and data held within the SIPRI Arms Transfers Project which is largely maintained by project members Cynthia Loo, Pieter D. Wezeman and Siemon T. Wezeman, all of whom are doing important work in a very professional manner in maintaining a unique and indispensable source. Vladimir Baranovsky, leader of the SIPRI Project on Russia's Security Agenda, made useful comments on the draft manuscript; John D. Hart, a SIPRI research assistant, made his considerable Russian-language skills available in helping identify and process some of the documents used in the study; Noemi Wallenius in SIPRI's library gave extensive help with language and terminology; and Boris Nevelev, an intern working at SIPRI during 1996–97, commented on and contributed valuable information to two of the chapters.

Eve Johansson edited and set the book in camera-ready format to SIPRI's customary high standards and Peter Rea prepared the index.

Several non-SIPRI experts have read and commented on part or all of the manuscript and SIPRI's appreciation is also due to them. I would particularly like to acknowledge Milton Leitenberg and Laure Després for the comments and information that they supplied.

Finally, I would like to thank the United States Institute for Peace for its generosity in providing financial support for the project.

Ian Anthony
SIPRI Project Leader
October 1997

# Acronyms

| | |
|---|---|
| AFV | Armoured fighting vehicle |
| ANC | African National Congress (South Africa) |
| AO | Joint-stock company (Aktsionernoye obshchestvo) |
| APC | Armoured personnel carrier |
| ASW | Anti-submarine warfare |
| AWAC | Airborne warning and control |
| $C^3I$ | Command, control and communications |
| CFE | Conventional Armed Forces in Europe (Treaty) |
| CIS | Commonwealth of Independent States |
| CMEA | Council for Mutual Economic Assistance |
| COCOM | Coordinating Committee for Multilateral Export Controls |
| CPSU | Communist Party of the Soviet Union |
| EPLF | Eritrean People's Liberation Front |
| FMLN | Farabundo Martí National Liberation Front (El Salvador) |
| FNLA | National Front for the Liberation of Angola |
| FRELIMO | National Liberation Front of Mozambique |
| FSLN | Sandinista National Liberation Front (Nicaragua) |
| GDP | Gross domestic product |
| GIU | Central Engineering Directorate (Glavnoye inzhenernoye upravleniye, USSR/Russian Federation) |
| GKI | State Committee for the Management of Property (Gosudarstvenny komitet po upravleniu imushchestva, Russian Federation) |
| GKVTP | State Committee on Military–Technical Policy (Gosudarstvenny komitet po voyenno-tekhnicheskoy politike, Russian Federation) |
| GNP | Gross national product |
| GTU | Central Technical Directorate (Glavnoye tekhnicheskoye upravleniye, USSR/Russian Federation) |
| GUMVS | Main Administration for International Military Cooperation (Glavnoye upravleniye mezhdunarodnogo voyennogo sotrudnichestva, Russian Federation) |
| GUSK | Central Directorate of Collaboration and Cooperation (Glavnoye upravleniye po sotrudnichestvu i kooperatsii, USSR/Russian Federation) |
| IAPO | Irkutsk Aviation Production Association |
| JAF | Joint Armed Forces (Commonwealth of Independent States) |
| KB | Design Bureau (Konstruktorskoye byuro) |
| KEKMO | Export Control Committee (Komitet eksportnogo kontrolya Ministerstva oborony, Russian Federation) |
| KGB | Committee of State Security (Komitet gosudarstvennoy bezopasnosti, USSR) |
| KMS | Interdepartmental Coordinating Council for Military–Technical Cooperation between the Russian Federation and Foreign States (Koordinatsionny mezhvedomstvenny sovet po voyenno- |

|          | tekhnicheskomy sotrudnichestvu Rossiyskoy Federatsii s inostrannymi gosudarstvami, Russian Federation) |
|----------|------------|
| KMSVTP | Interdepartmental Coordinating Council for Military–Technical Policy (Koordinatsionny mezhvedomstvenny sovet po voyenno-tekhnicheskoy politike, Russian Federation) |
| KPA | Korean People's Army |
| KVTS | Interdepartmental Commission on Military–Technical Cooperation between the Russian Federation and Foreign States (Komitet voyenno-tekhnicheskogo sotrudnichestva, Russian Federation) |
| LCA | Light Combat Aircraft (India) |
| MAPO | Moscow Aviation Production Organization |
| MCCH | Military Cooperation Coordination Headquarters (CIS) |
| MD | Military District (USSR/Russian Federation) |
| MFER | Ministry of Foreign Economic Relations (USSR/Russian Federation) |
| MPLA | Popular Movement for the Liberation of Angola |
| MTCR | Missile Technology Control Regime |
| NATO | North Atlantic Treaty Organization |
| NLFE | National Liberation Front of Eritrea |
| NPO | Scientific Production Association (Nauchno-proizvodstvennoye obedinenie) |
| OKB | Experimental Design Bureau (Opytno-konstruktorskoye byuro) |
| OSCE | Organization for Security and Co-operation in Europe |
| PLA | People's Liberation Army (China) |
| PLAAF | People's Liberation Army Air Force (China) |
| PLAN | People's Liberation Army Navy (China) |
| PLO | Palestine Liberation Organization |
| Polisario | People's Liberation Front of Sakiet Al-Khamra (Western Sahara) |
| R&D | Research and development |
| RENAMO | National Resistance of Mozambique |
| RSFSR | Russian Soviet Federal Socialist Republic |
| SPLA | Sudanese People's Liberation Army |
| SWAPO | South-West Africa People's Organization |
| TPLF | Tigray People's Liberation Front (Ethiopia) |
| TsKB | Central Design Bureau (Tsentralnoye konstruktorskoye byuro) |
| UAE | United Arab Emirates |
| UNITA | National Union for the Total Independence of Angola |
| VPK | Military–Industrial Commission (Voyenno-promyshlennaya komissiya, USSR) |
| V/STOL | Vertical/short take-off and landing |
| WTO | Warsaw Treaty Organization |
| ZANU | Zimbabwe African National Union |

## Conventions used in tables

| | |
|---|---|
| . . | Not available or not applicable |
| ( ) | SIPRI estimate |

# 1. Introduction

*Ian Anthony*

## I. New tendencies shaping international arms transfers

As with all other aspects of international security, the new political environment created by the end of the cold war has profoundly changed the pattern of the international arms trade. No country has been more affected by these changing conditions than the Russian Federation.

From the 1950s the arms trade was dominated from the supply side by the two superpowers—the United States and the Soviet Union—which used arms transfers to support their wider foreign and security policies in the framework of the cold war. The end of the cold war and the changes that accompanied it removed this competitive ideological dimension as a factor driving arms export decisions. The United States and the Soviet Union began to adopt a cooperative approach to the management of security both in Europe and in places such as Afghanistan, Angola and Central America. Moreover, as part of this new agenda former adversaries—including European countries—began to discuss the impact of arms transfers on international security in a serious way for the first time.[1] The twin catalysts for this multilateral dialogue (which is still in its early stages) were the invasion of Kuwait by Iraq and the gradual relaxation of the embargo operated by members of the former Coordinating Committee for Multilateral Export Controls (COCOM).

The fact that Iraq had built its conventional arsenal from imported equipment and technology brought into sharp relief the fact that arms transfers could, under some circumstances, have a negative impact on international security. Under these conditions the five permanent members of the UN Security Council (the P5) held talks in 1991 and 1992 on the impact of arms transfers on the stability and security of the Middle East.

The gradual relaxation of the COCOM embargo was a manifestation of the change in East–West relations and the growing preference for cooperation over confrontation. In 1993 the decision was taken to end the embargo entirely and replace it with a new forum in which states could exchange information and perspectives on the international arms trade. However, Russia was not involved in these discussions (although Russian officials were briefed about their progress). It was not until 1995 that Russia became a partner in the talks that led

---

[1] In the late 1970s a US initiative for bilateral Conventional Arms Transfer Talks (CATT) together with the Soviet Union failed. The CATT initiative and the reasons for its failure are examined in Spear, J., *Carter and Arms Sales: Implementing the Carter Administration's Arms Transfer Restraint Policy* (Macmillan: London, 1995).

to the establishment in 1996 of the Wassenaar Arrangement on Export Controls for Conventional Arms and Dual-Use Goods and Technologies.[2]

While the concept of collective security is gradually evolving into one of common or cooperative security in Europe, the same cannot be said elsewhere. After the cold war there will still be a large number of countries which see their armed forces as a central component in safeguarding their national security but which do not have comprehensive defence industrial capabilities. In these countries there will continue to be a demand for foreign-made arms.

In foreign policy terms first the Soviet Union and then Russia used arms transfer policy to send a clear signal to other countries—and particularly to the United States—that the changes under way were more than cosmetic.

From a strategic perspective the end of the cold war has also had an impact on the way in which major exporters view the international arms trade. During the cold war the strategic decisions of the superpowers in particular were driven by their central competition. This meant that they focused much of their attention on the main theatre of potential conflict—Europe. Arms transfers to allies were one important instrument by which the United States and the Soviet Union shaped the balance of forces in this theatre. Arms transfers were also an important element of military assistance programmes intended to achieve strategic objectives outside Europe in places such as the Persian Gulf and the Horn of Africa.

There is still uncertainty about how armed forces will be used in the new international environment and for what reasons. Under these conditions there is no clear picture of whether and how strategic arguments might underpin military assistance. Russia in particular lacks a new strategic framework in which to take decisions about its future force structure.

Traditional strategic arguments have lost some of their relevance. For example, military assistance is no longer needed to secure base rights, listening stations or other intelligence facilities in support of a global military strategy. However, these arguments might still be applied in new strategic arrangements, for example within the Commonwealth of Independent States (CIS).

At the same time, there are strategic arguments that would support a cautious approach to the spread of military technologies unless and until a clearer strategic picture emerges. Since the end of the cold war Russian armed forces have regularly found themselves deployed in conflict regions or engaged in military operations of various kinds.

At the same time as these important politico-military issues were emerging commercial and industrial factors were also exerting pressure on arms export policy. During the cold war their own armed forces provided by far the most important market for goods produced in the defence industries of the United

---

[2] Anthony, I. and Stock, T., 'Multilateral military-related export control measures', *SIPRI Yearbook 1996: Armaments, Disarmament and International Security* (Oxford University Press: Oxford, 1996), pp. 542–45; and Anthony, I., Eckstein, S. and Zanders, J. P., 'Multilateral military-related export control measures', *SIPRI Yearbook 1997: Armaments, Disarmament and International Security* (Oxford University Press: Oxford, 1997), pp. 345–48.

States and the Soviet Union—although even in this period exports were not unimportant. However, the rapid and deep cuts in military expenditure that followed the end of the cold war—in which spending on equipment was often reduced to a greater extent than other forms of military expenditure—created new pressures on industry to find new markets for its products.

As noted above, in a large number of states arms imports are a central element of arms procurement policy. However, the existence of demand for a product is not the same thing as the existence of a market. For a market to exist demand must be combined with a means of exchange. At the time when the superpowers were prepared to offer large-scale military assistance in pursuit of their political and strategic objectives this means of exchange could be political as well as financial. However, the changed international environment has largely removed the political incentives to subsidize arms exports. Therefore it is likely that in future a much higher proportion of arms transfers will have to be financed by the buyer.

## II. Russia in the new international arms trade system

The new political, economic and strategic features of the international arms market outlined above had an impact on all arms-exporting countries. However, the impact on those newly independent countries that succeeded the former Soviet Union was particularly dramatic.

The disruption to Russia's foreign relations brought about by the changes of the past 10 years has been more fundamental than the effects in most other countries. Changes in Soviet foreign policy initiated by President Mikhail Gorbachev and Foreign Minister Eduard Shevardnadze weakened bilateral relations with countries that were important clients for Soviet arms, for example, Afghanistan, Angola, Cuba, Iraq, North Korea, Libya, Nicaragua, Syria and Viet Nam. The dissolution of the Warsaw Treaty Organization (WTO)[3] and the Council for Mutual Economic Assistance (CMEA)[4] effectively ended the system of military–technical cooperation within the state socialist countries of East–Central Europe.[5] The subsequent dissolution of the Soviet Union itself broke many inter-enterprise ties within what had been an integrated production system.

The reductions in military expenditure have been deeper in Russia than in most other countries. Since 1992 Russia has been following a macroeconomic policy of controlling inflation through a combination of fiscal and monetary

[3] The WTO was formed in 1955 and disbanded in 1991. Its membership in 1991 consisted of Albania, Bulgaria, Czechoslovakia, the German Democratic Republic, Hungary, Poland, Romania and the USSR.

[4] In 1990 the membership of the CMEA consisted of Bulgaria, Czechoslovakia, the German Democratic Republic, Hungary, Poland, Romania, the USSR, Cuba, Mongolia and Viet Nam.

[5] In this book East–Central Europe after 1990 is defined as those non-Soviet countries that were members of the WTO—Bulgaria, the Czech Republic, Hungary, Poland, Romania and Slovakia—but excepting Albania.

measures.[6] Fiscal measures have been intended to ensure that money which is in circulation in Russia is primarily directed to the development of the private sector, rather than being channelled through state agencies. Among economists the verdict on this approach to economic reform is mixed. However, there is no doubt that one of its consequences has been reductions in military expenditure. While the available data are difficult to interpret, a survey in 1997 has concluded that the reduction in Russian military expenditure between 1992 and 1995 was between 40 and 50 per cent in real terms.[7]

Industrial dependence on revenues from arms sales (domestic and foreign) was also higher in Russia than in other countries and as a result the impact of the shrinking of the market has been greater. Enterprises have had to develop new strategies to manage the consequences of shrinking markets against the background of the transformation of the domestic legal, political and economic system away from a command economy and towards a market economy.

The combined impact of the domestic and international changes led to a collapse in the volume and value of Russian arms exports. According to SIPRI estimates, in 1987 the Soviet Union accounted for 38 per cent of the worldwide trade in major conventional weapons. By 1992 the Russian share had declined to 12 per cent of the world total and in 1994 (the lowest point of deliveries of major conventional weapons from Russia) to only 3 per cent.[8]

Since 1995 Russia has increased the level of its arms sales, whether measured in value or in volume terms. According to official Russian data from the state trading company Rosvooruzhenie (the State Corporation for Trade in Armaments and Military Technical Cooperation), the value of Russian arms exports rose from $1.7 billion in 1994 to $3.1 billion in 1995. According to the preliminary estimate for 1996, the value of Russian arms exports for that year would again be over $3 billion.[9]

This increase in sales reflects the fact that in 1995–97 Russia has managed to stabilize and consolidate its arms transfer relations with one important Soviet client, India, and to renew or open new bilateral arms transfer relationships with countries such as China, Iran and Malaysia.

Under these conditions Russia will be an important factor in the international arms trade during the coming years. However, the way in which Russia will align itself with the emerging pattern of the international arms trade is not at all clear. It is noteworthy that there was limited consensus among the group of contributors to this volume about the success of recent Russian policies and about what kind of policies would be appropriate for Russia to pursue in the future. Differences of view extend to fundamental issues such the role of arms transfers in Russia's foreign and security policy, whether or not arms transfers represent

---

[6] Volossov, I. V., 'The Russian economy: stabilization prospects and reform priorities', Paper delivered to the NATO Economic Colloquium 1996 on Economic Developments and Reforms in Cooperation Partner Countries, Brussels, 26–28 June 1996.

[7] Norberg, R., *Rysslands försvarsutgifter under perioden 1992–1997* [Russia's military spending during the period 1992–1997]. Unpublished manuscript, May 1997 (in Swedish).

[8] These shares are based on SIPRI trend-indicator values. For further discussion, see chapter 2.

[9] Tarasova, O., [Rosvooruzhenie calls for unity], *Segodnya*, 1 Nov. 1996 (in Russian).

a viable survival strategy for arms-producing enterprises, and the appropriate relationship between the Russian Government and the arms industry.

The chapters in this volume suggest that no single factor can explain the recent tendencies in Russia's arms export policy. Some analyses have tended to focus almost exclusively on the economic and industrial imperative for Russia to export arms. The balance in decision making between politico-military factors and economic and industrial factors has certainly changed since the Soviet period, with politico-military factors becoming relatively less important. However, a good deal of evidence is presented in this volume which suggests that Russia would still like to use arms transfers as an instrument of its foreign and security policy. Russia seems to derive little economic benefit from using arms transfers to develop its relations with fellow members of the CIS, for example, and Moscow may even be willing to pay an economic premium where the political rewards are considered high enough.

If it is accepted that Russian foreign and security policy will continue to play a role in shaping arms export policy, there are disagreements about how Russia should develop its foreign policy. One approach suggests that Russian national interests would be best served by flexibility and independence of action, after a brief experiment in trying to align its foreign policy closely with those of the group of Western states. However, there is evidence in the chapters in this volume that no firm Russian position has yet been formed on this question. Russia is participating actively in multilateral discussions of arms exports and has made decisions—for example, regarding arms transfers to Iran—that indicate a willingness to listen to the views of other countries.

All are agreed that economic and industrial factors will be a major influence shaping Russian arms export policy. However, there is no consensus about the extent of that influence or about the way in which government relations with industry are and should be managed. Some argue for a state monopoly over arms exports in which the interests of industry are taken into account but where there is no direct industry involvement in the management of arms sales. Others argue for a commercial approach in which large enterprises or industrial groups in particular would manage their own export sales. With this approach state involvement would be primarily through issuing export licences and other documents needed for sales to be made legally.

The authors of the chapters in this book suggest that few if any Russian industrialists expect to be able to depend on orders from the Russian armed forces as the main element in their enterprise strategy. Managers seem to fall into two broad categories: those who believe that their enterprise can benefit from a strategy based on arms exports and those who believe that they must find an alternative strategy based on non-military sales.

The arguments presented by the first group are that export contracts, when they can be won, are extremely profitable. If the Russian arms export system can be developed into an efficient and competitive mechanism, enterprises can expect to derive significant revenues from foreign sales. The current barriers to this strategy are the failings and inefficiencies of the current Russian authorities.

The second group argues that there are many external barriers to Russian success in winning a large share of the remaining global market for military equipment.

These arguments are not mutually exclusive since different enterprises have different prospects, depending on the nature of the market for their specific products. While arms exports could not support the entire defence industry inherited from the Soviet Union, individual enterprises or industrial groups may survive for a transitional period through foreign sales.[10]

It is true that in many cases external barriers to Russian sales exist and that these are outside the control of Russian authorities. For example, Russia cannot do much to change the fact that most of the countries which import large quantities of arms have close political and military ties with the United States. Equally, Russia has limited possibilities to address the problem that many countries are unable to pay for the equipment that they would like to buy.

The Russian authorities can do certain things to help in cases where opportunities to open new markets arise. For example, a willingness to license exports of production technologies may give competitive advantages in countries where the United States is unwilling to license this type of technology transfer in spite of close political and military ties.

The problem of finding financing in cases where countries cannot make payments using reserves is also being addressed through the growing involvement of Russian banks in the management of credit related to arms exports. Most of the new financial–industrial groups include large banks, some of which have enough capital to lend to foreign governments even in cases where large projects are being considered.

## III. Conducting research on Russian arms transfers

Against this background SIPRI initiated a study of Russian arms transfer policies and practices. In conducting the project it was decided to seek cooperation from Russian researchers and officials. The objective was to create a balanced group that included both official and non-government perspectives and individuals with expertise in military, foreign policy and industrial issues. It was also considered important to include not only Moscow-based analysts but also researchers from other centres of Russian arms production.

During the cold war an approach of this kind would have been impossible. It was impossible for an independent researcher to meet with responsible Soviet officials to discuss the issue of arms transfers. To the extent that any meetings could take place—for example, in the framework of United Nations confer-

---

[10] Long-term survival through exports alone is difficult to envisage since foreign sales cannot generate enough revenue to support the development of new generations of equipment unless prices are raised to a level that would make the systems offered uncompetitive.

ences—they were not productive. Soviet scholars and academicians themselves could make only a very limited contribution to the international discussion of the issue of arms transfers. Soviet researchers had no access to primary information and they could not publish independent evaluations.[11]

After the end of the cold war the conditions in Russia changed dramatically. It is now possible to get access to primary information from official Russian sources—although it is still prohibited to publish many types of information. Russian scholars are willing to participate in discussions and to offer an independent view of trends and developments. Moreover, it has become possible to cooperate with responsible officials. New sources of information exist. For example, not only government agencies but also industrial enterprises increasingly publish press releases describing their activities and have a press office to assist with outside enquiries. During the Soviet period the existence of these enterprises was often denied and the movements of enterprise employees were strictly monitored and controlled. Russia has produced active and inquisitive print and electronic media. A small cadre of journalists specialized in military matters has emerged. The Russian Parliament has several committees which have taken an active interest in military issues, including arms transfers.

While the level of disclosure remains far behind that in the United States, the level of transparency in Russian arms transfer policies and practices is now comparable with or greater than that in many European countries including, for example, France and the United Kingdom.

## IV. Unresolved issues in Russian arms exports

Although new research approaches are now possible, a stable domestic Russian environment was not expected or assumed when this volume was prepared. It is clear that it will be some time before a clear and consistent Russian policy towards conventional arms exports can be developed and implemented.

There continues to be uncertainty about basic issues and some very important questions remain unresolved.

### Principles guiding Russian arms export policy

In the late Soviet period the then Foreign Minister, Eduard Shevardnadze, promised that new primary legislation would be passed which would establish the basic principles by which arms export policy would be guided. The process of drafting the legislation was initiated in 1991. This legislation has not yet become law. On 20 June 1997 the State Duma (the lower chamber of the Russian Parliament) passed a draft law on principles of state policy on military–

---

[11] At a seminar organized by SIPRI with the Carnegie Moscow Center on 2 June 1997 it was confirmed that during the Soviet period officials and scholars rarely if ever met each other, let alone foreigners. Moreover, there was a low level of communication and cooperation between responsible officials within the Soviet Government.

technical cooperation with foreign states.[12] The law was passed by the Federation Council (the upper chamber of the parliament) on 4 July 1997. However, on 22 July 1997 President Yeltsin vetoed the legislation, arguing that it was inconsistent with the constitution.[13] A motion to overturn the veto was supported by only 200 votes in the Duma, which was insufficient for it to be carried. The draft law was then sent for consideration to a conciliation commission which was to consider whether and how it might be modified before being submitted to parliament again.[14]

Primary legislation defines the rights and responsibilities of the state and of legal citizens (which could be companies or citizens).

Secondary legislation (often called regulations) is an instrument for the executive branch to apply primary legislation. In this way the government can implement its own policy within the framework of the law. Typically these regulations will include guidelines for individual licensing decisions, control lists describing which items require export licences and differentiated lists of recipient countries to which special rules apply. These regulations might be proscriptive (for example, specifying countries under embargo) or they might be permissive (for example, specifying countries for which less demanding licensing procedures are applied).

Regulations can be expected to change at fairly regular intervals. For example, when one government succeeds another new licensing guidelines may be introduced in line with the policy of the new government. Control lists may be updated in response to technology change or the differentiated lists of destinations may be revised in response to changing events.

Through executive orders Russia has established secondary regulations which are comparable in their scope and structure with those used in, for example, West European countries. However, it is a more open question whether Russia has established adequate primary legislation.

There is no universal approach to drafting primary arms-export control legislation and the approach adopted by each country reflects its own legal system and definition of national interest. However, there are some general features that can be identified by surveying arms-exporting countries.[15] Only the United States has the relative luxury of allowing foreign policy interests to predominate in shaping its primary legislation.[16] In most countries primary legislation must command a broad base of support, including support among the different constituencies most affected by it. Through primary legislation the relative weights attached to the needs of three broad interests need to be established:

[12] ITAR-TASS, 20 June 1997 (in English) in Foreign Broadcast Information Service, *Daily Report–Central Eurasia* (hereafter FBIS-SOV), FBIS-SOV-97-171, 20 June 1997; and Bogatykh, M., 'Why not sell arms legally?', *Moscow News*, no. 27 (10–16 July 1997), p. 8.

[13] ITAR-TASS, 26 July 1997 (in English) in FBIS-SOV-97-207, 26 July 1997.

[14] Interfax, 5 Sep. 1997 (in English) in FBIS-SOV-97-248, 5 Sep. 1997.

[15] The basic elements of such a survey can be found in Anthony, I. (ed.), SIPRI, *Arms Export Regulations* (Oxford University Press: Oxford, 1991).

[16] The 1976 Arms Export Control Act and the 1961 Foreign Assistance Act together currently form the basis for US primary legislation.

foreign policy interests, strategic interests and economic interests. In addition, primary legislation should determine the rights and responsibilities of the executive and legislative branches of government (for example, the level of oversight, if any, that the legislative branch should exercise and the information, if any, that legislators should be entitled to receive). It is also necessary to decide on the role of the judiciary in arms export questions.

In Russia the executive branch of government has argued that presidential decrees should be considered as primary legislation, carrying greater weight than decisions by the prime minister (who is the chairman of the Council of Ministers). However, many representatives in the Russian Parliament argue that primary legislation should reflect the views of the legislative and executive branch and are not willing to accept that presidential decrees should be treated in the same way as legislation passed by a majority of elected representatives in parliament. Equally, many parliamentarians would like to see constraints placed on the ability of the president to govern by decree. This is partly for reasons of constitutional principle and partly for more practical reasons.

While the Russian Constitution contains some ambiguous provisions regarding the relative rights and responsibilities of different state authorities, it does give the president and the executive branch powerful arguments for saying that they should have a decisive voice in shaping and implementing arms export policy. Article 71 of the constitution states that the jurisdiction of the Russian Federation includes (among other matters) defence production, determining procedures for the sale and purchase of arms, ammunition, military hardware and other equipment. Article 80.3 of the constitution states that the president 'shall define the basic domestic and foreign policy guidelines of the state in accordance with the constitution and federal laws'. Article 115 establishes that the president and the government have the right to issue decrees and executive orders and that these are legally binding. Article 114.1 states that the government shall adopt measures to implement the defence, security and foreign policy of the Russian Federation.

At the same time, article 114.2 of the constitution says that the work of the executive shall be regulated by federal laws, which can be interpreted to mean that the president and the government should have a basic law setting the framework in which decrees and executive orders are issued.

The practical argument in favour of a single legal framework established for a longer period is that the tendency for the president to issue decrees that are not consistent with one another creates a climate of uncertainty for government agencies responsible for implementing policy and for industry.

In recent years the procedures for managing Russian arms transfers and military–technical cooperation have been changed regularly by decree. If there were a law passed by the legislature establishing the rights and responsibilities of different participants in the arms trade then these different actors would be able to adapt their activities in the knowledge that conditions were unlikely to change. To underline this point, since the draft law on principles of state policy on military and technical cooperation with foreign states was vetoed a new set of decrees has been issued by President Yeltsin.

On 28 July 1997 in decree number 792 'On measures to improve the system of management of military–technical cooperation with foreign countries' the prime minister was charged with supervising military–technical cooperation.[17] Then on 20 August 1997 in decree number 907 'On measures to strengthen state control of foreign trade activity in the field of military–technical cooperation of the Russian Federation with foreign countries', important changes were made to the procedures for coordination and management of arms exports.[18]

As a result of these decrees the central assumptions underpinning the coordination and management of arms exports were changed for the fourth time in the five years since 1992. Initially the assumption was that industry would play the leading role in coordination and management while the government exercised control through export licensing. Later it was assumed that organs of the state would have a full monopoly over coordination and management of arms exports. Later on still a mixed system was introduced in which some enterprises were permitted to choose either to pursue exports through the organs of the state or to act independently (although still subject to export licensing). The most recent change in 1997 appears to alter the status of the responsible state agency. Rosvooruzhenie was disbanded in the form of a state corporation and recreated by decree 910 of 20 August 1997, 'On the Federal State Unitary Enterprise the State Company Rosvooruzhenie', as an entity which, although not privatized, will behave like a company charged with the management of major, complex arms deals with foreign states.[19] At the same time, by decree no. 907, two new federal state unitary enterprises (Promexport and Rossiyskiye Tekhnologii) were created. Promexport will organize supplies of spare parts and components for Russian equipment that is being operated by foreign states as well as managing the disposal of equipment from the Russian armed forces. Rossiyskiye Tekhnologii will manage negotiations and deals involving sales of production licences and technology transfer.[20]

## The magnitude of Russian arms exports

Another basic question that has not been answered fully is related to the magnitude of Russian arms exports—whether measured by volume or by value. Rosvooruzhenie now publishes annual official data for the value of Russian arms exports. However, it is not clear what the basis for these data is or what exactly they measure.

---

[17] Reproduced in appendix 3 of this volume as document 25.

[18] Reproduced in appendix 3 of this volume as document 26.

[19] At the time of writing (Nov. 1997) aspects of the management of Rosvooruzhenie were still unclear. Decree no. 910 is reproduced in appendix 3 of this volume as document 28.

[20] Press Conference with Vice-Premier Yakov Urinson, 25 Aug. 1997, at URL <http://www.infoseek. com/Content?arn/ix.KMLN72363048&qt/federal+news+serv/ice&colIX&K/A&ak/industrynews>.

Four basic approaches can be taken to measuring the value of arms exports. First, information can be collected by government—either in negotiations with foreign governments or during the process of issuing export licences. Second, information can be obtained from customs authorities. Third, information can be obtained from industry. Fourth, an estimate can be made.

Rosvooruzhenie negotiates contracts on behalf of industry. The published data may represent the value of those contracts and agreements which Rosvooruzhenie has negotiated. This seems to be the most likely source. However, there is one enterprise (MiG-MAPO—the Moscow Aviation Production Organization) which prefers to discuss directly with foreign clients and it is not clear if MiG-MAPO data are included in Rosvooruzhenie data. Also, Rosvooruzhenie does not participate in managing or negotiating small-arms deals. There may therefore be individual transactions of relatively low dollar value that are not included in Rosvooruzenie data.

Most financial aspects of Russian arms transfers are handled by the Central Bank of Russia. Foreign governments make payments into an account set up for them at the Central Bank, and the Russian state authorities are then responsible for dividing this money among Russian enterprises. The data may be derived from reports from the bank about payments into these accounts.

In the process of licensing arms exports the responsible authorities ask for information about the value of the contract for which a licence is sought and the time during which the contract will be implemented. This information may be the basis for data on the value of arms exports. It would be held within the agency responsible for issuing licences which, in the case of Russia, has changed several times.

These data can provide an official trend indicator but do not reflect the specific arrangements in a contract. The schedule for deliveries may not correspond to the schedule of payments and the contract may also include provisions for special types of financing that affect the overall value (such as long-term credit, interest, currency arrangements, military assistance or counter-purchase). The contract may also be linked to offsets of various kinds.

Data derived from licences therefore probably cannot measure financial flows associated with arms sales.

Some governments task their customs service with collecting the documents that accompany physical shipments and recording the shipment values. The data collected by customs consist of the estimated value of the particular shipment as recorded by the shipper. Most customs services work with a harmonized set of international trade classifications in which one category is arms and ammunition. However, this category includes only armoured vehicles, artillery and lethal items such as ammunition, bombs and torpedoes. Many types of military equipment appear under other classifications. Moreover, some military equipment transfers are not subject to customs inspection. In Russia it is also the case that some points of exit are not manned by customs officers.

Customs data from Russia probably could not measure trends or financial flows accurately even if they were available.

It is possible that data on the value of foreign sales are also obtained by surveying arms manufacturers. Industry can provide data not only on the value of agreements but also on invoices issued and payments received. In many ways industry data provide the best measurement of financial flows associated with the arms trade. However, there are some problems in creating a data set of this kind. The selection of manufacturers is important in that data could reflect conditions in large enterprises responsible for final assembly of equipment or could be requested from a wider group that supply large enterprises with those things they need to manufacture major defence equipment. Surveys could be based on membership of a manufacturing association. In the case of Russia different approaches to the selection of enterprises could have a great impact on the information obtained. For example, enterprises affiliated with the former Ministry of Defence Industry included many that were engaged in both military and non-military production and represented many different levels of the production process. At the same time enterprises manufacturing equipment used in the production of military systems were affiliated with other state agencies—for example, the State Committee for Industry.

In cases where information is not known, official data can be supplemented by estimates. Rosvooruzhenie data may include estimates derived from applying a price index to the number of units supplied to a foreign country. This may be applied to include transfers that have no direct financial value (for example, deliveries offset against existing debts to East–Central European countries).

It seems likely that the basis for Rosvooruzhenie data is the value of contracts which the state trading company has negotiated. However, this may be supplemented by other data.

The magnitude of Russian arms exports can also be measured in terms of the number of items exported. In 1991 the United Nations established a Register of Conventional Arms[21] to which the Secretary-General asked each member state to report on the number of items in seven categories of equipment that were imported and exported in the past calendar year. Russia has submitted returns to the UN Register for each of the four years 1992–95.

The UN Register is a voluntary exercise which contains no procedures for verifying data submitted by member states. The question arises whether or not the Russian submission accurately reflects the quantities of items transferred that are eligible for reporting.

The Russian submission is prepared by the Ministry of Foreign Affairs on the basis of information provided to it by other state authorities, including Rosvooruzhenie and the Ministry of Defence. The Ministry of Foreign Affairs has no independent means of verifying the data provided. During 1997 it has emerged that title to and control over a large number of items which fall under the definition of conventional arms used by the UN may have been transferred from

---

[21] Wulf, H., 'The United Nations Register of Conventional Arms', *SIPRI Yearbook 1993: World Armaments and Disarmament* (Oxford University Press: Oxford, 1993), pp. 533–44 (appendix 10F); and Anthony, I., Wezeman, P. D. and Wezeman, S. T., 'The trade in major conventional weapons', *SIPRI Yearbook 1997* (note 2), pp. 281–91.

**Table 1.1.** Russian arms reportedly supplied to Armenia as a share of exports reported to the UN Register of Conventional Arms, 1993–96

| Category | A<br>No. of units trans-<br>ferred to Armenia | B<br>No. of units reported<br>to UN Register | A as % of B |
|---|---|---|---|
| Main battle tanks | 74 | 130 | 57 |
| Armoured combat vehicles | 54 | 998 | 5 |
| Large-calibre artillery | 90 | 353 | 25 |
| Missiles and missile launchers | 40 | 436 | 9 |

*Source*: Annual reports to the UN Register of Conventional Arms; and 'Summary of presentation by Lev Rokhlin to the State Duma session on violations in arms deliveries to the Republic of Armenia', *Sovetskaya Rossiya,* 3 Apr. 1997 (in Russian) in Foreign Broadcast Information Service, *Daily Report–Central Eurasia*, FBIS-SOV-97-067, 3 Apr. 1997, p. 3.

Russia to Armenia between 1993 and 1996 (see table 1.1) without being reported.

If the information (which was released by Lev Rokhlin, Chairman of the Defence Committee of the State Duma) is correct, this could mean that Russia's returns to the UN Register for the years 1993–95 understated the volume of Russian exports of tanks and large-calibre artillery pieces in particular. However, this need not necessarily be so. First, at least some of the equipment may have been delivered in calendar year 1996 and may be reported to the UN in the return for that year which will be submitted in 1997. Second, most of the equipment appears to have been transferred from the inventory of the Group of Russian Military Forces in the Caucasus rather than being supplied from the territory of Russia itself. The UN Register does not include a detailed definition of an arms transfer and there is national discretion in determining which items qualify to be reported. It may be that Russia decided that these items were not eligible to be reported.

While the level of transparency in Russian arms exports (whether measured in values or volumes) is far higher than it was during the Soviet period, it can be concluded that there are still serious shortcomings in the data collection mechanisms in use in Russia. It also seems that the low level of cooperation and trust between different state authorities is one of the main factors behind the data problems.

## The distribution of revenues from arms exports

A third issue which was not resolved by this SIPRI study is the question how the money obtained from foreign sales is distributed between state agencies and manufacturing enterprises.

According to Rosvooruzhenie, 7 per cent of payments from foreign customers are retained to cover its own costs in helping enterprises identify, negotiate and win contracts. The other 93 per cent are paid to enterprises. If it is the case that the official estimates of the value of arms exports reflect financial flows, this should mean that roughly $2.8–3 billion was returned to Russian industry in each of the years 1995 and 1996. However, most enterprises insist that they receive very small sums even where they have participated in foreign sales arranged by Rosvooruzhenie.

There are several possible explanations for this discrepancy. One is that the value data published by Rosvooruzhenie do not reflect the financial flows associated with the arms trade for the reasons suggested above. A second possibility is that the money is distributed among many enterprises so that very few receive substantial sums. A third is that payments are structured in such a way that large revenues will not be generated in the early years of a contract. A fourth possibility is that the revenues are received from foreign governments but that the money is kept either by Rosvooruzhenie or by other Russian state agencies (usually named are the Ministry of Defence and the Ministry of Finance) to plug holes in the Russian state budget.

## V. The future for Russia as an actor in the international arms trade

In spite of the continued uncertainties surrounding Russia as an actor in the international arms trade, it is possible to reach some tentative conclusions about likely future patterns.

Predictions frequently made that within a few years the value and volume of arms exports from Russia could reach levels similar to those recorded for the United States are not supported by the available evidence. These predictions are based on a forward extrapolation of statistical trends for 1995–97 using the official data published by the US and Russian governments. According to these data, during these years the value of new agreements by US arms suppliers has declined slightly while the value of new agreements from Russian suppliers has increased. If these trends continue, at some point they will cross. However, on the basis of the evidence presented in this volume there seem to be few additional customers that are likely to turn to Russia as a primary arms supplier. It does not seem likely that the level of demand for Russian equipment from existing customers will increase at the rate needed to sustain the growth levels of the past two or three years.

Nevertheless, it is clear that Russia does not intend to exit the scene as an arms supplier. On the contrary, the main thrust of government policy on this issue in the mid-1990s has been to take steps that can increase the competitiveness of Russian suppliers in world markets. This trend seems likely to continue.

A second and related conclusion is that the level and rate of growth of arms exports from Russia seem to be tied to the procurement behaviour of a small number of countries. Two recipients in particular—China and India—stand out as countries which might acquire significant quantities of Russian arms in future.[22] However, the defence budget in India is under severe pressure as a result of recent choices in its macroeconomic policy. There is greater uncertainty about the direction of future arms procurement choices by China and under certain scenarios a significant increase could occur in Chinese expenditure on imports of modern armaments.

A third conclusion is that the Russian Government still sees arms transfers as an important instrument in its foreign and security policy. In particular, as far as relations with newly independent states on the territory of the former Soviet Union are concerned, the role of arms transfers in Russian foreign and security policy seems to be growing.

A fourth conclusion is that in many cases Russian manufacturers of arms and military equipment see increased revenues from exports as an important element in their own strategies for transformation. However, it also seems that there is a growing awareness in industry—largely born of recent experience—that revenues from exports are very unlikely to be sufficient to ensure their survival.

A fifth and related conclusion is that Russian manufacturers need state support for export activities but see the Russian Ministry of Defence as a less and less trustworthy and reliable partner. The Russian defence budget does not offer sufficient funding for the research and development needed to provide new products. In addition, equipment procurement by the Russian armed forces and other state agencies is too low to sustain existing production capacities. At the same time, Russian arms exports collapsed in 1993–94—when state support for Russian arms exporters was at its lowest.

Russia has not yet found an effective mechanism for integrating the contributions of various government agencies and committees in a manner which produces a decision by consensus. Instead, these different agencies and committees tend to compete with each other for export-related rights and responsibilities. Under these circumstances the regular changes in Russian arms export policy and practice can be expected to continue for some considerable time.

---

[22] The arms procurement decision-making processes in these two countries are analysed in detail in Singh, R. P. (ed.), SIPRI, *Arms Procurement Decision Making, Vol. 1: China, India, Israel, Japan, the Republic of Korea and Thailand* (Oxford University Press: Oxford, forthcoming, 1998).

# 2. Trends in post-cold war international arms transfers

*Ian Anthony*

## I. The new international environment and the arms trade

During the cold war the existence of an international market for conventional arms was most often—and most convincingly—explained by reference to the structure of the international system itself. In a system composed of sovereign states each government reserves the exclusive right to make and implement policies that affect the military security of the state.[1] Each government reached the conclusion that armed forces should be raised under state control as a central element of security policy. However, the technological and industrial capabilities to develop and produce equipment for use by these armed forces are not distributed evenly across the international system. Many states lack the capacity to meet their perceived requirements from their own resources. These states have tended to seek the equipment they require from states with more highly developed technological and industrial capacities.

While this observation explains the demand for arms transfers in a general sense, it does not offer guidance as to the specific factors which shape the market. Why are some governments willing or able to buy arms from a particular supplier while others are not? Why are some countries willing or able to sell arms to a particular recipient while other potential suppliers are not?

In addressing these questions, four 'baskets' of issues and the interaction between them are important. The baskets consist of politico-military issues, economic issues, industrial issues and technological issues.

The first are politico-military issues. In analysing patterns of arms trade it is not sufficient to consider only those elements that shape markets for civilian goods—such as price, quality and availability. In recipient countries the armed forces and other state agencies are usually the only legitimate recipients of military end-items. From the supplier perspective the development and manufacture of and trade in military items are always under some form of state regulation—although the systems of regulation vary widely between countries. State intrusion into the arms transfer market therefore occurs throughout the entire cycle of development, production and trade.

From a supplier perspective, when the possibility of an arms transfer arises it will be asked whether or not it is desirable to increase the military capabilities

---

[1] The definition of military security in this context is the ability to deter the threat or use of military force and the control of military force.

of the recipient in question. Answering this question requires a consideration of political and strategic factors.

From the perspective of a recipient which wants to introduce a certain capability, the question will be raised whether obtaining it from a given supplier could have political or military disadvantages—for example, whether this bilateral linkage could lead to a deterioration of relations with third parties or whether supplies of necessary spare parts and support will be assured in all conditions.

Although the price of equipment is not sufficient to explain given arms transfers, it is not an irrelevant factor. Since military end-items are bought by state agencies, levels of military expenditure do set limits on the overall value of the arms trade. The limit may not be defined by the military expenditure of the recipient country, however, since suppliers may be prepared under some conditions to offer military assistance to offset some or all of the costs of the transfer.[2]

Industrial factors may also be relevant in determining the pattern of arms transfers. The supplier may take into account a range of industrial policy questions in evaluating whether or not to allow a given export. For example, the ability to recover the costs of research and development (R&D) of the equipment in question and to maintain capacity utilization (and employment) in the defence industry may be relevant factors. From a recipient perspective the extent to which a supplier is prepared to assist with technology transfers, place orders for goods produced in the recipient country or make various other kinds of investment may have a strong bearing on the decision whether or not to go ahead with a particular programme.

Manufacturing companies may also be the recipients of technology and component transfers that can, under certain conditions, be considered to be arms transfers. The idea of defining the arms trade to include industry-to-industry sales is a relatively new one and is related to the changing nature of the defence industrial base. In the past there were cases where suppliers of defence equipment transferred production technologies to enable the recipient to produce the end-item rather than buying an item manufactured in the supplier country 'off the shelf'. These transactions would have required licences at least from the supplier state and were, in that sense, comparable to the transfers of manufactured items.

Recent trends may make the defence industry less homogeneous as the range of equipment armed forces require to carry out their activities grows. The importance of goods and technologies which are not in themselves military in their application—for example, in areas such as data processing, telecommunications and sensors—has increased relative to that of more traditional types of military equipment. International sales and cooperation between companies

---

[2] This kind of military assistance to other countries is normally recorded as military expenditure by the aid donor but not by the aid recipient.

which do not consider themselves to be part of the defence industry may contribute to the military capabilities of recipient states.[3]

From this brief discussion it is evident that there is no clear and universal definition of what constitutes the conventional arms trade. A definition can be attempted using one or both of two elements.

First, the definition may be based on the end-user of the goods or technologies traded. The end-user could be defined as the ministry of defence or could include all of the armed forces of a state (therefore including units such as border guards, paramilitary police or coastguards even if they are outside the control of the ministry of defence). One disadvantage of this approach is that the armed forces buy many items that are non-military in the course of their activities (such as fuel, electricity or paint). Another disadvantage is that it may be desirable to take into account purchases by buyers outside state control (for example, private industry or non-state armed factions).

A second alternative would be to use a definition based on the technical characteristics of the items traded. As noted above, the boundary between military and civilian technologies is becoming increasingly difficult to draw. A third alternative—which is the preferred approach of most data collectors—is to combine both end-user and technical characteristics. However, the precise choices about which end-users and which types of goods and technologies to include differ as between different agencies and institutes (each of which has its own interests, objectives and information-gathering capacities).

This chapter uses SIPRI estimates of international transfers of major conventional weapons to try to assess how the volume and distribution of these transfers have changed in the 1990s. In the final section the chapter considers whether the four baskets of issues described above are still adequate to describe the trends in the international market, whether the relative importance of these issues has shifted and, if so, how.

The SIPRI estimates are based on a weighted index that can be used to measure changes in the volume and distribution of weapon flows. The index cannot be taken as a proxy for expenditure.[4] It is created by multiplying the number of units delivered in any calendar year by the trend-indicator value assigned to that unit. The trend-indicator value is the average programme unit cost in 1990 prices for those systems for which these data are available. In other cases the assigned value is based on comparisons of technical parameters (such as speed, range, weight and first year of production) between the system concerned and similar systems whose cost is known.[5]

---

[3] This is particularly true of small and medium-sized companies in high-technology sectors which may be unaware of or unable to verify the nature of the activities of foreign customers or partners.

[4] There is no data set which can measure aggregate revenue from arms exports or expenditure by governments on imported weapons. The data produced by the US Arms Control and Disarmament Agency (ACDA) are themselves a weighted index—although compiled according to criteria that are different from those of SIPRI. The data published by the United Nations in its COMTRADE database are supplied by national customs authorities. However, some important arms exporters and importers do not supply data to the UN and the most of the data are not disaggregated into military and civilian categories.

[5] A longer discussion is contained in 'Sources and Methods for SIPRI Research on Military Expenditure, Arms Transfers and Arms Production', SIPRI Fact Sheet, Jan. 1995, available from SIPRI.

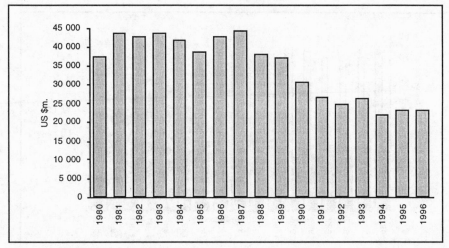

**Figure 2.1.** Deliveries of major conventional weapons, 1980–96

*Source:* SIPRI arms trade database.

If the aim is to evaluate the economic impact of the arms trade, this approach is inadequate. Since economic conditions are different in different countries, programme costs can also be expected to differ. In any specific transaction the purchase price will not only reflect production costs but also pricing methods (for example, whether or not R&D costs are recovered), the length of production runs, military aid, other forms of programme financing (for example, the use of credit or barter), excise taxes, and costs of transport and installation.

The following section uses the SIPRI estimates to describe changes since 1980 in international transfers of major conventional weapons.

## II. The general market trends

Looking at the period 1980–96, the overall trend in the volume of deliveries of major conventional weapons has been downwards since 1987—a peak year for deliveries (see figure 2.1). The level of deliveries recorded for 1996 represented 52 per cent of the level of deliveries in 1987.

The reductions in the market have not been distributed evenly between suppliers. During the cold war the United States and the Soviet Union dominated on the supply side. Their combined share averaged over 65 per cent in the period 1980–96. Since the end of the cold war there has been a large increase in the US share and for the five-year period 1992–96 the USA alone accounted for over 50 per cent of all deliveries. This increased share reflects not a large increase in the total volume of US deliveries (which increased in the period 1991–93, reflecting deliveries to countries around the Persian Gulf following the war against Iraq, but has been broadly stable across the whole period since 1980) but a constant volume of deliveries in a shrinking overall market.

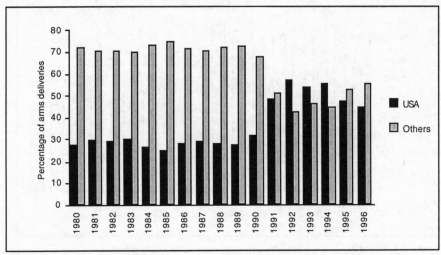

**Figure 2.2.** The US share of deliveries of major conventional weapons, 1980–96

*Source:* SIPRI arms trade database.

By contrast, the volume of Russian deliveries has fallen very significantly. In the period 1992–96 the average share of Russia in total deliveries was 13 per cent, compared with an average of 36 per cent recorded for the Soviet Union in the period 1980–91.

While the United States and the Soviet Union were the two most important suppliers for most of the period of the cold war, other suppliers taken together typically accounted for around 30 per cent of deliveries. However, within this group a small number of countries accounted for the overwhelming bulk of this 30 per cent. According to SIPRI estimates the 10 largest suppliers consistently account for 90 per cent of all deliveries.

Other than the United States and the USSR/Russia, the main arms suppliers since 1980 have been France, the United Kingdom, Germany and China. Others tend to occupy a market niche, with industries that specialize in making a narrower range of products and have a smaller customer base. At particular times countries such as Canada, the Netherlands and Sweden may increase their importance as suppliers on the basis of a single very large contract. Over time, however, they are relatively minor suppliers of major conventional weapons.

The United States seems certain to remain by far the most important supplier of major conventional weapons. A growing number of countries see it as the only external power able to offer a credible security guarantee.[6] The ability to work effectively alongside US armed forces is important not only because of

---

[6] This can often be true regardless of geography. For example, the Chief of the Directorate for European and North Atlantic Integration in the Romanian Ministry of Defence has observed that 'we can say a lot of things about equality and so on, but we are convinced that the trans-Atlantic link is the most important thing for security in the area'. Cody, E., 'Romania steps up efforts to secure spot in NATO', *Washington Post*, 26 Aug. 1997, p. A11.

possible defence cooperation but also because the USA has played a leading role in recent multinational operations. The extent to which other powers will participate in such operations is currently subject to review.[7] Between 1992 and 1996 the USA accounted for roughly 65 per cent of global military R&D expenditure.[8] In addition, the accumulated effect of high levels of military expenditure over a long period has created in the USA a very large and powerful defence industry capable of contributing independent resources to R&D. The USA also has many products which have reached the stage at which R&D costs have been fully amortized and can be subtracted from the purchaser price.

For these reasons, US companies will be very formidable opponents in most acquisition programmes that are opened to international competitive tender in countries that are not subject to export restrictions under US law. It is even likely that in many cases the two strongest competitors in any tender involving large, complex weapon platforms will both be US companies.

Countries in Western Europe have some disadvantages in comparison with the United States: individual West European countries have fewer resources to devote to defence than the USA.

Two countries—France and the UK—use arms exports as part of the military element of a foreign and security policy which, even if it is no longer global, has extra-European dimensions. For both permanent membership of the UN Security Council creates a requirement for intelligence-gathering capabilities and military-to-military contacts with states outside Europe that other European countries do not share. For example, both countries deploy forces in the Persian Gulf and a large share of their arms exports is directed at this subregion. Other European countries (with limited exceptions such as Italy) do not have either the capability or the ambition to play a major strategic role outside Europe, and some countries—such as Germany and Sweden—have national arms export policies which are incompatible with those of France and the UK. For example, neither Germany nor Sweden permits sales of lethal items to countries located around the Persian Gulf. As a result, there is no integrated West European arms export policy and no realistic prospect of creating one in the immediate future.

In that part of the defence industry which manufactures less complex equipment it is unlikely that any country (including the United States) will be able to gain a very large share of the world market. Apart from suppliers in Western Europe, a significant number of countries continue to invest in maintaining their defence industries. Countries such as Bulgaria, China, the Czech Republic, Israel, North Korea, Poland, Romania, Singapore, Slovakia and Ukraine have the capacity to make armoured vehicles, artillery of various kinds and infantry weapons together with the spare parts and ordnance that this equipment

---

[7] The French Government in 1997 decided to reduce the French military presence in Africa. In the United Kingdom, although the Labour Government elected in May 1997 has announced a full-scale defence review, the leadership has made clear that the British commitment to NATO and the maintenance of an independent nuclear force are not in question. Any reductions in commitment are therefore likely to fall on operations in other areas.

[8] Arnett, E., 'Military research and development', *SIPRI Yearbook 1997: Armaments, Disarmament and International Security* (Oxford University Press: Oxford, 1997), p. 219.

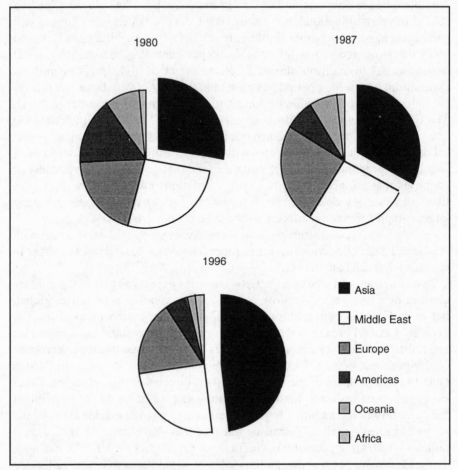

**Figure 2.3.** Regional shares of deliveries of major conventional weapons in 1980, 1987 and 1996

*Source:* SIPRI arms trade database.

requires. Several new suppliers of this type of equipment are beginning to enter the world market. South Africa, which has a significant defence industry, was excluded as a supplier by many countries for many years for diplomatic reasons. Other countries such as Egypt, South Korea and Turkey have developed capacities to produce equipment which they are offering for international sale.

### Recipient perspectives

The international demand for major conventional weapons has also been fairly concentrated. The most important recipients have been located mainly in Asia,

Europe or the Middle East, but the relative importance of these regions as markets has changed.

A comparison of the shares of deliveries on a regional basis in 1980, 1987 and 1996 illustrates the changing importance of Asia, Europe and the Middle East over the period. Figure 2.3 illustrates the growing importance of Asian countries as recipients of major conventional weapons in the 1990s. Their share in total deliveries has grown, particularly since 1993. Moreover, within Asia demand has been concentrated particularly in the subregion of North-East Asia.

By 1996 four recipients—China, Japan, South Korea and Taiwan—accounted for 33 per cent of all deliveries of major conventional weapons.

China depended on its own defence industry to meet the needs of the People's Liberation Army (PLA) for many years after the Sino-Soviet split, the Soviet Union having decided to withdraw from all cooperation in 1960. After 1990 China and the Soviet Union renewed their military–technical cooperation, a relationship that has been strengthened by Russia.

For many years Taiwan depended for its security mainly on the guarantees contained in bilateral security arrangements with the United States. After the Carter Administration unilaterally abrogated these arrangements, Taiwan invested heavily in its own defence industry. At a time when it was considered an international pariah, Taiwan's international defence cooperation was confined to a small group of states—notably Israel and South Africa—which also found themselves fairly isolated diplomatically. More recently, as Taiwan has gained acceptance in the international community, more states have been prepared to sell Taiwan defence equipment. By 1996 Taiwan was receiving large amounts of equipment from France and the United States in particular.

Although Europe is one of the centres of global defence production, many European countries import a lot of the equipment that their armed forces require. In the past the United States has been a particularly important supplier: in the 1950s and 1960s many West European countries rebuilt their armed forces with US equipment. In the 1990s even countries which have historically invested significant resources in military R&D have not been able to maintain a production base for all types of major defence equipment. In these conditions not only small and medium-sized European countries but also France and the United Kingdom have turned to the United States for off-the-shelf purchases as solutions to some of their military requirements.[9]

The Middle East, a region of continuous high-intensity conflict, has been an important centre of demand for major conventional weapons. Between 1980 and 1988 the Iraq–Iran War created a high demand for arms and equipment from Iraq in particular. After 1990, the decision by the UN Security Council to impose a mandatory embargo on Iraq following its invasion of Kuwait removed Iraq as a market for arms and equipment. Libya is also subject to a mandatory UN arms embargo. Nevertheless, there are some Middle Eastern states that are

[9] For example, the United Kingdom and France bought US E-3 airborne warning and control systems (AWACS); the UK and the Netherlands bought the US AH-64 Apache attack helicopter; and the UK bought the US BGM-109 Tomahawk cruise missile.

modernizing their armed forces. Saudi Arabia has been a large recipient of major equipment since the 1970s. Egypt and Kuwait have increased the volume of their major conventional weapon acquisitions in the 1990s.

Outside these three regions the demand for major conventional weapons has never been very high in Latin America and Oceania. In sub-Saharan Africa there have been periods when one country—for example, Angola in 1975 and Ethiopia in 1977–78—received relatively large amounts of equipment. These acquisition programmes have usually been associated with the conduct of high-intensity military operations. The low level of military expenditure and benign threat environment of Latin America makes it unlikely that this region will emerge as a large market for major conventional weapons. However, the fact that the economic and financial crises of the 1980s led to the delay or cancellation of many equipment programmes makes it likely that there will be some limited modernization of old equipment.

Africa's low level of military expenditure and low strategic salience from the perspective of major powers make it unlikely that this region will emerge as a large market for major conventional weapons. At the same time the persistence of conflicts in and between some African countries mean that both state and non-state actors in this region will continue to acquire simple, cheap weapons.

## III. Changing economic conditions

The impact of changing economic conditions on the arms trade is not easy to evaluate. While there has certainly been a reduction in worldwide demand for defence equipment, it does not automatically follow that this will be translated into a reduction in the economic value of international sales.

As noted above, there is no adequate measure of the economic value of the arms trade and it is therefore not possible to state the trend in value with full confidence. It could be that reductions in total demand mainly affect domestic producers rather than international sales. It could also be that international transfers have not declined but have changed their form—and so are no longer being measured effectively by existing statistical indexes.[10]

In 1996 a study by the Fiscal Affairs Department of the International Monetary Fund (IMF) concluded that military expenditure has declined significantly since 1985.[11] Among the 130 countries on the IMF World Economic Outlook database, it was estimated that military expenditure dropped from 3.6 per cent of the value of global gross domestic product (GDP) in 1990 to 2.4 per cent in 1995. Other data sets—for example, the military expenditure data published by SIPRI—suggested even greater declines. Although the pattern of global military

---

[10] Keeping instruments used for measurement in line with changing market practices is a general problem with foreign trade statistics and is not confined to defence equipment. However, the tendency of governments to treat transfers of defence equipment in a unique way makes the problem more difficult to address for these goods and services.

[11] Gupta, S., Schiff, J. and Clements, B., *Worldwide Military Spending, 1990–95*, IMF Working Paper WP/96/64 (International Monetary Fund: Washington, DC, June 1996).

expenditure is heavily influenced by spending in a small group of countries, the IMF found that 'the decline in military spending has been widespread both geographically and by level of development'.[12] While military expenditure fell in industrial countries, it was also the case that the developing countries as a group reduced their military expenditure by almost 50 per cent between 1990 and 1995. From a regional perspective, only one region—Asia—increased its military expenditure levels during this period (although spending by Latin American countries was stable).

For most countries very little information is available about how aggregate military expenditure is distributed between different functions. This reduction in military expenditure therefore does not necessarily mean that there was less money spent on equipment—which would be a more important finding from the perspective of the impact on the arms trade. One group of countries—the members of the North Atlantic Treaty Organization (NATO)—does publish highly disaggregated data on military expenditure. As noted above, many NATO countries are themselves significant arms importers, so that the development of expenditure on equipment within the alliance will have some bearing on the overall global pattern of arms transfers.

A recent survey of NATO military expenditure concluded that almost all the allies made substantial reductions in defence expenditure after the late 1980s. Moreover, in spite of the improvement in the general economic climate in many of the larger members of NATO, most nations have continued to project a further decline or at best levelling off in real defence expenditure.[13]

Looking specifically at expenditure on equipment, the overall trend within NATO has been downwards since 1985. In the period 1987–96, expenditure on equipment by the NATO members other than France (which does not provide NATO with this breakdown of data) fell from $121 billion to $81 billion. However, reductions have not been distributed evenly between the allies. Some countries—for example, Germany and the United Kingdom—have recorded steep reductions in expenditure on equipment while others—most notably Turkey—have recorded sharp increases.

As a result of this uneven distribution, the overall net reduction in demand for defence equipment may not translate directly into reductions in international arms transfers. Reductions in expenditure in countries with large defence industries (such as Germany and the UK) might primarily affect domestic manufacturers (and incidentally increase the pressures on companies to export). At the same time increases in expenditure by countries with less developed defence industries may have the effect of sucking in imports.

This dynamic may also contribute to the development of new types of transaction in the market for defence equipment in cases such as that of Turkey,

[12] Gupta, S., Schiff, J. and Clements, B., 'Drop in world military spending yields large dividends', *IMF Survey*, May 1996, p. 183.
[13] Presentation of Frank Boland, Head of Force Planning Analysis, NATO at the North Atlantic Cooperation Council Seminar on Economic Aspects of Defence Expenditures and Legislative Oversight of National Defence Budgets, Brussels, 14–15 Dec. 1995.

where the government has an ambition to develop its own defence industry. Imports may increasingly take the form of industrial cooperation rather than transfers of finished end-items.

There is some evidence of a growing tendency for countries to seek cooperation of this kind, known as offsets or counter-trade. Offsets are forms of industrial or commercial compensation required as a condition of purchase of defence articles and/or services, such as co-production, licensed production, sub-contractor production, overseas investment or technology transfer. Counter-trade is a generic term which embraces a number of trading arrangements by which some or all of the payments from arms buyers are compensated by the vendor purchasing goods and services and includes specific forms such as barter, counter-purchase or buy-back.

The practice of using offsets between industrialized countries began in the 1960s: offsets are as old as industrial collaboration in armaments production.[14] However, while 20 years ago only about 20 countries (mostly within NATO) had offset policies, in 1997 the figure is c. 130.[15] There is also evidence that industrialists in the United States, who were historically reluctant to engage in this kind of trade practice, are now more willing to consider technology transfers and licensing agreements.[16]

## The changing supplier base

An additional issue to consider is the changing nature of the supplier base for defence forces worldwide. The issue can be illustrated using the example of signals and communications. However, there are other examples where certain goods and services may in future be available from suppliers which are not specialized in defence sales.[17]

Military signals and communications have always been an important element of any armed force. However, after the year 2000 it is expected that several of the major powers will make a very large investment to expand their strategic and tactical communications systems. Moreover, these programmes are likely to be based on digital communications technologies: communications that in the past were discrete and separate can in theory be integrated, allowing large amounts of information to be sent almost instantly across very great distances. These new technical possibilities have come to play a very important role in the discussion of military doctrine and how to apply force in support of foreign and security policy objectives after the cold war.[18]

---

[14] Hammond, G. T., 'The role of offsets in arms collaboration', ed. E. Kapstein, *Global Arms Production: Policy Dilemmas for the 1990s* (Lanham Books: New York, 1992).

[15] Wood, D., *Australian Defence Offsets Program, Proceedings of Defence Offsets Seminar* (Australian Department of Defence: Canberra, 1992).

[16] See, e.g., the discussion of offsets by the US Aerospace Industry Association located at URL <http://www.access.digex.net/~aia/fp_tools.html>.

[17] The possibility that high-resolution satellite images could be available from commercial suppliers and in close to real-time is another such example.

[18] This discussion is building on thinking carried out in the United States in the 1970s and 1980s where the implications of these technological developments were already being discussed. E.g., *Discriminate*

The cost implications of transforming military organizations to take advantage of the new technical possibilities and the costs of investment in hardware and software needed to implement new programmes are difficult to assess. Communications programmes are spread between the different branches and agencies of the US defence establishment and some classified programmes in the budget almost certainly relate to communications. Moreover, a criticism of recent US budget decisions has been that too little attention has been paid to funding equipment programmes of this kind while too much funding has been concentrated on more traditional programmes—acquisition of platforms and weapons.[19]

These cost considerations are themselves likely to influence (and in fact are already influencing) defence industrial policies in both governments and companies in the larger countries of the Euro-Atlantic area.[20] Two possible consequences can be summarized.

First, an increase in the resources allocated to functions such as command, control and communications ($C^3I$), intelligence gathering and precision-strike weapons is implied by the organizational and doctrinal changes noted above. This is likely to squeeze even further the resources available to more traditional defence suppliers (which have already been shrinking), thus accelerating the current trends within the defence industry towards concentration and a search for new markets.[21] Second, acquisition strategies may be revised. This may lead ministries of defence to buy equipment from outside their traditional group of suppliers. Large companies such as Netscape or Sun Microsystems as well as many small and medium-sized companies which do not think of themselves as part of the defence industry may in future become very important suppliers to defence ministries.[22]

The paths of development taken by different countries' armed forces may increasingly diverge. While the United States seems to be about to step across a technological threshold, which will have a major impact on the operations and equipment of its armed forces, it is an open question how many countries will follow. Other NATO countries as well as US allies in Asia are likely to

*Deterrence*, Report of the Commission on Long-Term Integrated Strategy (US Department of Defense: Washington, DC, 1988).

[19] In some cases Congress has even allocated funds to acquire platforms that the Secretary of Defense has not requested.

[20] See, e.g., the article by an executive from the Raytheon company: Stein, R., 'US military technical requirements: views from the US defense industry', *Comparative Strategy*, vol. 13 (1994), pp. 93–100. Structural changes in industry, such as the decision by Siemens to create an integrated command, control and communications division in the company where both civil and military projects reside, are also relevant. *Siemens Annual Report 1995*.

[21] Sköns, E., 'Arms production', *SIPRI Yearbook 1997* (note 8), pp. 239–44.

[22] The Technology Reinvestment Program (TRP) in the United States is perhaps an early example of this kind of change. Under the programme the US military would enter into partnerships with private industry in which the same products that were developed to meet military requirements would also be marketable by private industry. This idea was praised by the Chairman of the Joint Chiefs of Staff, Gen. John Shalikashvili, as 'a great deal for the military and a great deal for the country'. For a discussion, see Lessure, C. A. and Krepinevich, A., *Technology Reinvestment Program: Potential Military Bargain* (Defense Budget Project, Washington, DC, 17 Feb. 1995). It is not yet known whether the types of idea sponsored under the TRP delivered the desired results.

incorporate some of the changes even if they will not be able to go as far as the USA. Most other armies seem likely to retain essentially the same structure that they currently have.

## IV. Patterns of Russian arms exports

After this brief review of what seem to be some of the more important current trends, this section considers the pattern of Russian arms transfers since 1980. The data used are again mostly taken from the SIPRI arms trade database.

After 1985, under the influence of President Mikhail Gorbachev and Foreign Minister Eduard Shevardnadze, the foreign relations of the USSR underwent significant revisions. German unification in 1990 and the dissolution of the WTO in 1991 disrupted what had been a Soviet-dominated integrated production system. Soviet collaboration with the USA to resolve ongoing conflicts in Central America, Southern Africa and South Asia and greater attention to the record of payment of recipients of Soviet arms transfers (some of whom had accumulated large debts for earlier deliveries) also had an impact on arms transfer relationships with important clients including Afghanistan, Angola, Cuba, Libya, Nicaragua, North Korea, Syria and Viet Nam.

In late 1991, before the implications of these ongoing changes had been digested, the Soviet Union itself went through a monumental convulsion which led to its dissolution in December of that year. Although Russia hosted the largest part of the Soviet defence industrial base, the breakup of the Soviet Union created a severe crisis for the defence industry, which now found relations between enterprises disrupted at a time of dramatic reduction in demand for its products.

Deep-seated changes have been set in motion in Russia which will have very important long-term implications for the defence industry. Factors that will be of particular importance include: the outcome of military reforms prompted by Russia's changed geopolitical circumstances; the shape of a new Russian foreign and security policy; the continued pursuit of macroeconomic objectives through, among other policies, a dramatic reduction in military expenditure; and the conviction that the defence industrial base contained elements that could reduce Russia's economic dependence on the sale of raw materials for foreign exchange.

The impact of these changes is examined in section V below.

Table 2.1 shows the changing percentage share of the 25 largest recipients of Soviet and Russian arms in the period 1980–96. It underlines the traditional importance of two groups of recipient—members of the WTO and Arab countries of the Middle East—in overall Soviet arms exports. Seven of the 10 largest recipients listed in the table belong to one or other of these groups. It also shows that since 1994 the 25 largest recipients have accounted for only around half of all deliveries from Russia. This reflects the emergence of clients for Russian arms—notably Algeria, Malaysia and the United Arab Emirates—which either are new or in the past received low volumes of Soviet equipment.

**Table 2.1.** Shares of the 25 largest recipients of major conventional weapons from the Soviet Union/Russia, 1980–96[a]

Figures are percentages.

| | 1980 | 1981 | 1982 | 1983 | 1984 | 1985 | 1986 | 1987 | 1988 | 1989 | 1990 | 1991 | 1992 | 1993 | 1994 | 1995 | 1996 |
|---|---|---|---|---|---|---|---|---|---|---|---|---|---|---|---|---|---|
| India | 9 | 9 | 11 | 14 | 7 | 10 | 16 | 21 | 24 | 23 | 17 | 29 | 35 | 11 | 13 | 21 | 24 |
| Iraq | 12 | 9 | 12 | 11 | 19 | 13 | 17 | 19 | 6 | 10 | 5 | 0 | 0 | 0 | 0 | 0 | 0 |
| Syria | 11 | 8 | 12 | 17 | 14 | 14 | 6 | 9 | 10 | 3 | 0 | 0 | 0 | 0 | 0 | 0 | 0 |
| Czech Republic[b] | 5 | 12 | 7 | 8 | 8 | 9 | 8 | 8 | 6 | 9 | 2 | 3 | 0 | 0 | 0 | 0 | 0 |
| German DR | 3 | 4 | 8 | 10 | 10 | 7 | 5 | 6 | 9 | 5 | 6 | 0 | 0 | 0 | 1 | 0 | 0 |
| Poland | 4 | 4 | 2 | 4 | 5 | 5 | 10 | 8 | 9 | 9 | 4 | 2 | 2 | 1 | 1 | 0 | 0 |
| Libya | 14 | 12 | 9 | 3 | 3 | 6 | 5 | 1 | 0 | 4 | 0 | 0 | 0 | 0 | 0 | 0 | 0 |
| Afghanistan | 3 | 3 | 2 | 1 | 1 | 1 | 3 | 3 | 6 | 14 | 20 | 26 | 0 | 0 | 0 | 1 | 2 |
| Bulgaria | 4 | 8 | 7 | 5 | 5 | 8 | 5 | 1 | 2 | 1 | 5 | 9 | 2 | 0 | 0 | 0 | 0 |
| Angola | 1 | 1 | 2 | 5 | 7 | 5 | 6 | 10 | 8 | 0 | 7 | 0 | 0 | 1 | 4 | 0 | 0 |
| Korea, North | 0 | 0 | 0 | 1 | 1 | 0 | 0 | 4 | 0 | 9 | 6 | 1 | 0 | 1 | 2 | 7 | 0 |
| Algeria | 7 | 4 | 6 | 3 | 2 | 0 | 0 | 4 | 9 | 4 | 3 | 9 | 0 | 0 | 18 | 11 | 40 |
| China | 0 | 0 | 0 | 0 | 0 | 0 | 0 | 0 | 0 | 0 | 0 | 3 | 37 | 32 | 0 | 0 | 0 |
| Cuba | 0 | 4 | 6 | 6 | 3 | 3 | 2 | 2 | 0 | 1 | 1 | 3 | 0 | 0 | 0 | 7 | 2 |
| Viet Nam | 5 | 7 | 0 | 5 | 3 | 1 | 2 | 3 | 4 | 0 | 2 | 0 | 0 | 0 | 0 | 0 | 0 |
| Yugoslavia[c] | 4 | 3 | 4 | 1 | 0 | 0 | 1 | 1 | 4 | 4 | 3 | 1 | 0 | 30 | 0 | 0 | 5 |
| Hungary | 2 | 2 | 2 | 2 | 2 | 0 | 0 | 0 | 0 | 0 | 0 | 0 | 0 | 0 | 0 | 0 | 0 |
| Romania | 2 | 2 | 2 | 2 | 1 | 0 | 0 | 0 | 0 | 0 | 6 | 0 | 3 | 0 | 6 | 0 | 0 |
| Iran | 0 | 0 | 0 | 0 | 0 | 0 | 0 | 0 | 0 | 0 | 5 | 11 | 6 | 12 | 6 | 0 | 4 |
| Yemen, South | 4 | 3 | 1 | 0 | 0 | 0 | 1 | 0 | 2 | 0 | 0 | 0 | 0 | 0 | 0 | 0 | 0 |
| Ethiopia | 1 | 1 | 3 | 0 | 0 | 1 | 1 | 0 | 1 | 1 | 0 | 0 | 0 | 0 | 0 | 0 | 0 |
| Jordan | 0 | 0 | 2 | 2 | 0 | 0 | 3 | 2 | 0 | 0 | 0 | 0 | 0 | 0 | 0 | 0 | 0 |
| Nicaragua | 0 | 0 | 0 | 0 | 1 | 1 | 1 | 1 | 1 | 1 | 2 | 0 | 0 | 0 | 0 | 0 | 1 |
| Peru | 0 | 1 | 0 | 0 | 0 | 1 | 0 | 0 | 0 | 0 | 1 | 0 | 1 | 0 | 0 | 0 | 0 |
| Yemen, North | 3 | 1 | 0 | 0 | 0 | 1 | 0 | 0 | 0 | 0 | 0 | 0 | 0 | 0 | 0 | 0 | 0 |
| Others | 4 | 3 | 1 | 1 | 3 | 3 | 1 | 1 | 1 | 1 | 4 | 4 | 13 | 12 | 50 | 52 | 22 |

[a] Countries are ranked by aggregate value of deliveries. [b] Prior to 1992 data refer to the former Czechoslovakia. [c] After Apr. 1992 the Federal Republic of Yugoslavia (Serbia and Montenegro).

**Table 2.2.** The 10 largest recipients of major conventional weapons from the Soviet Union/Russia, 1982–96

| Rank | 1982–86 | 1987–91 | 1992–96 |
|------|---------|---------|---------|
| 1 | Iraq | India | China |
| 2 | Syria | Afghanistan | India |
| 3 | India | Iraq | Hungary |
| 4 | Czechoslovakia | Poland | Iran |
| 5 | German Democratic Rep. | Czechoslovakia | Malaysia |
| 6 | Bulgaria | Korea, North | United Arab Emirates |
| 7 | Poland | Angola | Slovakia |
| 8 | Angola | Syria | Kazakhstan |
| 9 | Libya | German Democratic Rep. | Algeria |
| 10 | Cuba | Yugoslavia | Viet Nam |

*Source:* SIPRI arms trade database.

This impression is confirmed by table 2.2.

Only two countries—India and Slovakia (formerly as part of Czecho-slovakia)—appear in all three time-periods. Eight of the countries listed in the most recent period (1992–96) do not appear in either of the previous columns. In one case—Kazakhstan—this is because the country itself is new. However, the other cases reflect the reorientation of arms transfer policy during the late Soviet period and subsequently by Russia. In the most recent five-year periods (1987–91 and 1992–96) only five of the countries listed are WTO member states or Arab states. Moreover, the two Arab states listed for 1992–96 (Algeria and the United Arab Emirates) are not countries that historically received large quantities of Soviet weapons.

Table 2.3 illustrates the balance between different equipment categories of Soviet/Russian exports of major conventional weapons. As can be seen, two categories—aircraft and armoured vehicles—have consistently been the main-stays of exports. In the 1990s it is noticeable that exports of ships (a category which includes diesel-powered submarines) have made up a significant segment of Russian exports.

### India as a recipient of Soviet/Russian weapons

The bilateral arms relationship with India has been the single most stable element in Russia's military–technical cooperation. India had a close arms transfer relationship with the Soviet Union after the mid-1960s and its armed forces depend heavily on equipment of Soviet origin. It is the single largest customer for Russian arms measured by number of licences issued.[23] India had a strong interest in the maintenance of stability in the Soviet Union.

[23] See chapter 6 in this volume; and Sergounin, A. and Subbotin, S., *Russian Arms Transfers to India: Incentives, Patterns and Implications* (University of Nizhniy Novgorod Press: Nizhniy Novgorod, 1996).

**Table 2.3.** Shares of different weapon categories in Soviet and Russian exports of major conventional weapons, 1980–96
Figures are percentages.

| Category | 1980 | 1981 | 1982 | 1983 | 1984 | 1985 | 1986 | 1987 | 1988 | 1989 | 1990 | 1991 | 1992 | 1993 | 1994 | 1995 | 1996 |
|---|---|---|---|---|---|---|---|---|---|---|---|---|---|---|---|---|---|
| Aircraft | 58 | 53 | 50 | 48 | 49 | 50 | 53 | 52 | 44 | 58 | 56 | 33 | 52 | 60 | 46 | 64 | 48 |
| Armoured vehicles | 17 | 19 | 23 | 23 | 23 | 22 | 22 | 18 | 21 | 17 | 16 | 22 | 24 | 22 | 41 | 16 | 40 |
| Artillery | 5 | 4 | 5 | 4 | 3 | 3 | 3 | 4 | 4 | 2 | 2 | 4 | 0 | 0 | 5 | 1 | 0 |
| Guidance and radar | 9 | 13 | 8 | 11 | 13 | 10 | 5 | 7 | 4 | 2 | 1 | 4 | 1 | 4 | 4 | 0 | 1 |
| Missiles | 6 | 8 | 9 | 8 | 8 | 13 | 11 | 15 | 17 | 16 | 19 | 29 | 9 | 8 | 4 | 7 | 4 |
| Ships | 6 | 3 | 5 | 6 | 4 | 2 | 7 | 4 | 10 | 5 | 6 | 8 | 14 | 6 | 0 | 12 | 7 |

*Source:* SIPRI arms trade database.

The Indian Government was one of very few which sent messages of support to the group of conservative individuals that mounted a coup against President Gorbachev in August 1991 with the intention of restoring the pre-reform system of government of the Soviet Union. With the dissolution of the Soviet Union it was first necessary to establish what the nature of relations between India and the new Russian Federation would be. In spite of the poor start, the relationship has become extremely important to Russia both for foreign policy reasons and because of the changed nature of the payment system for weapons.[24]

On a pragmatic level, India needed reassurance that the repair and maintenance of equipment supplied to it under previous agreements would not be compromised by changes in Russia. As Air Vice-Marshal S. Krishnaswamy of India noted with some understatement, there was a 'hiccup' in supply relations during 1991–92.[25]

The issue of arms and technology transfers was raised at the highest level during Russian President Boris Yeltsin's visit to New Delhi in June 1994 when then Prime Minister Narasimha Rao apparently requested approval for the transfer to India of additional MiG-29 fighter aircraft.[26] In July 1994 Air Chief Marshal S. K. Kaul and his deputy, Air Marshal S. R. Deshpande, visited Russia for discussions while Defence Secretary K. A. Nambiar visited Russia twice in 1994.

India and Russia have agreed at the highest political level that future military–industrial cooperation is desirable and after 1994 reports emerged that new agreements had been signed for transfers of major systems.[27] In the event, not all reported agreements appear to have materialized as of 1997. In 1995 India purchased 10 MiG-29 fighter aircraft to replace aircraft of the Indian Air Force that had been damaged. The agreement covered eight single-seater and two twin-seater aircraft. India also purchased 12 Tunguska air defence systems in 1995.[28]

A long-expected decision by India to produce the MiG-29 fighter aircraft at facilities initially built to manufacture Soviet MiG-27 aircraft under licence has never materialized. This programme was expected to provide aircraft to replace MiG-21 fighter aircraft in the air defence role in the Indian Air Force. A decision to replace Vijayanta tanks with the Russian T-80 main battle tank also

---

[24] This economic dimension of the Russo-Indian arms transfer relationship is discussed in chapter 4 in this volume.

[25] For example, of 122 fighter aircraft engines sent to CIS countries for repair between July 1990 and Jan. 1992, only 79 were returned to India by June 1992. *Aviation Week & Space Technology*, 25 July 1994, pp. 49–50.

[26] *Defense News*, 27 June–3 July 1994, p. 28.

[27] Systems other than the MiG-29 and T-80 which figure consistently in press reports are the Su-35 fighter aircraft; the Ka-50, Ka-52, Mi-35 and Mi-28 attack helicopters; and 152-mm calibre self-propelled howitzers. *Aviation Week & Space Technology*, 25 July 1994, pp. 58–59; *Jane's Defence Weekly*, 30 July 1994, p. 4; *Defense News*, 3–9 Oct. 1994, pp. 1, 36; *Defense News*, 17–23 Oct. 1994, p. 58; Foreign Broadcast Information Service, *Daily Report—Central Eurasia* (hereafter FBIS-SOV) FBIS-SOV-94-205, 24 Oct. 1994, p. 15; FBIS-SOV-94-207, 26 Oct. 1994, p. 12; and *Jane's Defence Weekly*, 5 Nov. 1994, p. 1.

[28] *Military Technology*, Apr. 1995, p. 64; FBIS-SOV-95-205-S, 24 Oct. 1995, p. 50; *The Hindu*, 27 Nov. 1995, p. 1; *The Hindu*, 21 Feb. 1996, p. 14; and *Moscow News*, 31 Mar.–6 Apr. 1996, p. 5.

has not been taken. In each of these cases the Indian Government is waiting for an indigenous programme (the Light Combat Aircraft [LCA] and Arjun tank) to meet the requirement of the air force and army, respectively.[29]

Discussions continued throughout 1995 regarding a programme to modernize 125 of India's fleet of MiG-21 aircraft as an interim measure, pending the production of the LCA. MiG has developed a retrofit package, the MiG-21-93, that involves installing new navigation and target-acquisition systems. This includes a radar which permits the aircraft to fire long-range air-to-air missiles which would be included in the package to India. It was reported that this agreement was signed on 1 March 1996.[30]

In 1997 India has purchased 40 Su-30M Flanker fighter aircraft.[31] It is also reported that India has ordered two additional diesel-powered submarines of Type 636 (the improved Kilo Class).[32]

It is clear that Russia is willing to meet India's conventional arms requirements, subject to agreement on terms. The fact that not all anticipated new agreements have materialized appears to reflect the existence of obstacles at the Indian end of the relationship. These stem from India's approach to public expenditure and defence budgeting, the process of setting priorities between the requirements of different branches of the Indian armed forces and the balance between imports and domestic production in India's overall arms procurement.[33]

# V. An assessment of Russia's future in the world arms market

From this survey of the dominant trends in the international market for conventional arms and the pattern of Russian exports, what conclusions can be drawn about the future prospects for Russia in this market?

---

[29] Thomas, R. G. C., 'Arms procurement in India: military self-reliance versus technological self-sufficiency', ed. E. Arnett, SIPRI, *Military Capacity and the Risk of War: China, India, Pakistan and Iran* (Oxford University Press: Oxford 1997), in particular pp. 119–23.

[30] *New Europe*, 24–30 Mar. 1996, p. 9; and *Times of India*, 8 Oct. 1996, p. 13. The MiG-21-93 can also include an extensive re-build of the airframe and installation of a new engine, the RD-93. It is not clear if India has bought this full re-fit. *International Defense Review*, May 1994, p. 16; and *Defense News*, 6–12 June 1994, p. 12. It has subsequently been reported that the MiG-21 upgrade programme has been suspended for budgetary reasons. *Defense News*, 9–15 June 1997, p. 34.

[31] The specific aircraft ordered by India exists currently only as a flight test prototype. The 8 aircraft delivered almost immediately to India by the Irkutsk Aircraft Production Organization (IAPO) were of the 'K' version. Under the terms of the agreement, these aircraft will subsequently be updated to 'M' versions. *Air Force Monthly*, Sep. 1997, pp. 14–15.

[32] This appears to have been a framework agreement between the two governments or a memorandum of understanding rather than a negotiated contract. *Defense News*, 6–12 Jan. 1997, p. 22; *Times of India*, 8 Jan. 1997, p. 9; and *International Herald Tribune*, 9 Jan. 1997, p. 4.

[33] Over the next few years production of several systems assembled in India under Soviet licences will end and it is unclear whether production assets built up in India (such as the MiG-27 production line in Bangalore, the T-72 production line in Avadi and the BMP-2 production line in Shankarpally, Andhra Pradesh) will remain idle, close down, switch to production of equipment of Indian design or begin production of follow-on Russian equipment types. Recent discussions of Indian arms procurement programmes include Arnett, E., 'Military technology: the case of India', *SIPRI Yearbook 1994* (Oxford University Press: Oxford, 1994), pp. 343–65; and Arnett (note 29), pp. 253–57.

First, it is clear that Russia still has a range of large weapon-delivery plat-forms and weapons for which there is international demand. Russia continues to export equipment of a type that was traditionally the mainstay of Soviet arms exports: aircraft (including military helicopters), armoured vehicles and war-ships (including submarines).

There have been several cases of Russian suppliers teaming up with foreign companies in joint ventures aimed at sales to third parties.[34] In these pro-grammes the role of non-Russian partners seems to be to supply advanced avionics (particularly navigation and communications systems) for integration into Russian weapon platforms. Since such systems rely heavily on data pro-cessing, this may point to a weakness in the Russian computer industry (both hardware and software).

Second, Russia's recent success in opening up new markets for this equip-ment suggests that potential recipients are satisfied with guarantees from the Russian state about the medium-term capacity of Russia's defence industry to continue to supply the spare parts and maintenance assistance necessary to keep this equipment in service.

As Russia has an inventory of mature designs that have proved their effec-tiveness and a stable group of clients that form the basis for its military–technical cooperation, it can be predicted that it will continue to be a significant arms supplier, probably at a level at least equivalent to the larger West Euro-pean countries, over the medium term.

Long-term success as an arms supplier seems less certain because of the diffi-culty of predicting the outcome of domestic processes under way in Russia. The background processes driving Russian policy are still incomplete. There is uncertainty about the path of military reform, foreign and security policy, eco-nomic policy and defence industrial policy.

**Military reforms**

During the cold war the basis for Soviet strategy was the confrontation in East–Central Europe. Large numbers of Soviet forces were deployed far forward in support of prevailing doctrine. By the end of 1995 Russia had withdrawn over 700 000 personnel and 45 000 pieces of equipment from the Baltic states and East–Central Europe. Nevertheless, a large number of Russian troops and infra-structure to support them remain stationed outside the territory of Russia. Each of the major branches of the Russian armed forces (Ground Force, Air Force, Air Defence Force, Navy and Strategic Rocket Force) has drawn up a develop-ment plan that sets objectives for fundamental reform and reorganization that should be achieved by the early years of the next century. The government, notably the Defence Council (which is part of the presidential apparatus) and the staff of the Defence Minister, have been active in developing alternative

---

[34] For example, MiG-MAPO teamed up with the British company GEC Marconi to meet the Malaysian fighter aircraft requirement and with French company Thomson-CSF to meet the Indian requirement for upgraded MiG-21 fighter aircraft.

options for military reform. On 19 June 1997 President Yeltsin signed a plan for military reform. Subsequently, in late July, he issued a series of decrees which he described as the 'first steps' in a process of reform in Russia's overall force structure as well as related 'power-wielding departments' of the government.[35]

As a result, it is not possible to state with any certainty what the main priorities and requirements of the Russian armed forces will be in the medium and long term.

## Russian foreign policy

In the period between 1989 and 1993 the development of cooperative relations between the Soviet Union/Russia and Western countries (in particular the USA) was the central focus of foreign policy in Moscow. During this period far-reaching demilitarization and conversion, in particular in Europe, were also elements of Russian declaratory policy. In 1994–95 a more balanced view of Russian national interest began to predominate. While relations with the United States and Western Europe remain very important, greater weight is now given to other regions and actors. Particular attention has been focused on rebuilding relations with the countries of the CIS and to a lesser degree of East–Central Europe, consolidating relations with India and opening new relationships with states that were hostile to the Soviet Union such as China, Iran and Turkey. It is also seen as important for Russia to establish normal relations with as many states as possible, particularly in regions such as South-East Asia and Latin America, which are seen to offer important economic opportunities. Moreover, a by-product of the evolving debate about European security has been the emergence of a consensus in Russia that future foreign and security policy for Europe will include a significant military component.

## Economic policy

After 1992 the Russian armed forces decisively lost the 'battle of the budget' as the Ministry of Finance successfully argued that the overriding priority of budget policy was control over inflation. With the dramatic reductions in the volume of state orders, the relative importance of arms exports to the defence industry has increased. This is true in spite of direct and indirect government support to the defence industry.[36]

There is growing competition for those funds which are received by the Ministry of Defence. Maintaining manpower levels and training, implementing the reforms sketched above and ongoing operations of different kinds in Azerbaijan, Moldova, Tajikistan and Chechnya all have more immediate claims on

---

[35] Informatsionnoye Agentstvo Ekho Moskvy, 25 July 1997 (in Russian) in FBIS-SOV-97-217, 5 Aug. 1997; and Voice of Russia World Service, 22 Aug. 1997 (in English) in FBIS-SOV-97-236, 24 Aug. 1997.

[36] E.g., defence-related enterprises appear to have received significant direct and indirect subsidies under different budget headings. For a discussion, see George, P. *et al.*, 'World military expenditure', *SIPRI Yearbook 1995: Armaments, Disarmament and International Security* (Oxford University Press: Oxford, 1995), pp. 399–408.

Ministry of Defence expenditure than the procurement of equipment. As a result, the armed forces were unable to place any orders in 1996 for some categories of equipment.

It is difficult to establish a clear picture of the pattern and trajectory of military R&D in Russia. Data returned to the UN by Russia in 1995 suggested that within Russian military R&D there is a continuing high priority on strategic forces and that new weapons are being developed for all three legs of the nuclear 'triad'.[37] Expenditure by the Russian Ministry of Defence on conventional arms (including the development of new platforms and tactical weapons) appears to be very low. There are reports that the ability of Russian design bureaux to develop new systems depends very heavily on how much revenue they receive from exports. The distribution of export revenues between design bureaux and production associations is itself contentious, with designers complaining that they do not receive adequate compensation for their contribution to exports. The attempt to create new industrial entities that combine research and production functions in financial–industrial groups is in part an effort by the Russian Government to address this problem. However, recent efforts by producers (such as the production associations in Irkutsk and Komsomolsk-na-Amure associated with the Sukhoi Design Bureau) to resist integration into such a group underline that this policy of state-mandated industrial integration has not been fully implemented and does not enjoy consensus support.[38]

## Restructuring the defence industrial base

It is now widely accepted that the Russian defence industry will have to be fundamentally restructured in the face of the dramatic decline in the demand for its products. However, how this restructuring will take place has been a subject of fierce disputes between different agencies of government in Russia. The State Committee on Defence Industries (Goskomoboronprom) was a successor to the Soviet sectoral ministries that had responsibility for defence production. It argued that it should have not just an executive function, carrying out policies determined elsewhere, but also a say in the fate of enterprises that fell under its umbrella. The prospect of Goskomoboronprom achieving this enhanced role seemed to improve when it was transformed into the Ministry of Defence Industry (Minoboronprom) in 1996. This meant that the minister, Zinoviy Pak, would have the status of a member of the cabinet and so participate directly in some of the most important decisions related to drafting the budget. However, in March 1997 the Ministry of Defence Industry was dissolved and most of its functions transferred to the Ministry of the Economy. The minister now

---

[37] Arnett (note 8).
[38] This issue is discussed in chapter 8 in this volume and in Kogan, E., *Are FIGs good for you? Russian Financial Industrial Groups and their Impact on the Aerospace Industry*, FOA Scientific Report R–97–00465–170–SE (Försvarets forskningsanstalt [Swedish National Defence Research Establishment]: Stockholm, Mar. 1997).

speaking on behalf of the defence industry in government discussions is not believed to be as sympathetic to its plight as Pak.

The State Committee for the Management of Property (Gosudarstvenny komitet po upravleniu imushchestva, GKI) has overall responsibility for privatization and has argued that the defence sector should not be exempt from its programme. Individuals, including Anatoliy Chubais, who played a prominent role in the activities of the GKI have since taken positions of great importance in the Office of the President and in the government.

The Ministry of Defence also has a strong interest in the fate of at least parts of the defence industry. First Deputy Defence Minister Andrey Kokoshin has argued that relations with the defence industry should be regulated by contracts. This would effectively give the Ministry of Defence control over defence industrial policy through its power to award contracts, which would almost certainly be used to sustain those elements of the defence industry expected to contribute to modern and effective armed forces after the year 2000. However, as noted above, exactly what the expression 'modern and effective armed forces' means to the Russian Government in the present context is not yet clear.[39]

In the last years of the Soviet Union and immediately after the creation of the Russian Federation a great deal of attention was paid to theoretical aspects of the conversion of the defence industry. However, even though a Law on Conversion was passed in 1992 and several state conversion plans were elaborated, relatively little was done to implement these documents. In July 1997 Russian Deputy Minister of Economics Vladimir Salo noted that, between 1995 and 1998, 4.2 trillion roubles had been allocated to conversion programmes in the budget but only around 10 per cent of these funds had actually been paid. As of July none of the funds contained in the 1997 budget for conversion had been released.[40] In 1997 there has been a revival of discussions of conversion. In July 1997 the State Duma drafted revisions modifying the existing Law on Conversion.[41] However, whether these new initiatives will lead to any more effective measures than previous efforts is unknown.

For all these reasons it is not possible to make a confident prediction of how Russia's role as an arms supplier will evolve. Chapters 3–11 offer a more detailed picture of the pattern of Soviet and Russian arms exports.

---

[39] For an overview of the background to this issue, see Kile, S., 'Military doctrine in transition', ed. I. Anthony, *The Future of the Defence Industries in Central and Eastern Europe*, SIPRI Research Report no. 7 (Oxford University Press: Oxford, 1994).

[40] *Rossiyskaya Gazeta*, 26 July 1997, p. 9 (in Russian) in FBIS-SOV-97-209, 28 July 1997. By the mid-1990s it was noticeable in discussions with Russian specialists (and in particular with industrialists) that the use of the word 'conversion' was treated with scepticism and sometimes even hostility.

[41] ITAR-TASS, 2 Aug. 1997 (in English) in FBIS-SOV-97-214, 2 Aug. 1997.

# 3. Conventional arms transfers during the Soviet period

*Yuriy Kirshin*

## I. Introduction

During the cold war the main feature of international relations was an active confrontation between two opposing social and economic systems: the capitalist and the socialist. The United States and the Soviet Union were the leading actors. The two systems confronted each other in the economic, diplomatic, ideological and military areas. The military confrontation assumed a variety of forms. These included threats and demonstrations of force, the arms race, and competitive military research and development and intelligence operations.

The cold war left few states unaffected. The disposition of political forces on the international scene changed further with the collapse of the colonial system and the formation of many new states. A number of states remained within the capitalist system even after winning independence and liberating themselves from colonial oppression. However, there were other states which adhered to a neutral status or leaned towards supporting the socialist system. Some states in Africa, the Middle East and Latin America took a further step and proclaimed their determination to build socialism. This process combined with the East–West competition set in train a struggle for a new division of the world, and the policies of the United States and the Soviet Union were intended to support the vitality of their allies and friendly regimes in various regions.

The cold war also involved military assistance to local wars and military confrontations, most often in the developing countries. Several regions in the world became areas of tension or 'hot spots'. These hostile activities pervaded the whole structure of international life.

Against this background, each of the opposing systems attributed great importance to the supply of armaments and military equipment as an instrument of policy.

The economic and military potential of the Soviet Union was strong enough to produce armaments in sufficient quantity to meet the needs not only of its own armed forces but also of many other countries. The nature of Soviet military supplies derived from both domestic and foreign policy imperatives. Due account was paid to the military and political situation in the world and in various regions and to the military policy of the Western countries.

The pattern of conventional arms exports can be best examined if divided into three parts: (*a*) supplies to Soviet allies—members of the WTO and other socialist states; (*b*) supplies to those developing countries which adopted a socialist orientation or pursued an anti-imperialist policy; and (c) supplies to various non-governmental political forces engaged in internal armed struggles against dictatorships and pro-imperialist states. These non-governmental forces included armed opposition groups struggling for power, national liberation movements and organizations fighting for independence and self-government.

The above classification is not perfect since in some cases the Soviet Union also supplied conventional armaments to forces involved in complicated conflicts that combined more than one of the features mentioned above. These different types of recipient are described in sections IV–VI below.

It is will be useful first to describe the general features of Soviet military supplies.

## II. General principles guiding Soviet military supplies

The factors guiding the USSR's conventional arms exports and the balance between different priorities were not fixed over the entire period of the cold war. Priorities changed depending on the military–political situation in the arms market, market conditions and the operational performance of particular models during different military conflicts. Special attention was devoted to the specific choice of a military–technical cooperation partner.

However, certain hard and fast general principles did apply throughout the period. The Soviet Union used conventional arms supplies to try to achieve its own political, military–strategic and economic goals. Political goals were the dominant factor when the decision to export conventional arms was taken.

### Political factors

Account was taken of the following considerations: (a) the socio-political system of the customer state; (b) the coalition of states to which the customer belonged; (c) the purposes for which the conventional arms were sought; (d) the commitment of the customer state to maintain a certain political regime in the country; (e) the desire of the country to draw closer to the socialist system; and (f) the possibility of aggressive action by that country against other countries of the socialist system, those friendly to the Soviet Union or those tied to it by peace treaties. It was necessary to avoid any risk of being drawn into aggressive actions against friendly states.

The political dimension was analysed most thoroughly when a country requested conventional arms from the Soviet Union for the first time. The analysis covered not just the state of the country at that particular time but also the near-term and more distant prospects. In particular, it was considered whether the country would continue to adhere to a political course that satisfied the Soviet Union and the countries of the socialist community. In other words, the customer countries were evaluated from the standpoint of their political orientation. Special attention was focused on countries going through profound socio-economic change. The requests of these countries for conventional arms supplies were considered with regard to the role, place and standing of the country in the overall world political process.

In decisions about conventional arms exports, preference was given to countries which adopted a socialist orientation, took an anti-imperialist attitude or were struggling for political and economic independence and the overthrow of dictatorships. Military supplies were of major importance for penetrating the political and ideological structures of many countries, winning new political allies and, in this way, providing support to the Soviet Union in the United Nations and other international organizations. As new states emerged in various regions as a result of the disintegration of the colonial system, the Soviet Union tried to fill vacant niches and used military supplies to maintain peace in regions of vital interest.

Dependence on Soviet weapons prevented a customer country from rapidly changing its policy and starting to buy arms from other countries. Many countries in Africa and the Middle East which equipped their armies with Soviet weapons now find that they cannot change to an arms supplier other than one of the members of the CIS.

As a rule the Soviet Union prevented arms from being exported to countries which could take aggressive actions that would destabilize the situation in a region. Arms were exported for defensive tasks. If there were sufficient negative aspects in evaluating these factors, requests for arms exports were rejected even if profitable. For example, when Syrian troops entered Lebanon in 1976—an action which was neither approved nor supported by the Soviet Union—this had an immediate effect on arms exports to Syria. Export supplies were temporarily suspended in spite of their great profitability for the Soviet Union and the number of military specialists in Syria was reduced. Profits were high but political and strategic reasons were decisive. Unfortunately, in some cases political errors were made.

Although pursuing economic interest was not the main objective of the Soviet Union—as has been stated, the emphasis was placed on political goals—many newly independent countries were rich in raw materials. The Middle East became a major export market for Soviet arms and appeared profitable from a financial point of view. Many countries of that region paid for weapons in the year of delivery and in hard currency, although in other cases even oil-producing states received arms against credit. In the late 1980s Czechoslovakia put forward an initiative to make economic relations with customers from the developing countries more fair. The Soviet Union opposed that initiative.

Even though conventional arms exports grew during the cold war period, the restraint shown was evident. In spite of requests by certain countries, the export of nuclear weapons and their components, components of chemical weapons and other types of weapons of mass destruction was never allowed. Moreover, the Soviet Union did not export strategic missiles, military space equipment, weapons based on new physical principles, advanced missile technologies, long-range air-defence fighter–interceptor aircraft to countries with small territories, nuclear-powered cruisers and submarines, heavy-calibre artillery, tactical missile systems and many other items.

There was also a list of states to which the export of weapons was forbidden which was periodically revised, depending on military–political conditions in one or another region of the world.

The restraint shown in conventional arms exports and the refusal to satisfy all requests were of major importance in preventing the cold war from escalating into a hot war. This restraint also prevented the development of large-scale arms production by other countries. Similarly, restraint in non-conventional technology transfer made it possible to stop the proliferation of nuclear weapons and other weapons of mass destruction.

The arms exported included both modern models and updated but essentially obsolete models of earlier generations. The ratio between modern and obsolete models varied at different times between 50 : 50 and 70 : 30. However, obsolete models were often re-examined before transfer and updated to match systems adopted in the armies of possible aggressors for another 20–30 years.

The newest types of arms tended to be supplied in emergency situations and to countries playing an extremely important political and strategic role in particular regions. At the same time, the Soviet leadership recognized the possibility of the most modern weapons being seized and the USA acquiring them.

Holding the government monopoly over arms exports, the Soviet leadership took the greatest care of Soviet state security. Adequate steps were taken to provide for secrecy where improvements to the technical and combat characteristics of exported weapons and their combat capabilities were concerned.

Soviet arms supplies were kept strictly secret with no exceptions. At the same time, Soviet leaders spread propaganda concerning the peaceful nature of the socialist state through all available channels.

Weapon systems intended for export were produced with due regard to the need to oppose the weapons of enemy forces. While the performance of some types of arms supplied to the Soviet armed forces was better than that of the arms exported, the performance and combat capabilities of exported arms were often made public.

Decisions to reveal or conceal the characteristics of weapons had both advantages and disadvantages. This fact was taken into account in the arms export concept. The Soviet Union did not participate in arms exhibitions or trade fairs. As data on weapons were mostly secret, it was easy to maintain doubts about their actual performance characteristics. Opponents of Soviet weapon exports identified and exaggerated the shortcomings of weapons used in military conflicts, thus reducing their competitive strength in the world market and beating down their prices. That was the case with T-72C tanks in the 1980–88 Iraq–Iran War and the anti-aircraft missile systems used in the 1967, 1970 and 1973 Arab–Israeli wars. It was also the case in the 1991 Persian Gulf War when much was made of the US MIM-104 Patriot anti-aircraft missile system while its Soviet counterpart—the S-300 MMY-1 (or SA-10)—was disparaged. Many other examples may be found. In the cold war period Soviet arms and equipment were regularly presented by the West as the worst in the world with the purpose of forcing the Soviet Union out of the arms market. However, there were some fields in which the Soviet Union lagged behind some other states. For example, it was seriously backward in electronics, miniaturized optical systems and some radio-electronic control systems.

At the same time the analysis of the impact of excessive secrecy on arms exports urged a fundamental revision of the arms export policy concept in the 1980s and 1990s. First the Soviet Union and then Russia became regular participants at military exhibitions, demonstrating the performance characteristics and describing the means of employment of Soviet weapons. Recently Russia has participated actively in arms exhibitions in Chile, China, Turkey, the United Arab Emirates and other countries. This openness has made it possible to clear away suspicions about the quality of arms being exported and, under those conditions, Russia's chances of success in the struggle for markets have become equal to or even better than those of other countries. Many types of arms and military equipment have become known worldwide and won praise in many countries.

The superiority of Soviet weapons was proved by their successful use in military conflicts at various levels and times. The performance of Soviet military equipment in some wars surpassed that of foreign counterparts. Examples include the Kalashnikov and other assault rifles, the ZSU-23 Shilka anti-aircraft system, T-62 and T-72C tanks, some types of artillery, the MiG family of combat aircraft, anti-aircraft missile systems such as the S-75 Dvina (SA-2) and S-200 Angara (SA-5), and several types of diesel-powered submarines and surface ships. Soviet arms proved to be simple in use and highly reliable yet had high performance characteristics.

During the cold war Soviet weapons showed their worth in local wars in Korea and Viet Nam and in the Arab–Israeli wars. Apart from Soviet allies in Europe, North Korea, North Viet Nam and some Arab states were the principal purchasers of Soviet weapons. As a result of close military–technical cooperation, 70–90 per cent of the weapons in service in these countries were Soviet-made. In the cold war years developing countries, particularly in the Arab world, occupied a leading position as markets for Soviet arms exports because they could produce extra profits for the Soviet Union. It is notable that the same countries remain the principal debtors of Russia as the legal successor of the Soviet Union.[1]

Local wars and military conflicts did more than give impulse to an expansion of the arms market. A side-effect was the re-export of Soviet arms to third countries. Using both legal and illegal means, some countries went ahead with re-export of arms, thus gaining political, economic and financial dividends. As

[1] See chapter 4, table 4.5 in this volume. According to Vladimir Belskiy, spokesman for the Africa Department of the Ministry of Foreign Affairs, African countries owe Russia about $20 billion. Researchers at the Russian Academy of Science have estimated that most of this sum is owed for weapons. Klomegah, K., 'Moscow wants back Soviet loans to Africa', St Petersburg Times, 15 Nov. 1996 (in English).

a result, some kinds of weapon—particularly small arms and artillery—appeared in countries banned as destinations for Soviet arms exports. The most intensive and widespread activities of this kind took place during military conflicts and local wars.

Some countries resold weapons exported from the Soviet Union to opposition forces in other countries. For example, China exported Soviet weapons to the Zimbabwe African National Union (ZANU), the National Liberation Front of Eritrea (NLFE) and the Tigray People's Liberation Front (TPLF) in Ethiopia. The opposition forces of Sudan received weapons from Ethiopia. The opposition in Chad bought arms from Libya and Sudan. The Arab countries supplied weapons to Sudan. As recently as 1995 Ecuador used Soviet weapons bought from Nicaragua in its war with Peru.

The cold war brought about the militarization of many countries. Some countries of Africa, Western Europe and the Middle East that had no armed forces of any significance became involved in modernization during the cold war, formed national armies and provided them with modern arms and military equipment. Military conflicts and local wars that were unleashed in various regions of the world subsequently accelerated the process. During the 1960s and 1970s entire regions became heavily militarized—for example, in Europe, the Middle East and parts of the Asia–Pacific region. Many of these countries bought their military equipment from the Soviet Union.

The intensive militarization of countries and the creation of militarized zones through Soviet arms exports was not a one-sided phenomenon. The scale of arms export activity was proportional to the threat that a particular state faced as estimated by specialists. For example, Viet Nam armed itself in response to US aggression while the militarization of the Arab states was a response to the build-up of armed forces in Israel.

**Strategic factors**

The suppliers of arms took account of military–strategic factors. For example, another important reason for militarization in the Middle East was the fact that the Arab and in particular the Persian Gulf states were the main source of oil and natural gas. The struggle for control over sources of raw materials, including threats of and even the actual use of force, intensified there.

The main and decisive consideration in this field was to prevent the security of the Soviet Union and socialist countries from being undermined. Requests for Soviet arms were thoroughly studied and analysed at the Ministry of Defence so as to avoid concentrations of weapons near the borders of the WTO and to rule out the possibility of aggression against Soviet allies. Each type of arms could be exported strictly on the condition that the requesting country was included on the list of regions allowed for the export of a given system.

Arms were also exported with due regard to the interests of the WTO countries, including their security, the impact on their military potential and relations with third countries that wished to purchase Soviet weapons. In some cases the

Soviet Union asked an ally for advice on a possible arms export to a certain country. If the ally disapproved of the export and saw it as a threat to its security, the advice not to export was followed.

In the 1970s and 1980s wars in various regions became an integral part of the world military–political situation, mostly in Africa, Asia and the Middle East. Wars in Central and Latin America were less frequent. The Soviet Union provided military aid to many countries engaged in wars since its military policies and those of the United States were directed at expanding their spheres of influence and creating a favourable political and military–strategic situation.

Supplies of military equipment were intended to provide favourable conditions in various theatres of war in case a world war was set off or to produce favourable military conditions in a local war should the forces of the Soviet Union or other socialist countries become involved. The equipment needs of states with an important strategic and geopolitical position had high priority. Military cooperation with armies of other countries (including training of their personnel) promoted favourable conditions for the development of the Soviet armed forces and for improving arms and military equipment.

The cold war years were marked by an active struggle by the Soviet Union and the United States for military and political superiority both at the global level and in particular theatres of potential conflict. The performance of military equipment supplied to one or other country by the Soviet Union was compared with the performance of analogous systems used in the armies of potential enemies. An important intelligence priority was therefore to know more about the types of weapon used by the customer's enemies. As the need arose, measures were taken to supply new arms of better quality and higher performance or to improve and modernize existing equipment.

The Soviet military leaders were well informed about the military doctrines and military potential of the customer countries as well as those of potential enemies. In the course of official talks the Soviet side might recommend the type and quantity of arms that were most advantageous to the customer. For example, it was often recommended that arms to counter specific capabilities— anti-aircraft, anti-tank and other systems—should be purchased. Requests did not always correspond to recipient countries' strategic needs, and in these cases Soviet military leaders had to persuade the military representatives of customers to take a more rational and well thought-out position.

There were rare cases when high-ranking Soviet officers effectively forced other countries to buy what they recommended. For example, when South Yemen on one occasion requested the supply of a gunship the Soviet military leaders instead forced it to buy a command version of a large landing ship although it did not fit in with the military doctrine of South Yemen.

Usually, however, the customer took the final decision after detailed consideration of all the political, military and economic factors. It was not unusual for the customer to advance arguments to reduce the prices for weapons. To help in the correct selection of arms for export, the weapons were demonstrated to a delegation of representatives from the customer. Such demonstrations allowed the customer to evaluate the performance of the equipment including advantages and limitations, service conditions in various types of climate, and principles and methods of its use in combat. As the need arose (at the request of the customer) the Soviet side demonstrated the functions of the equipment, for instance, by live firing or launching of missiles. Representatives of the community of socialist countries were invited to attend numerous live-fire exercises and manoeuvres.

Experts of both the sellers and the buyers studied the following aspects of any transfer thoroughly: the need for a certain type of arms; the conditions for maintenance; and combat use in the given region or theatre of war. Modifications and improvements were introduced when needed. If the purchasing country was unable to provide long-term service and repair of the arms delivered, the Soviet Union also provided the necessary military–technical assistance. Systems were chosen to ensure that the repair of equipment exported was not too complicated, could be done quickly and required only small quantities of spare parts. Exports also involved exporting the means to repair, test and adjust equipment. Some countries were helped with building repair bases for medium and major repairs of the arms purchased after the expiry of the overhaul period stated in the agreement. In considering the abilities of the recipient properly to master and use arms supplied one factor was the possibility of training personnel either in educational establishments in-country or in the Soviet Union.

Arms for export were selected with regard to the capabilities of the customer to standardize and support them. In particular, attempts were made to maintain consistency with the armament and ordnance of the customer's armed forces.

Circumstances are particularly favourable for selling arms after one side is defeated in a war and takes measures to strengthen its military power by, among other things, buying new arms and military equipment. The Soviet Union and the USA took that fact into account when supplying arms to other countries.

## Economic factors

The arms and equipment that were in most demand by foreign recipients were also prominent in Soviet state orders to manufacturers. In this way arms exports supported state defence orders which, in turn, made it possible to maintain highly skilled personnel and develop new weapon models. The powerful military–industrial complex of the Soviet Union had an overall positive effect on the economy of Russia. Its achievements in the military industrial field were promptly introduced into heavy industry. Winning markets for armaments also allowed the Soviet Union access to sources of strategic raw materials.

Estimates of the financial solvency of the partner were closely connected with political and military–strategic considerations. The prices for transfers which could bring political benefit to the Soviet Union were reduced. However, this was compensated for by prices charged to partners which were not considered so important. Sales opportunities were studied thoroughly. The financial reserves, raw material base and other material resources of the customer were estimated. In evaluating the transfer one thing to be found out was whether the partner could pay for military articles promptly when the deal was made or whether a period of grace was needed.

The decisive consideration in the economic estimation was to maximize profits and to cover the costs of manufacturing, expanding production and transport and other costs, regardless of the socio-political system of the customer. This approach could be modified if there were particular reason to take another view. Prices for exported equipment were based on world prices for that kind of product. Some commercial methods were used to maximize profits, such as the granting of credit, barter and trade through third countries or through agents. Sometimes obsolete arms and military equipment were transferred at no charge to the acquiring country, but subsequent supplies of spare parts for these items were subject to payment, thus compensating over the long term for the cost of the original material. The desire to obtain maximum profits determined a number of terms and conditions.

In exceptional cases, decisions were made by the Central Committee of the Communist Party of the Soviet Union (CPSU) and the Council of Ministers to export arms at reduced prices or hand them over free of charge. These were minor shipments of arms, most frequently small arms and spare parts. The costs of such supplies were compensated for by increasing the prices of and taking additional profits from exports of more complicated and up-to-date arms. Sometimes compensation came through the state budget. Most frequently, the arms supplied free of charge were destined for armed groups struggling for power and those representing national liberation movements.

As manufacturing enterprises belonged to the state, priority was given to state interests in allocating revenue. Profits from exports were put into the state budget and made up one part of state revenues but a small percentage was used to cover the costs incurred by state organizations and ministries during the process of managing the export: for example, the costs of modifying weapons for export, carrying out servicing and repair, transporting weapons, and providing escort and protection could be directly recovered. Orders were allocated with regard to enterprises' workload and the employment of labour resources.

Particular attention was paid to obtaining profits to be used for scientific research and experimental designs. These and other economic factors were thoroughly analysed and evaluated prior to signing a contract.

The trading organizations that were directly involved in arms trade (which are described more fully in section III of this chapter) were interested in selling equipment at higher prices. They received interest from earnings to compensate for their expenses during the conclusion of contracts and as a result they were interested in larger volumes and values of supplies.

The manufacturers of export-oriented military items and their supervising ministries were not directly engaged in the arms trade. They received a state order to produce arms and military equipment and finance from the government to fulfil it. It was not important for the manufacturer or the ministry whether the article produced was to be used by the Soviet armed forces or exported abroad.

Military articles manufactured were always paid for by the state in Soviet roubles. Neither the manufacturers nor the ministries had any role in looking for cooperation partners.

The interest of the manufacturer was in timely production of high-quality articles in order to receive stable payments from the state. The volume of military orders to the manufacturer determined the level of payments: more money was given for larger orders and, correspondingly, minor orders meant less money and lower profits. Manufacturers producing military articles, particularly those making finished items, were always overloaded with orders. This was the socialist method of managing production, with its positive and negative sides.

The assertion that the Soviet Union supplied a substantial or even a greater part of its military equipment free of charge—in particular to the socialist countries or to countries which declared their determination to build socialism—is not true. The fact that the majority of deals were not free of charge is demonstrated by the tens of billions of dollars which customers still owe Russia as the legal successor of the Soviet Union. Some countries did, however, receive free of charge some samples of arms or individual weapons given as presents.

The WTO countries were not supplied with weapons free of charge. Each package supplied was backed by a contract that stipulated all the conditions of the deal. However, it is true that the prices established for the weapons were only to recover the primary costs and that cash payment in foreign currency was rare. To a great extent those deals were of a barter nature, weapons being exchanged for industrial and agricultural goods. It is also the case that debts were supposed to be paid over long periods of time and in some cases were remitted.

The debts that Russia now owes to the countries of the former WTO are good evidence of the fact that all supplies were backed by contracts. At the time when the WTO was dissolved not all weapon supplies for which payment had been made in advance were complete. Upon the dissolution of the WTO those countries which had not received deliveries declared that Russia, being the legal successor of the USSR, had to repay the outstanding debts. Under the same system of accounting some former members of the WTO owe debts to Russia.

As far as other socialist countries are concerned, most of the supplies were of a credit nature and aimed at developing international economic cooperation. In addition the credits were used as a tool of economic and political influence on the debtors. However, some transactions based on financial compensation also took place with these states.

Commercial credit was the basic means of settling the accounts for weapons and combat *matériel* supplies to Cuba. Credits were granted to Cuba for periods that allowed repayment over 10–15 years and even on occasion 20 years with interest at 1 or 2 per cent per annum and with a price discount. There were certainly favourable conditions in the trading of weapons, but these weapons were not given to Cuba free of charge. Barter was widely used as well. Soviet ships carried military products to Cuba and took back Cuban sugar, citrus fruits, coffee and other goods.

The presence of Soviet military advisers and specialists in Cuba and the training of Cuban military personnel in military institutions of the Soviet Ministry of Defence were also governed by contracts. This kind of assistance was paid for either in hard currency or by supplies of consumer goods from Cuba.

Another of the socialist countries, North Korea, paid for Soviet military supplies. During certain periods of the Korean War supplies were sent with a 50 per cent discount and repayment was by instalment.

The prices of weapons supplied to North Korea were not constant over time and during the period 1947–50 were changed several times. The prices for small arms and artillery were set in March 1947 but new prices were introduced in January 1949. Prices were doubled for mortars, increasing by 180 per cent for 76-mm calibre towed guns and 190 per cent for 76-mm calibre self-propelled guns. In 1950, the following prices were in effect: for a 120-mm calibre mortar, 10 200 roubles; for a 76-mm calibre towed gun, 23 050 roubles; for a 122-mm calibre howitzer, 54 100 roubles; for a 95-mm calibre anti-aircraft gun, 93 600 roubles; for a 7.62-mm calibre rifle, 266 roubles; for a 7.62-mm calibre carbine, 289 roubles; for a projectile for a 76-mm calibre gun, 103 roubles; for one round for a 122-mm calibre mortar, 221 roubles; and for a round for the main gun of a T-34 tank, 224 roubles.

In the period 1947–49 the prices for aircraft changed three times. In 1949, they were: for a Il-10, 641 500 roubles; for a Yak-18, 183 500 roubles; and for a PO-2, 56 726 roubles. In 1950 the cost of one T-34 tank ranged from 142 000 to 197 000 roubles depending on where it was manufactured.

North Korea paid for part of its weapon supplies in gold. It also supplied scrap lead and lead concentrate to the Soviet Union. On 26 March 1951, Soviet Prime Minister Joseph Stalin received a coded cable from Kim Il Sung, head of the North Korean communist administration, with the information that 1710 tonnes of lead had already been sent to the Soviet Union—210 tonnes more than planned under the agreement. In addition, Kim Il Sung assured Stalin that 5500 tonnes of lead would be dispatched to the Soviet Union by August 1951.[2]

At the present time no debt-repayment formula has been found for some countries and several are trying by every possible means to avoid paying their debts to Russia.

*Forms of payment*

The form of payment for equipment exported also depended on conclusions made in the course of analysing the political and military–strategic factors. The specific forms of payment varied.

The forms of payment for export supplies were: (*a*) cash payment in US dollars promptly on delivery; (*b*) payment by instalments; (*c*) payment by barter; (*d*) part payment in cash, either in dollars or in the local currency of the buyer, on delivery and part payment by barter; (*e*) payment against credit;

---

[2] Letter of Kim Il Sung to Joseph Stalin, 26 June 1951, reproduced in *Cold War History Project Bulletin* (Woodrow Wilson Center for Scholars: Washington, DC, winter 1995/96), p. 63.

(f) part payment by instalments in cash, part on credit and part payment by barter; and (g) various combinations of these forms of payment depending on what the purchasing country could afford and how profitable the deal was for the USSR.

The form of payment was chosen to achieve a number of goals. Political, military and economic factors were an essential basis for this choice and included: (a) the level of development of the customer country; (b) the prospect for a long-term partnership; and (c) the aims of the customer in buying military equipment.

Each deal or contract took account of the direct profit and was normally analysed by economic experts to this end. If the results of such an analysis were unsatisfactory, the deal was rejected in its existing form and appropriate changes and additional terms were introduced into the contract by agreement between the parties.

The starting-point for setting prices for exported arms was the total cost of manufacturing, demonstration and transport and the profit required to replenish the state budget. The cost of each item to be supplied was negotiated with the military cooperation partner and entered in the contract. Where arms and spare parts were obtained from the Soviet Ministry of Defence, the price included the costs of preparing them for export, adapting them for export, transport and delivery.

## III. Decision-making procedures in the Soviet Union

In the Soviet Union during the cold war period the monopoly in trade in weapons belonged to the state. All questions and transactions related to weapon exports were handled only by state organizations. The decision-making system was intended to ensure that decisions strictly reflected the policy and ideology of the Soviet state. The system forbade trading organizations from exporting weapons independently.

At the same time, the practical measures for arms exports had to be flexible enough to be easily controlled given the large number of applications for Russian armaments.

The majority of applications and requests from recipient states were received by the Political Bureau (Politburo) of the Central Committee of the CPSU, which exercised direct rule over all state bodies of the country. It was stipulated that in the countries of the socialist commonwealth political authority also belonged to the communist or workers' parties. Hence it was natural for their leaderships to address such requests, as a rule, to the Politburo. Some applications were received directly by one or other individual member of the Politburo. These were considered as official and as far as possible were satisfied. Finally, applications were sometimes received by parts of the government—the Council of Ministers of the USSR—such as the Ministry of Foreign Affairs and state organizations which traded in weapons. The Ministry of Defence took a special place in the reception of applications.

Delegations from foreign countries, especially developing countries, as a rule brought with them applications for purchase of weapons. Requests for purchase of weapons were almost always addressed to the Minister of Defence, who had no rights to export weapons but transmitted the requests to the government or to trading organizations.

There was a strict order for decision making on exports of weapons. The Politburo decided the list of countries to which export of weapons was authorized and the categories of weapons permitted for export. Separate decisions were required from the Politburo concerning countries not already on the list.

Each specific agreement with a given country was issued either in the form of a decree of the Central Committee of the CPSU and the Council of Ministers together or, under the instruction of the Central Committee, by the Council of Ministers alone. All decrees were first coordinated in an inter-agency body including the State Planning Commission (Gosudarstvennaya planovaya komissiya, Gosplan), the Ministry of Foreign Affairs, the Ministry of Defence and trading organizations. The Minister of Foreign Affairs and Minister of Defence were as a rule members of the Politburo.

The Military Industrial Commission of Gosplan participated directly in preparing the decrees of the Central Committee and of the Council of Ministers on exports of arms and military equipment. As Gosplan directly supervised defence industries, it decided the timetable and terms for manufacturing and delivering arms and military equipment for export.

In the period between 1950 and 1970, and especially in the years of maximum export deliveries to socialist and developing countries, the state order for industry was planned by Gosplan and the Ministry of Defence together. The order consisted of two parts. The first and main part of the order was to support the needs of the Soviet armed forces and to create stocks that could be required in case of war. This part of the order was financed from the state budget using the resources allocated for defence. The other part consisted of arms and military equipment for export. The share of export deliveries in the overall state order was between 3 and 12 per cent at different times. This part of the order was partly financed from the state budget and partly from the budget of trading organizations (at different times, the Central Engineering Directorate (Glavnoye inzhenernoye upravleniye, GIU), the Central Technical Directorate (Glavnoye tekhnicheskoye upravleniye, GTU) and the Central Directorate of Collaboration and Cooperation (Glavnoye upravleniye po sotrudnichestvu i kooperatsii, GUSK). Using the arms allocated under the state order, trading organizations accumulated stocks of arms and military equipment at warehouses from which items were delivered for export. The trading organizations were obliged to repay to the state budget the equivalent of the resources used in the manufacture of arms after the completion of a sale.

In cases where it was impossible to manufacture arms for export by the time fixed, equipment could be taken from the stocks of the Ministry of Defence by agreement between the ministry and Gosplan. These items were compensated for by new production from industry—consisting of the latest models.

Spare parts, fittings and some particularly sophisticated weapon systems were delivered by the Ministry of Defence through trading organizations. From the profit on such transactions 1–5 per cent was deducted to meet the costs of the trading organizations. Of the rest, the Ministry of Defence was obliged to sell 50 per cent of any currency received to the state at the appropriate rouble exchange rate. The money received by the Ministry of Defence—either in foreign currency or the rouble equivalent—was spent essentially on improvement of the living conditions of personnel, the construction of housing, and the development of repair and workshop facilities in the main and central directorates of the armed forces.

The Ministry of Foreign Affairs had information about the foreign and security policy intentions of the state purchasing the armaments and drafted conclusions about the course of Soviet foreign policy concerning that country.

The Ministry of Defence participated in the decision with the intention of preventing the creation of force groupings in various regions or countries which could threaten the military security of the Soviet Union and other countries of the socialist commonwealth. The ministry compared the tactical and technical characteristics of the arms exported with the weapons and equipment of the Soviet armed forces: they should not be better than those of weapons adopted for service in the Soviet armed forces. The export of the most modern models of equipment was therefore not, as a rule, permitted. The Ministry of Defence also ensured that equipment exported to one end-user was not re-exported to countries that had not been authorized to receive them. For this purpose an end-user certificate was required from buyers.

The decree or decision adopted usually specified which organizations, departments and ministries were responsible for different aspects of its fulfilment.

Only such a decree or decision could be the basis for foreign trade activities involving the export of arms and military equipment. Without one, orders from trading organizations for the transport and support of equipment were not accepted, the Ministry of Defence would not issue an end-user certificate, and weapons would not be allowed to cross the border by customs and the border security services. In this way there was a harmonious and clear system for taking decisions which excluded any unauthorized deliveries by trading organizations, factories and manufacturers or any other organization, department or ministry.

Trading organizations concluded contracts with the buyer countries based on the decrees and decisions reached by the Central Committee and the Council of Ministers. These agreements could be either standard contracts or more complex contracts. Standard contracts referred to a single act of purchase and sale without additional obligations or conditions.

Most standard contracts were concluded with countries of the socialist commonwealth. Many military experts and officers from these countries had received training in the educational system administered by the Soviet Ministry of Defence and were familiar with many terms and concepts. This also eased the use of Soviet troops on the territory of allies to help master arms and military equipment. To help allies master and use effectively complexes of arms and equipment such as rocket battalions or anti-aircraft missile regiments, the Ministry of Defence, on the basis of a decision of the government, established educational centres and training grounds with their own equipment on the territory of the USSR. Military units from socialist countries held exercises and live firing.

In this way standard contracts were transformed into a system of managing mutual relations in the field of military–technical cooperation with allies. This permitted not only standardization of equipment systems but also common practices for supporting arms and equipment. This, in turn, simplified the system of combat maintenance and supply and promoted the accumulation of stocks of war *matériel* necessary for defensive and offensive operations.

Complex contracts for deliveries of arms and military equipment were concluded with developing countries. These included contractual obligations for maintenance in combat-ready condition, the supply of spare parts over a long time-period, the sending of military advisers and experts to the buyer, the training of personnel in the military colleges of the Soviet Ministry of Defence, the transfer of special and technical literature, and the development of basic repair facilities.

To maintain arms and military equipment in combat-ready condition it was necessary to supply ammunition, petroleum, oil and lubricants, and test and tuning equipment. In order for the receiving country to be able to master the weapons quickly, Soviet military advisers and experts were sent who were able to teach a domestic cadre to exploit, use and maintain arms and military equipment directly in combat units. In a more long-term perspective, these advisers helped create training units and training colleges.

A special part of the complex contracts regulated deliveries of spare parts. Spare parts were delivered as part of a repair complex along with weapons. However, conditions for deliveries over a long period were stipulated in the contract. The dependence on spare parts 'attached' the buyer country to the Soviet Union in its military–technical cooperation. Deliveries of spare parts were complicated because production was not as profitable for industry. These difficulties were overcome and spare parts were delivered regularly in accordance with orders.

Complex contracts were expedient as they allowed profits to be received directly at the moment of delivery and also against future production.

Industrial enterprises, design bureaux and the Ministry of Defence Industry, which all received allocations from the state budget, played no role in trading in weapons and were forbidden to do so. In the case of design bureaux there were more rigid regulations specifically for them. Many designers had no right to go abroad, in order to prevent any outflow of this type of information.

The management of work at enterprises which produced arms and military equipment was carried out by ministries and departments which, taken together, constituted the military–industrial complex. Nine ministries were principally involved: the Ministry of Air Industry, Ministry of General Mechanical Engineering (Minobshchemash), Ministry of Defence Industry (Minoboronprom), Ministry of Mechanical Engineering (Minmash), Ministry of Radio Industry (Minradioprom), Ministry of Communication Industry (Minpromsvyaz), Ministry of Electronic Industry (Minelektroprom), Ministry of Shipbuilding Industry (Minsudprom) and Ministry of Light Mechanical Engineering (Minsredmash). Some military production was also undertaken by enterprises subordinated to the Ministry of Heavy Transport Mechanical Engineering, Ministry of Road and Municipal Mechanical Engineering, Ministry of Tractor and Agricultural Mechanical Engineering and Ministry of Electrotechnical Industry. These ministries executed the state order for industry, which they received from Gosplan as far as it affected military production through their management of enterprises. They were also responsible for coordinating orders placed with other ministries for parts used in manufacturing military equipment such as electric motors, storage batteries or measuring and control devices.

### The evolution of administrative arrangements for arms exports

The direct administration of Soviet weapon exports was carried out by specially created trading organizations. These had their own history of development, and during the cold war they were powerful organizations in their own right.

The first special division for military–technical cooperation was developed in 1921—the Department of External Orders of the National Commissariat on Military and Naval Affairs. It directed the activity of engineering departments that were attached to the trade agencies of the Russian Soviet Federal Socialist Republic (RSFSR) abroad. These departments carried out purchases of military and other property on behalf of the Soviet Government. In 1939 the Department of External Orders was brought within the structure of the National Commissariat of Foreign Trade under the general title 'Engineering'. During World War II the Engineering Department supervised lend-lease deliveries of military equipment from Canada, the UK and the USA; later it administered the credits given to the Soviet Union by its allies within the anti-Hitler coalition. In 1942 it became the Engineering Directorate and its size and resources were increased as required by the increasing amount of work.

During the cold war the volume of export deliveries of arms and military equipment, the number of different types of equipment exported and the quantity of services rendered to foreign countries for military purposes increased greatly. In this context, by an order of the Council of Ministers of the USSR of 8 May 1953, the GIU was created on the basis of the Engineering Directorate of the Ministry of Internal and Foreign Trade. The GIU consisted of specialized divisions which could decide complex questions of military–technical cooperation professionally. In 1955 it was included in the structure of the newly formed Central Directorate of Economic Cooperation with Countries of Socialist Democracy. In 1957 it became a division of the State Committee for External Economic Cooperation of the Council of Ministers.

After World War II the GIU executed the reparations required of Germany and deliveries of military property, arms and military equipment to the armed forces of the European socialist countries.

In 1950–60 construction for military purposes in foreign countries was considerably expanded. Airfields, training grounds, educational centres and repair factories were usually lacking and were built with the help of Soviet experts and equipment. In 1968 the GTU was created on the basis of some of the divisions of the GIU in order to manage this work. Also in 1968, GUSK was set up to improve cooperative links with the members of the WTO on questions of joint manufacture of arms and military equipment and their standardization.

In 1988, within the framework of the Ministry of Foreign Trade and the State Committee on External Economic Relations of the Council of Ministers, a new body was created, the Ministry of Foreign Economic Relations (MFER), incorporating the GIU, the GTU and GUSK with all their rights and responsibilities.

## IV. Military supplies to socialist countries

The socialist countries, especially the WTO countries, were the leading recipients of Soviet weapons and combat *matériel*. They enjoyed a privileged place in the military market of the Soviet Union. This could be explained by the political basis of the coalition—the protection of common interests in the event of combat action through joint efforts. An aggression against any member of the WTO was considered to be an aggression against all its members. The WTO member states were equipped with this in mind and, in addition, a considerable part of their forces and means was allocated to a joint command which would conduct a military operation. Large strategic formations consisted of units from different countries of the treaty organization. All member states had standard weapons and combat *matériel*, which simplified the procedures of command, logistics, maintenance and manufacture of spare parts.

All states participating in the manufacture of weapons and combat *matériel* adhered strictly to standardized production based on cooperation in manufacture. Assembly plants in member states were supplied with sub-assemblies and units (such as case blanks, engines, weapon systems and communication facilities) from other treaty partners. The weapons were mostly Soviet designs.

As a rule, cooperatively produced weapons were exported to other socialist countries. Export to a developing country required the exporter to get an export licence from the original manufacturer. The countries licensed to export Soviet-made weapons were Bulgaria, Czechoslovakia, the German Democratic Republic, Romania and, to a lesser extent, Hungary. Some of these states also re-exported surplus weapons and combat *matériel* bought from the Soviet Union.

Production, co-production, supply, export and re-export were integrated in a common military–industrial complex. Policy for this complex where weapons and combat *matériel* were concerned was worked out in the Headquarters of the

Joint Armed Forces of the WTO. The profits earned through exports were spent on the national needs of the exporter.

Cooperation in manufacturing enabled WTO member states to develop new types of weapon and replace one generation of weapons with another quickly and reliably. Although they enjoyed a priority position, the socialist countries were never supplied with weapons of mass destruction. They were, however, in possession of the means of delivery—missile complexes and aircraft—of tactical nuclear weapons. Nuclear ammunition was in the custody of Soviet armed forces stores and was to be issued only in extraordinary circumstances with special permission and approval from the highest political level.

Exports of weapons and combat *matériel* to socialist countries outside the WTO were handled in almost the same way. However, with these countries there was no cooperation in the manufacturing of weapons and less attention to standardization.

Newly independent socialist countries received large shipments of all types of weapon and combat *matériel* from the Soviet Union.

The Korean People's Army (KPA) was founded in 1946 and the Soviet Union then began deliveries of weapons and combat *matériel*. In 1947, 17 362 rifles and carbines, 5816 sub-machine-guns, 268 mortars and 234 artillery pieces were supplied to North Korea. On 1 June 1949 the KPA had 36 622 rifles and carbines, 345 mortars, 352 artillery pieces, 64 tanks and 48 combat aircraft. On 1 January 1950 these numbers had increased to 43 371 rifles and carbines, 442 mortars, 515 artillery pieces, 151 tanks and 89 combat aircraft. The Soviet Union also helped North Korea establish a small fleet which included 3 hunter-killers, 5 torpedo boats, 3 minesweepers, 6 patrol boats, and 60 schooners and launches. By March 1950 the ambassador of the Soviet Union to North Korea reported to Moscow that the KPA had been fully equipped with Soviet weapons and combat *matériel*.

The Soviet Union also played a great role in creating the Chinese People's Liberation Army (PLA). In 1949 the PLA received 360 anti-aircraft guns (sufficient to equip 10 anti-aircraft artillery regiments), 332 aircraft, 32 radio stations of different types, 130 telephone sets and 196 parachutes. In February 1950, Mao Zedong, Chairman of the Chinese Communist Party, sent a letter to Stalin in which he asked for further supplies of weapons and combat *matériel*. On 27 February 1950, Marshal of the Soviet Union Nikolay Bulganin discussed this request with Chinese Prime Minister Zhou Enlai and, in March 1950, it was decided to supply China with naval *matériel* in the third quarter of the year. This included 4 minesweepers, 52 patrol boats and naval aircraft worth a total of 460 million roubles. Chinese air divisions were equipped with Soviet-made aircraft, specifically the MiG-15, La-9, Il-10, Il-28 and Tu-2. Also in March 1950 the decision was taken to send China 450 aircraft (a single delivery of 184 aircraft was made in June 1950). In May 1950, 235 railway wagons loaded with ammunition and 92 with spare parts were sent to the PLA.

A great deal of assistance was given by the Soviet Union to countries of the socialist community during wars and military conflicts in which they were engaged. North Korea and Viet Nam are examples. In both cases the Soviet Union provided not only arms but also other forms of military assistance, including the direct participation of Soviet forces in combat roles. In exchange it received information about the performance of its own systems in combat (which was valuable for improving their capabilities). It was also very interested to receive examples of US equipment captured during military operations.[3]

On 25 June 1950, North Korea attacked South Korea and the Korean War began. It lasted until July 1953. Documents now available show that the war efforts of North Korea and of the Chinese forces which later assisted in combat operations were almost entirely underwritten by the Soviet Union. North Korea fought with Soviet-made weapons. The quantity and types of weapons and combat *matériel* supplied corresponded to the objectives of the war and to the specific requirements of the combat actions anticipated. Military supplies were also needed to compensate for the great losses suffered by the KPA and materials for maintenance of equipment in the field, such as oil and lubricants, were also sent from the Soviet Union.[4]

During the first stage of the war (25 June–24 September 1950) North Korean units carried out successful offensive operations, seizing Seoul, the capital of South Korea, and inflicting a heavy defeat on US and South Korean forces in the process of reaching Pusan. In this period of the war intensive supplies of Soviet combat *matériel* began. On the 10th day of the war the decision was taken to send North Korea 32 self-propelled guns, 310 mortars, 248 artillery pieces, 84 anti-aircraft guns, 50 000 rifles and carbines, 705 sub-machine-guns, 68 000 mortar bombs, 82 000 rounds of artillery ammunition, 15 000 rounds of tank ammunition and 128 radio systems. On 29 July 1950, 124 aircraft were sent to North Korea and 130 tanks were sent under a directive issued on 4 July. The armed forces of North Korea received equipment of the first rank including the latest model T-34 tank.[5]

Despite the success of the offensive, the KPA was suffering heavy losses, amounting to 40 per cent of its artillery and 50 per cent of its tanks. On 22–24 August 1950 the Soviet Council of Ministers decided to supply weapons and combat *matériel* to North Korea as a matter of urgency. Within the framework of that decision North Korea was sent 110 aircraft, 150 tanks, 100 self-propelled guns, 480 mortars, 674 artillery pieces and 53 anti-aircraft guns.

---

[3] According to a recent volume based on archives of the CPSU. Gaiduk, I. V., *The Soviet Union and the Vietnam War* (Ivan R. Dee: Chicago, Ill., 1996).

[4] By early 1951 tens of thousands of tonnes of petrol, diesel lubricant, brake-oil and grease were being transferred to Chinese and Korean forces each month. Ciphered telegram from Zhou Enlai to Stalin, 16 Nov. 1950, reproduced in *Cold War History Project Bulletin* (note 2), p. 49.

[5] Chang-Il Ohn, 'Military objectives and strategies of two Koreas in the Korean War', Paper prepared for *The Korean War: An Assessment of the Historical Record* (Georgetown University: Washington, DC, 24–25 July 1995), p. 10. [Editor's note. This was a contrast with the equipment initially supplied to the People's Republic of China, most of which was surplus equipment that had been produced during World War II. Only rocket artillery was of the latest generation. I would like to acknowledge Milton Leitenberg and the Cold War International History Project of the Woodrow Wilson Center, Washington, DC, for much detailed information and primary documents related to the Korean War.]

**Table 3.1.** The value of Soviet military aid to North Korea, 1949–51[a]

Figures are in thousand current roubles.

| Year | General | Air Force | Armoured force | Chief Artillery Department |
|------|---------|-----------|----------------|----------------------------|
| 1949 | 249 962 | 195 293 | .. | 51 388 |
| 1950 | 869 677 | 347 757 | 1 238 | 383 164 |
| 1951 | 2 612 822 | 1 182 044 | 179 253 | 881 585 |

[a] Total military aid including aid to the air force, the armoured forces, artillery forces and unspecified recipients.

*Source:* Goncharov, S. N., Lewis, J. W. and Xue Litai, *Uncertain Partners: Stalin, Mao and the Korean War* (Stanford University Press: Stanford, Calif., 1993), p. 147.

During the second stage of the war (25 September–24 October 1950) there was an urgent need to form new divisions quickly and equip them with aircraft, tanks and artillery after US and South Korean forces counter-attacked and reached northern areas of North Korea near the Chinese border. The KPA left 200 tanks behind in South Korea during its retreat. On 27 September the Soviet Minister of Defence, Marshal A. M. Vasilevskiy, contacted Stalin with a proposal to form and equip six new infantry divisions urgently. Stalin granted permission and 600 artillery pieces, 630 mortars, 40 000 rifles and 12 000 sub-machine-guns were dispatched to North Korea. In September–October 1950, 80 fighter aircraft, 20 000 anti-tank mines, 40 000 anti-personnel mines and 100 000 overcoats were supplied.

During the third stage of the war (25 October 1950–July 1953) Chinese volunteers joined combat operations. US and South Korean forces retreated from the territory of North Korea and the armed forces of the warring parties conducted combat operations in areas close to the 38th parallel. In November 1950, the North Korean air forces received 24 Yak-9 fighter aircraft and 15 PO-2 aircraft (used for night-time missions).[6] In the same month Stalin approved the creation of a North Korean air division equipped with MiG-15 fighters and a bomber regiment equipped with Tu-2 bombers.[7] In December 1950, Stalin gave an order to send weapons and combat *matériel* sufficient for nine new infantry divisions. The supplies included 940 mortars, 900 artillery pieces, 59 000 rifles and carbines. In May and June 1951 additional shipments were approved.[8]

[6] Ciphered telegram from the Soviet Military Representative in Beijing to Stalin, 2 Nov. 1950, reproduced in *Cold War History Project Bulletin* (note 2), p. 48.

[7] Ciphered telegram from Stalin to Kim Il Sung, 20 Nov. 1950, reproduced in *Cold War History Project Bulletin* (note 2), pp. 50–51.

[8] The May shipment included 25 000 rifles, 5000 sub-machine-guns, 1200 light machine-guns, 550 medium machine-guns, 275 heavy machine-guns, 500 anti-tank rifles, 700 82-mm calibre mortars and 125 120-mm calibre mortars. Ciphered telegram from Stalin to Kim Il Sung, 29 May 1951, reproduced in *Cold War History Project Bulletin* (note 2), p. 59. The June agreement included equipment sufficient for 16 divisions as well as a commitment that 8 Soviet fighter aviation divisions would be made available to support Korean operations. Ciphered telegram from Stalin to Gao Gang and Kim Il Sung, 13 June 1951, reproduced in *Cold War History Project Bulletin* (note 2), p. 60.

After Chinese volunteers joined in combat actions, the Soviet Union also supplied China with a great deal of weapons and combat *matériel* to support these operations. In November 1950, it sent China 214 railway wagons of small arms and ammunition, 37 loaded with aviation equipment and 1400 with petrol, oil and lubricants.[9]

During the war China suffered heavy losses in aviation and the Soviet Union supplied both aircraft and aircraft engines to compensate. In December 1950, China received 257 aircraft and 360 aircraft engines and in February 1951 an additional 190 aircraft engines. At the end of June 1951 Stalin approved the release of the MiG-15 fighter aircraft to the PLA Air Force (previously the MiG-9 had been the main combat aircraft supplied) and Soviet instructors began retraining Chinese pilots from three fighter aviation divisions to fly these aircraft.[10] During 1951, 13 air divisions, 3 artillery divisions, 2 divisions of rocket artillery, 2 anti-tank divisions, 8 artillery regiments, 3 tank divisions and 59 anti-aircraft artillery battalions were formed on the basis of Soviet combat *matériel*. The Chinese volunteers fought using Soviet tanks and armoured vehicles. At the beginning of the war the Chinese had no tanks, but by the end of it they were in possession of 316 tanks and 75 self-propelled guns. In 1953, the Soviet Government issued China with a licence to manufacture 76-mm calibre guns and 122-mm calibre howitzers.

By 1 August 1952, the equipment of the PLA was dominated by Soviet-made items. In 1952, 1056 artillery pieces of all types were delivered to the Chinese forces. All combat aircraft, 85 per cent of their anti-aircraft guns, 80 per cent of their heavy machine-guns, 60 per cent of their tanks and self-propelled guns, 40 per cent of their mortars and 40 per cent of their anti-tank guns were of Soviet origin.

Throughout 1953 equipment continued to be provided to the Chinese forces. In that year equipment sufficient for 20 infantry divisions was scheduled for delivery along with 1652 artillery pieces of various types (including some self-propelled guns).[11]

After the end of the Korean War, Soviet military advisers helped the KPA to develop a new organizational structure. It required large amounts of weapons and *matériel,* which were supplied from the Soviet Union and Poland. In 1954 North Korean purchases from the Soviet Union included 124 12-mm calibre howitzers, 166 76-mm calibre guns, 16 57-mm calibre guns and 24 120-mm calibre mortars. Later, in 1955–56 a further 192 122-mm calibre howitzers, 144 76-mm calibre guns, 112 57-mm calibre guns, 136 anti-aircraft guns and 160 000 rifles and carbines were supplied by the Soviet Union.

---

[9] The shipment included 140 000 rifles, 26 000 sub-machine-guns, 7000 light machine-guns and 2000 heavy machine-guns along with over 250 million rounds of ammunition. Ciphered telegram from Mao Zedong to Stalin, 8 Nov. 1950, reproduced in *Cold War History Project Bulletin* (note 2), p. 48.

[10] Weathersby, K., 'Stalin and a negotiated settlement in Korea, 1950–53', Paper prepared for the conference on The Cold War in Asia, University of Hong Kong, 9–12 Jan. 1996, p. 23.

[11] Telegram from Stalin to Mao Zedong, 27 Dec. 1952, reproduced in *Cold War History Project Bulletin* (note 2), pp. 79–80.

After the Korean War China also received a large amount of additional combat *matériel* from the 64th Air Corps, which had been deployed on Chinese territory and participated in the Korean War. By 10 December 1954, 296 MiG-15 fighter aircraft and 302 anti-aircraft guns had been transferred.

## Arms transfer decision making during the Korean War

In principle, decisions on military and technical assistance during the Korean War were made in a way that was similar to those in peacetime, by the Politburo—although in practice it was Stalin who had the final word. After the Politburo had made its decisions, decrees were issued by the Council of Ministers, signed by Stalin as the Chairman of the Council of Ministers, and sent to the ministries for implementation.

The decision that North Korea would unleash war against South Korea was taken in March 1950, after negotiations between Stalin, Mao Zedong and Kim Il Sung. Also in March, several decrees of the Council of Ministers pertaining to massive supplies of weapons and *matériel* to North Korea were adopted.

Marshal Bulganin supervised all military matters, including those involving the Korean War, in the Politburo and Council of Ministers. Decisions on several matters seem to have been taken by Bulganin himself. In January 1948, for example, Kim Il Sung sent a request to dispatch ships, tools and training aids for a navy school. The resolution of Bulganin says: 'Allow implementation through the Ministry of Foreign Trade'.

Dealing with military and technical assistance to North Korea were the Ministry of Defence, Ministry of the Navy, Ministry of Foreign Affairs, Ministry of Armaments, Ministry of Foreign Trade, Ministry of Communication, Ministry of Health and others. Each ministry had a deputy minister dealing specifically with supplies of weapons, combat *matériel* and equipment.

Within the Ministry of Defence the Minister of Defence (Marshal of the Soviet Union Alexander Vasilevskiy), Army General Shtemenko, the Chief of the General Staff, the commanders of the various branches of the armed forces and the heads of the main departments of the Ministry of Defence were all involved in dealing specifically with matters of military and technical assistance. However, the greater part of this work was assigned to the General Staff and the office of the Chief of General Staff in particular. Smaller structures dealing with some aspects of weapons and combat *matériel* supplies were established in many different bodies of the Ministry of Defence.

Decrees of the Council of Ministers provided guidance for the General Staff to determine what types of weapon and *matériel* should be supplied to North Korea and from which military districts as well as the terms, procedures and routes of supplies.

Weapons and *matériel* for North Korea were taken from many military districts and plants but, as it shared the border with North Korea, the Soviet Maritime Territory (Primorskiy Kray) Military District (MD) played a leading role. Soviet officers would hand over weapons and *matériel* to the North Koreans at the border points. For example, on 22 September 1950 Lieutenant-Colonel Yuriy Pavlovich Maksimov handed over 47 aircraft of different types to North Korean representative Li Phar in the city of Vozdvizhenka while 32 577 anti-tank mines and anti-personnel mines were handed over by Major Bryantsev to North Korean representative Pak Sobon.

Kim Il Sung usually delivered his requests for weapons and *matériel* via the Ambassador of the Soviet Union and the Chief Military Adviser in North Korea and sometimes addressed his requests directly to Stalin, Bulganin, Andrey Vyshinskiy (Vice-Minister of Foreign Affairs), Andrey Gromyko (First Deputy Minister in the Ministry of Foreign Affairs), and Deputy Minister of Trade Yeremin. Most often, however, requests were addressed to General Shtemenko.

## V. Soviet military assistance to developing countries

The collapse of the colonial system was largely completed during the 1960s and 1970s. The result was the emergence of a large number of independent and sovereign states. During the cold war these countries did not become a 'neutral zone' between the two social systems but rather became the theatre of that war.

The Soviet Union and the United States regarded developing countries as the scene on which the confrontation of capitalist and socialist models of social development could be played out. Political, economic and ideological contradictions between newly independent countries were often the cause of military conflicts and wars. Political orientation in favour of capitalism or socialism was also at the root of some civil wars. In addition to the cold war and decolonization, there were other explanations for why developing countries went to war— competition over resources, materials and territory, and ethnic and religious conflicts.

In any event, many of the hot spots and areas of tension in the world were found among the developing countries. The sharpest conflict was probably the Middle East conflict, which has lasted for more than 40 years. While the hostility between Israel and the Arab countries formed the main axis of conflict it was not the only one. Arab forces were also occasionally used against one another—as in the case of the crisis in Jordan in 1970, for example.

To wage war against one another and to ensure their security, the newly independent states required armed forces, armaments and combat equipment. However, they lacked the resource base, the industrial base and the experience and know-how to produce their own arms and combat equipment. Consequently, those that made their choice in favour of socialism requested the Soviet Union to supply them with arms. Others asked France, the UK or the USA. Some states requested arms from both the Soviet Union and the West. As the number of newly independent states increased, arms deliveries increased. In some of the wars between developing countries both belligerents were using Soviet weapons, for example, in the wars between Algeria and Morocco (1963),

India and Pakistan (1971), Iran and Iraq (1980–88), Ethiopia and Somalia (1978), and North Yemen and South Yemen (1994).[12]

In the years of the cold war East–West hostility was an important factor. If, in some country, a pro-Western orientation emerged, the Soviet Union tried to find a counterbalance among the neighbouring countries. The USA and the other Western countries were pursuing their own political objectives—supporting counter-revolution in countries with a socialist orientation and supporting countries which were oriented to the West.

Having become an importer of Soviet arms, a developing country became 'tied' to the USSR in important ways. Arms deliveries were followed by the expansion of military assistance. In some cases the Soviet Union also sent military advisers and military units to advise the national military command on the national defence or military construction, and advisers were sometimes present even during the repelling of an aggression. The USSR had relations of this type with Algeria, Ethiopia, the United Arab Republic, North Yemen and later Syria. Countries were also forced to set up economic relations to manage payment. Arms exports from the Soviet Union were sometimes cancelled if the armed forces of the importer threatened or waged war against countries friendly to the Soviet Union or might become a threat to the Soviet Union itself.

The Soviet Union was forced to take account of military and political factors—according to or against its wishes—since the collapse of the colonial system was not peaceful but was the source of a number of local wars and acts of aggression. It cannot be claimed that the Soviet Union always had a purposeful, well-thought out foreign and military policy concerning the newly independent states and the conflicts between them. Often political and military activity took place in confused circumstances immediately after liberation from colonial dependence. Often war began by surprise and was recognized too late for policy to be established. Sometimes military action was opportunistic—the simple seizure of available territory. Sometimes subjective factors were decisive: for example, Soviet General Secretary Leonid Brezhnev liked the Ethiopian leader Lieutenant-Colonel Haile Mariam Mengistu and wanted to make him a great revolutionary 'African Castro'.

Military–technical cooperation with the developing countries could also have negative political aspects. Developing countries which used fine rhetoric about building socialism were often insincere. Public opinion was being deceived, the idea of socialism was being discredited and countries of the socialist community ignored these facts or published without criticism information known to be untrue. For example, during the immediate post-colonial time the Central Committee of the CPSU announced that Algeria, Congo, Egypt, Ethiopia, Guinea-Bissau, Libya, Madagascar, Mali, Somalia, Syria, Tanzania, Zambia and other countries were of a socialist orientation. These were never true socialist states.

---

[12] Since the end of the Soviet Union, in the recent war between Peru and Ecuador in 1995 both parties used weapons of Russian origin. However, while Peru was supplied directly by Russia, Ecuador received its weapons through third countries.

Events also changed the political context in which arms transfers took place. On occasion, the Soviet Union was supplying a country with weapons when the recipient changed its political course.

This happened more than once in East Africa. In 1967 the Soviet Union and Czechoslovakia began to supply arms to the government forces of Sudan. In 1969 the reactionary dictatorship in power there was toppled by the Revolutionary Command Council. When the new leaders of Sudan asked the Soviet Union for military assistance, it delivered tanks and BM-21 multiple-rocket launch systems to the Sudanese land forces as well as MiG-19, MiG-21, MiG-23 and An-24 fixed-wing aircraft and Mi-1 and Mi-8 helicopters to the air forces. In addition, S-75 Dvina (SA-2) anti-aircraft systems were provided for air defence.

In 1971 Sudan changed its political orientation once again and pursued a pro-Western course until 1985. During that period the USSR stopped its military assistance. In February 1974 there was a revolution in Ethiopia and the monarchy was overthrown. However, the relationship between Ethiopia and Sudan remained antagonistic. At this time the Soviet Union started to supply Ethiopia with weapons. In all it supplied Ethiopia with more than 1000 T-55 and T-62 tanks, about 1000 anti-tank guided missiles, about 100 MiG-21 and MiG-23 fighter aircraft, 3500 artillery pieces and mortars, about 400 BM-21 multiple-rocket launch systems, 25 ships of different kinds, and more than 10 S-125 Pechora (SA-3) and S-75 Volga (SA-2) anti-aircraft missile complexes.

In October 1969 Siad Barre came to power in Somalia through a military *coup d'état* and declared the Somalian Democratic Republic. Between 1969 and 1977 the Soviet Union supplied the Somali armed forces with equipment including tanks, armoured personnel carriers (APCs), heavy artillery, anti-aircraft artillery, combat aircraft and a range of combat and support ships.

In July 1977 Somalia began military operations against Ethiopia in order to seize control of the vast Ogaden region through which the border between the countries passed. It was intended to incorporate all of the lands where Somali-speaking tribes were living and roaming into a 'Greater Somalia'—an idea that had been maturing in Mogadishu since the 1960s. By accelerating its deliveries of arms and combat equipment from the Soviet Union and with the active help of Cuba, Ethiopia was able to resist Somalia.[13] On 9 March 1978 the Somali Government announced the withdrawal of troops from the Ogaden region.

[13] The governments of Ethiopia and the USSR signed an agreement on arms transfers and military cooperation in Dec. 1976. However, this agreement covered only a limited volume of small arms. Ethiopia requested additional Soviet arms and military assistance in Mar. 1977. Having observed increased military activity along the border with Somalia, it also approached the USA for assistance. The Ethiopian forces relied heavily on US equipment. The USA did not approve exports of spare parts for this equipment, and in late Apr. Ethiopia abrogated the bilateral agreement with the USA on the preservation of mutual security. As a result, Ethiopia approached both North Viet Nam (for equipment and spare parts of US origin) and the USSR (for equipment of Soviet origin). Memorandum of Conversation between Soviet Acting Chargé d'Affaires in Ethiopia, S. Sinitsin, and Ethiopian official Berhanu Bayeh, 18 Mar. 1977; and Memorandum of Conversation between Soviet Acting Chargé d'Affaires in Ethiopia, S. Sinitsin, and Political Counselor of the US Embassy in Ethiopia, Herbert Malin, 9 May 1977, reproduced in *Cold War History Project Bulletin,* issue 8–9 (Woodrow Wilson Center for Scholars: Washington, DC, winter 1996/97), pp. 56–57, 61–62.

Large-scale conflicts as well as small ones were taking place between developing countries, involving all kinds of weapons—aviation, rocket forces, armour and artillery—supplied to the belligerents by China, France, the UK, the USA, the USSR and other countries. In smaller conflicts the warring parties were using, as a rule, small arms, artillery and mortars, and helicopters.

The USSR rarely supplied the least developed countries with complex or expensive weapons and combat equipment. As a rule exports to them involved small arms, light artillery, anti-aircraft missile systems (including light anti-aircraft systems) and aircraft with a small radius of operation. The weapons delivered were simple because their military personnel were not well educated.

## Economic dimensions

In each case the method of payment for the weapons and combat equipment was determined by agreement between governments and specified in a contract.

The prevalent system of payment for weapons by developing countries was with hard currency—usually US dollars. Some countries—Iraq, for example—were both paying some hard currency and supplying the USSR with petroleum. Such orders provided for part payment in order to produce the weapons while the main payment was put into effect on delivery.

Immediately before and in the course of an armed conflict no requests to produce weapons were transmitted to industry because time was needed to produce them. Weapons were delivered from the warehouses or taken from the Soviet Ministry of Defence and later replaced from new production. Because weapon deliveries in war often had an immediate operational task, air and sea bridges were thrown to deliver them. In these cases the price for weapons to be delivered was a little higher than usual.

A second main method of payment was instalments against credit. The deferment of payment for different countries varied and fluctuated between 7 and 10–15 years. The annual rates of interest for deferment also differed and were between 1.5 and 2 per cent. Depending on the rate of interest and the length of payment, credit terms could increase the costs of armaments significantly. For example, weapons were delivered to Syria on the basis of a 50 per cent cost increase over 10 years with interest at 2 per cent per annum.

This method was widely used and credit had a number of positive features. While getting credit from the USSR for military cooperation, developing countries often applied in other areas of cooperation as well. Credit arrangements also created one more condition binding the buyer into the Soviet weapon mar-

The USSR agreed to give logistical support to North Yemen, which began supplying arms and military equipment to Ethiopia from around Apr. 1977. In addition, the USSR agreed that Czechoslovakia, Hungary and Poland would supply military equipment to Ethiopia. Memorandum of Conversation between Soviet Ambassador to Ethiopia A. N. Ratanov and Cuban military official Arnaldo Ochoa, 17 July 1977, reproduced in *Cold War History Project Bulletin* (winter 1996/97), pp. 65–66.

The decision to supply large amounts of heavy weapons to Ethiopia was taken in Oct. 1977 during the visit to Moscow of President Mengistu. Background Report on Soviet–Ethiopian Relations, 3 Apr. 1978, reproduced in *Cold War History Project Bulletin* (winter 1996/97), pp. 90–93.

ket. Finally, a country which bought armaments and combat equipment on credit was usually supportive of socialism and generally favoured the consolidation of the socialist system and the authority of the Soviet Union.

A third form of financing was full or part payment in goods—that is, barter. Barter deals were used not only with countries of the socialist community but with developing countries as well. This allowed the Soviet Union to get scarce goods in exchange for weapons. Sometimes the Soviet Union would also help developing countries to sell goods in other markets as part of a financing arrangement associated with weapon transfers.

As a consequence of difficulties in the political and economic spheres and a deterioration of relations in the military sphere, Russia has suffered economic and financial costs. It suffered a great deal of damage as a result of the disintegration of the Soviet Union because debts which were being remitted were questioned by many developing states. For example, Ethiopia still owes Russia $6.2 billion, Libya $3 billion, Syria about $4.2 billion and Zambia $300 million.

Under current conditions a number of developing countries are refusing to pay their debts or resorting to tricks in order to evade payment, justifying this by referring to current relations in the area of military–technical cooperation or by questioning the formula used in the calculations. Some do not want to continue cooperation with Russia but are applying to other weapons markets.

There has been a sharp reduction in the number of military advisers and specialists sent on official trips to countries which buy Russian weapons. Neither educational and technical materials, services for repair and maintenance nor bases to carry them out are being created any longer. The number of foreign persons studying in the colleges of the Russian Ministry of Defence has been reduced. All of these were formerly sources of revenue both for Russia and for its Ministry of Defence.

## The main recipients of Soviet arms

Large deliveries of weapons and combat equipment were sent from the Soviet Union to Angola, Egypt, Ethiopia, India, Iraq, Libya, Somalia, Syria, and North and South Yemen.

During the crisis which preceded the 1956 war between Egypt and British, French and Israeli forces, the Soviet Union made an emergency delivery of weapons to Egypt.[14] After the war Egypt paid increased attention to the creation of national armed forces and sought help from the Soviet Union. Its land forces were supplied with T-54, T-55 and T-62 tanks as well as the Luna-M (FROG-7) tactical rocket system, BM-21 and BM-24 multiple-rocket launch systems, and Malutka portable anti-tank missile systems. The Kvadrat anti-aircraft missile system and the Shilka self-propelled anti-aircraft gun were provided for air defence. The air force received Tu-16 bombers, MiG-17 and MiG-23 fighter aircraft as well as An-12 and Il-14 transport aircraft and Mi-4 Hound transport

[14] Engelmann, B., *The Weapons Merchants* (Elek Books: London, 1968), pp. 174–75.

helicopters. The air defence forces were supplied with S-75 and S-125 anti-aircraft missile systems and portable Strela-2 anti-aircraft missiles. The Egyptian Navy was supplied with diesel-powered submarines, destroyers, landing ships, motor torpedo boats and patrol boats. In all, Egypt received more than 2000 tanks, 5000 APCs, 21 tactical rocket systems and 14 submarines.

The USSR also supplied Syrian land forces with a wide range of equipment including tactical missiles, tanks, artillery and mortars, APCs and Strela-1 (SA-7) portable anti-aircraft missiles. The Syrian air forces and air defence forces were armed with Soviet aircraft, helicopters, anti-aircraft artillery and missile systems while the naval forces received warships, fast patrol boats, mobile coastal defence missile systems and helicopters equipped to attack ships. By 1992 Syria had been supplied with about 5000 tanks, more than 1000 aircraft, 4000 artillery pieces and mortars, and 70 combat and support ships.

Soviet arms deliveries to Egypt and Syria were at their height in 1973, when the war against Israel was going on. When Syria suffered heavy losses, the USSR helped to restore and update its forces by sending military advisers and specialists and delivering weapons and combat equipment in 1974. The Syrian armed forces were not only restored but increased in numbers and quality: for example, Syria was supplied with new T-62 tanks and Su-7 fighter aircraft.

Libya bought many modern weapons in the USSR, including about 200 tactical missiles, 4000 tanks and about 600 aircraft. A large number of air defence systems were supplied including S-200 Angara (SA-5), Kub (SA-6) and Osa (SA-8) missile complexes and six submarines. The proportion of Soviet weapons in the armed forces of Libya is close to 95 per cent.

The USSR supplied South Yemen with weapons for political and military–strategic reasons. South Yemen was always among the most faithful supporters of Soviet positions in the UN—even though this was not always in its own interests—and had strategic significance for the USSR as it made available port facilities for Soviet warships.

Military relations between the USSR and both South Yemen and North Yemen were established in the 1960s. When the leaders of North Yemen declared that they would fight actively against imperialism, the Soviet Union began to supply the country with aircraft, APCs, artillery pieces and mortars, anti-tank guided missiles and small motor boats. Military cooperation with South Yemen started in 1969. The USSR was simultaneously supplying both North and South Yemen with weapons. In 1979 South Yemen began a war against North Yemen with the aim of uniting both parts into one country and Soviet weapons were used in that war. The Soviet Union approved the war and Soviet military advisers remained in both South and North Yemen. Some third countries were reselling Soviet weapons to North Yemen—notably Egypt.

In its 1980–1988 war with Iran, Iraq relied heavily on Soviet weapons. It received far greater supplies than Iran from the Soviet Union and the most modern weapons, including T-72 tanks. However, Iran did receive T-54 and T-55 tanks and other armoured vehicles. Later, after the war was over, MiG-29 and Su-24 combat aircraft and air defence systems were also delivered.

The Soviet Union gave significant military assistance to all branches of the armed forces of India to strengthen its security. Of the developing countries, India received most assistance with arms production. The Soviet Union helped it to build factories where MiG-21 and MiG-23/27 fighter aircraft were first assembled and then produced under Soviet licences. It also helped to build a tank factory designed to repair T-72 tanks.

## VI. Military supplies to political movements and organizations

During the cold war internal wars and conflicts based on clashes of economic, territorial, ethnic, religious and ideological interests between different political forces were not unknown. Self-determination and political power were the main goals of the conflicting parties.

The Soviet Union's aim was to bring to power pro-Soviet, pro-communist, anti-imperialist forces. These forces were supported not only by political means but also through military assistance and in particular by supplies of arms and military equipment. This equipment was used in *coups d'état*, guerrilla campaigns and acts of terrorism.

During the cold war the Soviet Union, the United States and their respective allies were involved—indirectly or directly—in internal conflicts and wars in Africa, Asia, Latin America and the Middle East. Both sides in the cold war assigned a very important role to military instruments of policy.

The Soviet Union supplied arms and military equipment to national liberation movements and organizations that carried on the struggle for national independence and to military opposition groups that shared a position ideologically close to that of Moscow and carried on a struggle for power inside their state.

In the period 1960–80 the stubborn struggle for national independence went on in many colonial possessions in Africa, Asia and the Near East. In several cases it took the form of armed conflict. Western countries that possessed colonies sought to suppress national liberation movements while the Soviet Union did its best to help them.

In the framework of this policy, the Soviet Union supplied weapons to many national liberation movements and political organizations. As noted above, it was often these kinds of supplies that were free of charge.

The International Department of the Central Committee of the CPSU and the Committee of State Security (Komitet gosudarstvennoy bezopasnosti, KGB) scrutinized the policy of ruling regimes in the Middle East, Asia, Africa and Latin America as well as opposition movements, which were divided into communist, social democratic, religious and nationalist tendencies. These analyses were to detect political forces that followed an anti-imperialist tendency and desired cooperation with the Soviet Union and with the Central Committee in particular.

The leadership of the Central Committee exaggerated the extent of revolutionary processes taking place in countries of Africa, Asia, Latin America and the Near East in order to enhance the prestige of the CPSU (especially in resolutions of party congresses). Under pressure from the Central Committee some movements were declared to be revolutionary although they were not. This was so in Afghanistan in particular. Nevertheless, it seems that the Central Committee did not give birth to revolutionary processes but scrutinized processes already taking place in national liberation movements and began to act only when positive developments were identified.

This can be illustrated by some examples.

In 1958, in accordance with a UN General Assembly resolution, the Federation of Ethiopia and Eritrea was proclaimed. Eritrea was to have its own parliament and autonomous administrative organs. The Emperor of Ethiopia, Haile Selassie, in defiance of the decision of the international community, began to curtail the rights of the Eritrean population. In 1962 the deputies, under pressure from the emperor, decided on a full merger of Eritrea with Ethiopia. Opposition groupings appeared in Eritrea. Christians and Muslims were united by the struggle against the central government. Opposition to the leadership was combined under the NLFE and the Eritrean People's Liberation Front (EPLF). Before the overthrow of Selassie the USSR supplied arms to the opposition forces through Egypt, Sudan and South Yemen.

As early as the late 1950s movements were formed in Angola to fight colonialism, the largest being the National Front for the Liberation of Angola (FNLA), the Popular Movement for the Liberation of Angola (MPLA) and the National Union for the Total Independence of Angola (UNITA). The leading role in the struggle against Portuguese colonialists belonged to the MPLA, which was the strongest and best organized group. From 1964 the Soviet Union began to supply weapons to that movement.[15] In January 1975 a transitional government was formed with the participation of the MPLA, the FNLA and UNITA. However, the FNLA and UNITA—with the support of the USA and the Republic of South Africa—attempted to deprive the MPLA of participation in the government. In the resultant outbreak of armed conflict the MPLA took control of the capital, Luanda, and then of the central and eastern regions of the country. Finally it gained control of the main ports on the Atlantic coast.

The emergence of a pro-Soviet group as the dominant force in Angola did not suit the USA, South Africa or neighbouring Zaire. In the summer of 1975 South African armed forces opposed to the MPLA entered Angola from Namibia while forces from Zaire invaded from the north. The MPLA had meanwhile formed a single-party government and the Soviet Union had increased the volume of military supplies considerably.[16] This Soviet military support and

[15] The main period of Soviet assistance was 1964–72. Message from Cuban Military Adviser Raul Diaz Arguelles to Armed Forces Minister Raul Castro, 11 Aug. 1975, reproduced in *Cold War History Project Bulletin* (note 13), p. 14. The USSR reduced its support as the liberation movement in Angola divided into factions in the period 1971–73. Westad, O. A., 'Moscow and the Angolan crisis, 1974–76: a new pattern of intervention', *Cold War History Project Bulletin* (note 13), pp. 21–32.

[16] The decision to increase arms transfers and military assistance to the Angolan Government was taken in Oct. 1975, after Cuba had decided, in Aug. 1975, to send military advisers to Angola. Gleijeses, P., 'Havana's policy in Africa, 1959–76: new evidence from the Cuban archives', *Cold War History Project*

numerous Cuban military forces made a decisive contribution to defending the MPLA regime. After repelling the initial external aggression, the Soviet Union continued to supply weapons to the MPLA (now the government) as it carried on a struggle against UNITA (still supported by the USA and South Africa).

The territory of Western Sahara was a colonial possession of Spain. After the proclamation of independence of Morocco in 1956 the activities of the national liberation movement in Western Sahara accelerated. However, Morocco had its own claim on the territory of Western Sahara, which is rich in phosphates. In May 1973 with the support of Algeria the People's Liberation Front of Sakiet Al-Khamra (the Polisario Front) was set up and headed the national liberation struggle. On 27 February 1976, on territory controlled by Polisario detachments, the Arab Democratic Republic of Sahara was proclaimed. The Soviet Union supplied weapons to Polisario through Algeria.

In the 1960s a conflict flared up in Namibia (called South-West Africa until 1968). In 1920 the League of Nations handed a mandate to govern South-West Africa to the Union of South Africa. After World War II the government of the new Republic of South Africa refused to return that mandate. In early 1960 the leadership of South Africa decided to divide Namibia into semi-autonomous administrative entities governed by tribal chiefs. Namibian nationalists began to protest against being placed under South African administration. In 1960 the South-West African People's Organization (SWAPO) was founded to pursue the goal of a unified state for all tribes and nations of Namibia. Meeting with repression from South Africa, from 1961 SWAPO initiated an armed struggle for national liberation and organized military formations—the People's Liberation Army of Namibia. The Soviet Union began to supply arms and military equipment to SWAPO. In 1990 the independence of Namibia was proclaimed.

In 1975 Mozambique became independent. The struggle against Portuguese colonialists had been headed by the Front for the Liberation of Mozambique (FRELIMO) and Soviet arms and military equipment were supplied to this organization as well. Here as in Angola the Soviet Union found itself supplying arms to the government of an independent state that was itself opposed by internal forces. From 1978 RENAMO (the National Resistance of Mozambique), headed by Afonso Dhlakama, began to act as an anti-government armed grouping, its activity being made possible because the USA and South Africa gave military and economic assistance.

In May 1961 apartheid, expressed through the adoption of a number of laws that infringed the rights of the native non-white population, was raised to the level of the official state policy of South Africa. Black and coloured peoples set

*Bulletin* (note 13), pp. 5–18. In discussions of documents released by the International Department of the Central Committee of the CPSU in 1994, Russian specialists suggested that the Soviet Union was not enthusiastic about increasing arms deliveries to Angola. However, the presence of Cuban troops created a new dimension to the decision. As one former Soviet official commented, 'we could not let them die there, be killed there, without helping them, sending our weapons'. Brutents, F. and Kornienko, G. (former Deputy Head and First Deputy Head, International Department, Central Committee of the CPSU, respectively), 'US–Soviet relations and Soviet foreign policy toward the Middle East in the 1970s', Transcript from a workshop at Lysebu, 1–3 Oct. 1994, Norwegian Nobel Institute, Oslo, 1995, pp. 43–47.

up a number of organizations to fight for their rights, the African National Congress (ANC) being the main one. The ANC had been proscribed as early as 1960 and it therefore began to act under ground. Its main forces were compelled to make their base on the territories of countries neighbouring South Africa. From early 1960 the Soviet Union began to supply the ANC with small arms.

From 1964 the Palestinian Liberation Organization (PLO) carried on a struggle against Israel, uniting the different organizations that made up the Palestinian resistance movement. At different times the PLO maintained armed formations in Algeria, Iraq, Jordan, Lebanon, Sudan, Syria and Tunisia. These formations were equipped with Soviet-made weapons including main battle tanks, multiple-rocket launch systems, anti-aircraft artillery, mortars and field artillery. Soviet-made arms and equipment were supplied to the PLO through Egypt, Iraq, Lebanon, South Yemen, Syria and other countries.

During the period 1975–92 civil war raged in Lebanon. Nationalist and right-wing Christian forces were fighting national patriotic forces supported through Syria and also the PLO. By 1988 the PLO had in various locations more than 100 T-54 and T-55 main battle tanks, 50 Grad multiple-rocket launch systems, over 200 guns and mortars, nearly 100 APCs and armoured fighting vehicles and over 400 machine-guns. In 1982 right-wing Christian forces were equipped with a small number of T-54 and T-55 main battle tanks as well as four Grad multiple-rocket launchers received from Israel, which transferred them from captured stocks seized from the PLO during the invasion of Lebanon in 1982.

The large amounts of arms in the possession of countries like Iraq, Libya, South Yemen and Syria inhibited them from carrying on any armed struggle against neighbouring countries. However, the supplies of weapons to the PLO had some negative impact on internal stability in some countries of the Middle East. Non-Palestinian armed revolutionary formations also existed on the territory of some of the states of the region and these armed opposition groups could cooperate with Palestinian formations in order to purchase weapons to use in civil wars. This feature contributed to the instability of Ethiopia, Lebanon, Somalia and South Yemen.

The presence of revolutionary forces in countries where they had military camps and bases often led to their being actively used by the ruling regimes. For example, Libya sent members of the PLO to carry out terrorist acts in Egypt, and Egypt sent members of the same organization to carry out subversive acts in Libya.

In the 1980s there was some amalgamation (both ideological and organizational) of left-wing militant organizations with nationalist organizations and Islamic fundamentalists. This led in some cases to the strengthening of Islamic opposition movements in countries such as Algeria, Egypt and Lebanon and to some extent also Syria.

Soviet arms and equipment were used in the internal military conflict in Sudan between the ruling regime and the Sudanese People's Liberation Army (SPLA), which was defending the interests of three southern non-Muslim provinces. The President of Sudan, Gafaar Mohammed Numeiri, returned to an open policy of Islamicization of southern Sudan in early 1980. In September 1983, when he introduced Islamic law, the SPLA, headed by Colonel John Garang, refused to lay down arms until these laws in the southern territories were lifted. As a result armed struggle resumed. Colonel Garang was supported by the regimes in Ethiopia, Kenya and Zaire and later Egypt. There are SPLA military bases in Ethiopia and its military formations are equipped with some Soviet-made weapons—small arms, mortars and anti-aircraft guns—supplied through Ethiopia.

In Iran, Iraq, Syria and Turkey and in several post-Soviet successor states lives a Kurdish population which is demanding national autonomy or the foundation of an independent Kurdish state and is waging a stubborn struggle using both peaceful and military means. During the cold war the Soviet Union supplied arms and military equipment to some Kurdish movements and organizations.

The Soviet Union had no role in instigating the revolutions that took place in Latin America in the 1950s—including the Cuban revolution. However, after that revolution Cuba was supported by Soviet First Secretary Nikita Sergeyevich Khrushchev, who regarded it as a bridgehead for future socialist revolutions. Small arms and then artillery, main battle tanks and military aircraft were sent to Cuba. Cuban leaders supplied these Soviet-made arms and equipment to other revolutionary organizations of Latin America without permission from the Soviet Union, which was against an export of revolutions. It was in this way that Soviet arms were supplied to the revolutionary forces in Guatemala.

The Central Committee of the CPSU had no purposeful revolutionary strategy for Latin America, which was far away from the Soviet Union and where the United States had a very strong influence.

From 1979 to 1981, during the Nicaraguan conflict, Soviet arms were transferred to the Sandinista National Liberation Front (FSLN) through Cuba and through the revolutionary forces of Panama. In July 1979 Sandinist forces succeeded in overthrowing the Somoza regime and from the spring of 1981 the Soviet Union began to supply arms and military equipment directly to the new regime rather than through intermediaries.

From 1979 to 1991, during the conflict between the ruling regime and the Farabundo Martí National Liberation Front (FMLN) in El Salvador, the Soviet Union supplied arms to the FMLN via the revolutionary organizations in Honduras and Panama.

# 4. Economic dimensions of Soviet and Russian arms exports

*Ian Anthony*

## I. Introduction

This chapter examines the economic dimensions of Soviet (and now Russian) arms transfers. From the discussion in chapter 3, it is clear that the primary determinants of Soviet arms transfer decisions were political and strategic rather than economic considerations. However, it is also clear from chapter 3 that the Soviet Union was not indifferent to economic returns from the arms trade. Since the dissolution of the Soviet Union the defence industry has been plunged into a deep, at times seemingly existential, crisis which is described in more detail in chapter 8. Under these conditions it is widely believed that economic motivations have become more important as a causal explanation of Russian arms export behaviour. However, many questions remain unanswered about the economic dimensions of Soviet and now Russian arms transfers.

Differences of view about the historical importance of economic factors in Soviet arms export behaviour are reflected among the Russian authors who have contributed to this book. For example, in chapter 5 Sergey Kortunov writes: 'for decades the Soviet military–industrial complex received guaranteed payments from the government for arms manufactured for export. A significant portion of this military equipment was either sold at concessional rates to foreign countries or, on occasion, given away'. This would suggest that Soviet arms transfers may have represented a net loss to the economy. In chapter 3 Yuriy Kirshin writes that 'the prices for transfers which could bring political benefit to the Soviet Union were reduced. However, this was compensated for by prices charged to partners which were not considered so important'. Kirshin suggests that the overall economic impact of Soviet arms exports was either neutral or made a net contribution to Soviet finances.

At the level of manufacturing enterprises it is also unclear how far Soviet and now Russian exports were and are beneficial to producers and how far revenues were or are retained by the state, either within the state trading companies or within the responsible ministries. In chapter 8 Elena Denezhkina writes that given a choice some enterprises in St Petersburg prefer foreign sales over sales to the Russian Government, which has become known as an unreliable customer. In chapter 11 Alexander Sergounin reports on the disappointment of enterprise managers in Nizhniy Novgorod that success in winning orders in China has produced such limited financial benefits for their enterprises.

**Table 4.1.** Official estimates of the value of arms exports, 1988–94
Figures are in current US $b.

|  | 1988 | 1989 | 1990 | 1991 | 1992 | 1993 | 1994 |
|---|---|---|---|---|---|---|---|
| State Committee | 12.00 | .. | 6.05 | .. | 4.00 | 2.15 | 2.80 |
|   on Defence Industries | .. | .. | .. | .. | .. | 4.00 | .. |
| Ministry of Foreign | .. | .. | 7.10 | 3.00 | 0.61 | 0.54 | .. |
|   Economic Relations |  |  |  |  |  |  |  |
| Oleg Davydov | .. | .. | .. | .. | 2.30 | 1.20 | .. |

*Source:* Després, L., 'Financing the conversion of the military industrial complex in Russia: problems of data', *Communist Economies and Economic Transformation,* vol. 7, no. 3 (1995), pp. 335–51.

These different perspectives give rise to two general questions. Were arms exports profitable to the Soviet (and now Russian) economy? Did arms exports yield hard currency and, if so, how much?

There is no single or simple answer to either question. However, this chapter attempts to shed some light on this aspect of arms transfers.

## II. Aggregate data on the value of arms exports

Several sets of data try to capture the volume, value and pattern of Soviet and now Russian arms exports. However, none of them is truly satisfactory.

During the final years of the Soviet Union officials began to make occasional statements about the value of Soviet arms exports. In 1991 I. S. Belousov, Chair of the Soviet Military–Industrial Commission (Voyenno-promyshlennaya komissiya, VPK), stated that the average annual value of the Soviet foreign trade in weapons was 11.7 billion transferable roubles in the period 1986–90.[1]

Between 1992 and 1994 Russian spokesmen made various statements, many of them contradictory, about the value of Soviet and Russian arms exports.

This reflected the general confusion within industry and within the state apparatus during these years. As explained in chapters 3 and 5, responsibility for the management of arms transfers was not centralized in one agency during this period, and cooperation and coordination between existing agencies were far from ideal. Between 1992 and 1994, according to correspondence between the author and the deputy chairman of the then State Committee on Defence Industries (Goskomoboronprom), central industrial organizations found it impossible to collect information either from individual enterprises or from regional industrial associations.[2]

---

[1] Quoted in Albrecht, U., *The Soviet Armaments Industry* (Harwood Academic Publishers: Chur, 1993), p. 290.
[2] Author's correspondence with G. G. Yanpolskiy, 2 Mar. 1994. Yanpolskiy cited both technical problems associated with economic changes (such as the high rate of inflation and finding a representative currency exchange rate) and the general difficulties of effecting plant-level transformation in the industrial sector as reasons for the difficulty in collecting usable statistics from enterprises and regional offices.

**Table 4.2.** Export of military and civilian products from enterprises under the State Committee on Defence Industries, 1988–93

Figures are in current US $b.

| | 1988 | 1989 | 1990 | 1991 | 1992 | 1993 |
|---|---|---|---|---|---|---|
| Exports of military output | 12.00 | .. | 6.05 | .. | 4.00 | 2.15 |
| Exports of civilian output | .. | .. | 2.00 | .. | 0.61 | 0.54 |

*Source:* Després, L., 'Financing the conversion of the military industrial complex in Russia: problems of data', *Communist Economies and Economic Transformation,* vol. 7, no. 3 (1995), pp. 335–51.

Table 4.1 illustrates the range of official data for the late Soviet and early Russian period. In some years widely differing estimates were produced by the same agency. In 1990 the Ministry of Foreign Economic Relations (MFER) released both $7.1 billion and $1.55 billion as values for arms exports, while in 1994 the State Committee on Defence Industries offered both $2.8 billion and $4 billion.[3] To add to the confusion, the Minister for Foreign Economic Relations, Oleg Davydov, released additional estimates in 1994 for the years 1992 and 1993.[4]

In 1994 the State Committee on Defence Industries released data on the value of exports from enterprises falling under its umbrella (see table 4.2). For some years these data were divided into the value of military items and the value of sales of civilian items and were published in US dollars.

In 1996 aggregated data on the value of arms exports covering the period 1985–96 were presented for the first time in public by the state trading company Rosvooruzhenie. These data are presented in figure 4.1 and suggest that the annual value of Soviet arms exports was in the region of $20 billion during the second half of the 1980s—close to the values estimated by Western government agencies such as the US Arms Control and Disarmament Agency (ACDA). For comparative purposes, figure 4.2 shows the value of Soviet arms exports as estimated by ACDA for a similar period. The similarity between the time series is surprising given all that has been published about the inadequacies of Soviet statistics. In the context of foreign trade, dollar-denominated Soviet statistics are said to be of limited value because agreements were denominated in foreign trade or 'convertible' roubles which were converted into dollars at an official exchange rate which was meaningless.[5]

The process by which ACDA estimated the constant dollar value of arms exports from the Soviet Union remains somewhat obscure. It publishes estimates of the value of goods delivered in a calendar year which it receives from

[3] The most likely explanation of the differences is that MFER data are based on the value of licences issued while the State Committee data are based on reporting by enterprises.

[4] *International Defense Review,* May 1994, p. 54.

[5] Information provided in author's correspondence with Prof. Laure Després, University of Nantes, 27 Feb. 1997. See also Tabata, S., 'The anatomy of Russian foreign trade statistics', *Post-Soviet Geography,* vol. 35, no. 8 (1994).

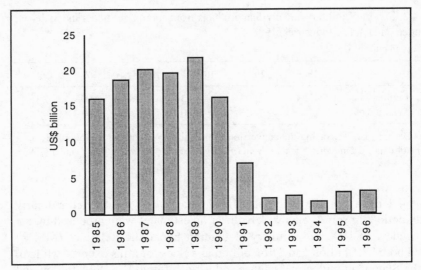

**Figure 4.1.** Trends in Soviet/Russian arms exports according to Rosvooruzhenie, 1985–96

*Source:* Tarasova, O., [Rosvooruzhenie calls for unity], *Segodnya*, 1 Nov. 1996 (in Russian).

other US government agencies. These estimates are already denominated in US dollars when ACDA receives them and are then deflated using a gross national product (GNP) index.

During the cold war most dollar estimates of Soviet arms exports were generated in Western government agencies and research institutes using volume indexes rather than estimates of the value of arms sales. However, there were also efforts to identify arms exports in Soviet foreign trade statistics. These estimates produced dollar values very different from those contained in the data released by Rosvooruzhenie.

These data were estimated by eliminating from Soviet foreign trade statistics all categories which were clearly non-military and assuming that most of the remaining exports were for military end-users. The resulting data were converted from roubles into dollars using the prevailing official exchange rate. Comparing the value for 1980 contained in table 4.3 ($5.6 current billion) with the value for 1980 given by ACDA ($8.8 billion), and allowing for the fact that the data in table 4.3 exclude trade within the WTO, there appears to be rough comparability. According to residual foreign trade data the average annual value of Soviet arms exports to developing countries was $3.2 billion (in current dollars) between 1971 and 1980. Looking at ACDA estimates for the same period, the average annual value is $5.6 billion.

These data should not be interpreted as hard currency earnings. The official rouble/dollar exchange rate was unable to capture the relative value of the two currencies because the foreign trade and 'convertible' roubles were not an

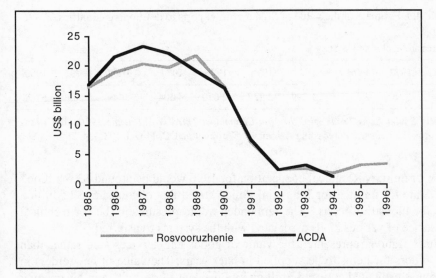

**Figure 4.2.** Rosvooruzhenie and ACDA data on Soviet/Russian arms exports, 1985–96

*Source:* US Arms Control and Disarmament Agency, *World Military Expenditures and Arms Transfers 1995* (US Government Printing Office: Washington, DC, 1996); and Tarasova, O., [Rosvooruzhenie calls for unity], *Segodnya*, 1 Nov. 1996 (in Russian).

accepted form of exchange. Moreover, the data do not reflect the impact of military aid or the different forms of financing (such as barter) that the Soviet Union employed in managing its foreign trade. At best they reflect the broad trends in foreign trade, although longitudinal analysis of Soviet economic activity is made more difficult by the difficulty of measuring the influence of inflation.

In 1996 and 1997 a great deal of international attention was paid to estimates by non-Russian analysts which indicated that Russia had achieved a market share comparable to that of the larger West European arms exporting countries, France, Germany and the United Kingdom.[6] Using official government data (which are not strictly comparable but which give a broad indication of the relative value of arms exports) in 1995 Russia exported arms and military assistance worth $3.1 billion compared with $3.8 billion from France, $3.3 billion from the UK and $1.2 billion from Germany.[7] A preliminary estimate by the General Director of Rosvooruzhenie, Alexander Kotelkin, suggested that the

[6] Nikolayev, A., 'Russia comes second in arms sales', *Power in Russia*, vol. 4, no. 56 (5 Feb. 1997), Internet edition translated by RIA Novosti and distributed by John Pike, Federation of American Scientists.

[7] Anthony, I., Wezeman, P. D. and Wezeman, S. T., 'The trade in major conventional weapons', *SIPRI Yearbook 1997: World Armaments, Disarmament and International Security* (Oxford University Press: Oxford, 1997), table 9.2, p. 270.

**Table 4.3.** Estimate of the value of Soviet arms exports to developing countries, 1971–80
Figures are in current US $m.

| 1971 | 1972 | 1973 | 1974 | 1975 | 1976 | 1977 | 1978 | 1979 | 1980 |
| --- | --- | --- | --- | --- | --- | --- | --- | --- | --- |
| 780 | 1 155 | 2 331 | 1 299 | 2 227 | 2 670 | 4 504 | 5 364 | 5 585 | 5 628 |

*Source: Soviet Arms Trade with the Non-Communist Third World in the 1970s and 1980s* (Wharton Econometric Forecasting Associates: Washington, DC, 11 Oct. 1983), p. 26.

value of military–technical cooperation for 1996 would be around $3.6 billion.[8] The State Committee for Industrial Policy stated that of this sum $2.5 billion was for industrial goods.[9] The remainder would presumably be for technical assistance of various kinds associated with the systems transferred.

These figures represented the value of goods and services sold rather than new orders for items to be supplied in later years. The value of new orders in 1995 was estimated at over $7 billion by President Boris Yeltsin in his opening statement to a conference of defence industry workers in Moscow in May 1996.[10]

According to Oleg Soskovets, at the time First Deputy Prime Minister and with overall responsibility for Russian military–technical cooperation with foreign countries, around 75 per cent of the arms trade business of Russia in 1995 involved hard currency payment.[11]

Country and regional data for some of the principal recipients of Russian arms have also begun to be published in the past few years. In 1991 the MFER published data which had been used in discussions between the five permanent members of the UN Security Council on approaches to arms transfer control.[12] These data are reproduced in table 4.4 and underline the importance of Asia, Europe and the Middle East as markets for Soviet arms.

According to then Prime Minister Yegor Gaidar, Russia concluded arms agreements worth $2.2 billion with China, India and Iran in 1992. Of this sum China accounted for $1000 million, India $650 million and Iran $600 million.[13] According to an article in *Rossiyskaya Gazeta*, China accounted for $2.1 billion of the estimated $3.6 billion sales in 1996.[14]

---

[8] Interfax (in English) in Foreign Broadcast Information Service, *Daily Report–Central Eurasia* (hereafter FBIS-SOV), FBIS-SOV-96-236, 5 Dec. 1996; and *Jane's Intelligence Review & Jane's Sentinel Pointer*, Jan. 1997, p. 2.

[9] *Atlantic News*, no. 2797 (6 Mar. 1996), p. 4.

[10] Interfax, 29 May 1996 (in English) in FBIS-SOV-96-105, 30 May 1996. Earlier, in Mar. 1996, Rosvooruzhenie had given an estimate of $6 billion for the value of orders in 1995. *Komsomolskaya Pravda*, 30 Mar. 1996 (in Russian) in FBIS-SOV-96-064, 2 Apr. 1996, p. 47.

[11] Interfax, 5 Mar. 1996 (in English) in FBIS-SOV-96-045, 6 Mar. 1996, p. 31; and *Financial Times*, 6 Mar. 1996, p. 2.

[12] See chapter 5 in this volume.

[13] *Defense News*, 7–13 Dec. 1992, p. 3.

[14] *Jane's Defence Weekly*, 6 Nov. 1996, p. 19.

**Table 4.4.** Regional distribution of deliveries of arms and military equipment by the former Soviet Union, 1991

Figures are percentages.

| Region | Share |
| --- | --- |
| Middle East | 61 |
| Asia | 17 |
| Europe | 12 |
| Near East | 8 |
| Africa | 1 |
| Latin America | 1 |

*Source: Nezavisimaya Gazeta,* 29 Sep. 1992.

# III. Managing foreign trade with different recipient groups

Soviet arms transfers can be divided into five categories for the purpose of evaluating their economic impact: (*a*) equipment provided for non-economic forms of payment such as political influence or strategic assistance (including basing rights and shore support for the Soviet Navy), corresponding to grant military aid; (*b*) equipment provided to socialist countries in the framework of the CMEA arrangements; (*c*) equipment provided to non-CMEA countries which did not reimburse the USSR in hard currency and with which bilateral clearing arrangements were used; (*d*) equipment provided against hard currency payments; and (*e*) equipment provided against delivery of commodities.

A comprehensive accounting of the economic benefits derived from arms transfers would require data for each type of transaction which are not available. However, it is possible to examine each type of financial arrangement in general terms.

## Grant military aid

Equipment transferred as grant aid was assigned a book value for accounting purposes but no financial transfers took place.

When the United States completed its military operations in Grenada in 1983, a large number of documents were recovered detailing the relations between Grenada and the Soviet Union. The documents included the agreements on deliveries of arms and military equipment to Grenada. In October 1980 the two countries agreed that the USSR would 'ensure in 1980–1981 free of charge the delivery to the Government of Grenada of special and other equipment in nomenclature and quantity according to the Annex to the present agreement to the amount of 4 400 000 roubles'.[15] In a subsequent protocol the value of the

[15] Document 13, 'Agreement between the Government of Grenada and the Government of the Union of Soviet Socialist Republics on deliveries from the Union of Soviet Socialist Republics to Grenada of

goods to be shipped was raised to 5 000 000 roubles.[16] Under another agreement of July 1982, the Soviet Union was to transfer special equipment worth 10 000 000 roubles in the period 1982–85.[17]

Under the terms of these agreements the Soviet Union also provided technical assistance and documentation free of charge. Deliveries of these were made via Cuba, which also performed some training and maintenance tasks. Separate agreements regulated this assistance provided to Grenada by Cuban specialists. Cuba was paid for its assistance by Grenada in US dollars on a per-person per-day basis.

## Arms transfers within the CMEA

The membership of the CMEA included all the members of the WTO. Within the WTO an integrated military–technical policy included transfers of equipment and technology between partners.

The CMEA was founded in January 1949 with the objective of integrating its members with the Soviet economy on the basis of specialization of trade and production among member countries. It was a planning mechanism which operated at several levels. Annual plans established quotas for cross-border trade between members in goods that were classified according to nine broad categories and many specific sub-categories.[18]

Beginning in the late 1950s CMEA members attempted to develop multilateral trade relations rather than acting as an umbrella organization managing a series of bilateral relations. To establish these multilateral plans an accounting unit (the 'transferable' or 'convertible' rouble) was invented to compensate for the fact that none of the local currencies in CMEA countries could be exchanged at a market-determined rate. However, according to one analyst the effort to develop a system of prices for trade between CMEA members that was independent of prices in the world market largely failed.[19] Consequently, by the mid-1970s the CMEA conducted annual reviews of prices and adjusted them according to data collected on prices in the wider global economy. If this is correct, then it is likely that the starting-point for establishing prices for arms traded between CMEA members was data collected on the prices of Western equipment sold internationally.

special and other equipment, 27 Oct. 1980', *Grenada Documents: An Overview and Selection*, Released by the Department of State and Department of Defense, Sep. 1984.

[16] Document 15, 'Protocol to the Agreement between the Government of Grenada and the Government of the Union of Soviet Socialist Republics on deliveries from the Union of Soviet Socialist Republics to Grenada of special and other equipment, 9 Feb. 1981', *Grenada Documents: An Overview and Selection* (note 15).

[17] Document 14, 'Agreement between the Government of Grenada and the Government of the Union of Soviet Socialist Republics on deliveries from the Union of Soviet Socialist Republics to Grenada of special and other equipment, 27 July 1982', *Grenada Documents: An Overview and Selection* (note 15).

[18] On the operation of the CMEA, see 'Trading patterns and trading policies', *Quarterly Review* (European Bank for Reconstruction and Development), 30 Sep. 1992, pp. 4–7.

[19] Knirsch, P., 'Economic relations between the CMEA states and the influence of trade with the West', ed. I. Oldberg, *Unity and Conflict in the Warsaw Pact*, Proceedings of a Symposium organized by the Swedish National Defence Research Agency, Stockholm, 18–19 Nov. 1982.

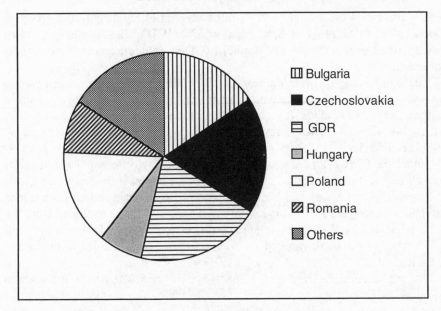

**Figure 4.3.** Shares of arms exports to socialist countries by the Soviet Union, 1980–83

*Note:* 'Others' include Cuba, Mongolia, North Korea, Viet Nam and Yugoslavia.

*Source:* Calculated from Vanous, J., 'Developments in Soviet arms exports and imports', *Centrally Planned Economies Current Analysis* (Wharton Econometric Forecasting Associates), vol. iv, no. 62 (15 Aug. 1984), p. 4.

Within the CMEA, each producing enterprise dealt only with its national authorities. These national authorities had already decided the scale of production for export and the schedule of interstate payments during their negotiations with the state authorities of other participating countries.

There have been several efforts to quantify the scale of intra-CMEA arms sales and military–technical cooperation using estimates derived from comparing published trade data from the CMEA and the Soviet Union. According to Jan Vanous, the Soviet Union deleted all arms trade data from the trade statistics supplied to the CMEA secretariat but included them in the Soviet foreign trade statistics. By comparing the Soviet *Foreign Trade Yearbooks* with CMEA foreign trade yearbooks, Vanous estimated total Soviet arms exports. According to his estimates, in 1983 the Soviet Union exported arms worth *c.* 9 billion roubles or 13.4 per cent of the value of total Soviet exports in that year.[20] In 1983 arms represented the second largest export category, although significantly smaller than oil and oil products, which accounted for over 41 per cent of Soviet exports. Vanous went on to disaggregate Soviet arms exports to WTO allies using the data on exports by commodity group contained in the Soviet foreign trade statistics. He estimated that in the years 1980–83 the Soviet Union

---

[20] Vanous, J., 'Developments in Soviet arms exports and imports, 1980–83', *Centrally Planned Economies: Current Analysis*, vol. iv, no. 62 (15 Aug. 1984), p. 2.

exported arms worth roughly 10.7 billion convertible roubles to the group of socialist countries, of which 83 per cent were for WTO allies. The largest shares went to the German Democratic Republic (GDR), Bulgaria and Poland, in order of magnitude.

The main value of these data is that they indicate the magnitude of arms exports relative to other commodity categories and show a rough distribution of arms sales to WTO allies. It seems likely that these data, converted into dollars using the official exchange rate, are contained in those recently released by Rosvooruzhenie.

Under the CMEA arrangements there was probably differential treatment of developed and non-developed members. For example, it is likely that Cuba, Mongolia, North Korea and Viet Nam received significant military assistance and also some grant aid.[21] This military assistance would be excluded from the data presented by Vanous, which are confined to the value of exports. It is not clear whether the book value of military assistance was included in the aggregate trade data.

The multilateral clearing arrangements within the CMEA were intended to produce balanced trade for any given year. In reality this was not achieved and when the CMEA was dissolved the Soviet Union owed significant outstanding debts to some of the other participating states for military equipment paid for in advance but not yet delivered. In recent years arms transfers and military–technical cooperation have been used as a way of clearing some of the debts to Bulgaria, Hungary and Slovakia assumed by Russia.[22]

### Soviet arms transfers to non-socialist countries financed through clearing arrangements

The Soviet Union maintained bilateral agreements specifying financial aspects of trade arrangements with roughly 20 Asian, African and Latin American countries which imported large quantities of Soviet arms, including Algeria, Egypt, India and Syria. These agreements defined the trends and structure in trade between the Soviet Union and partner countries.[23]

They also specified arrangements for making payments to clear specific trade deals. Unlike the multilateral arrangement in the CMEA, bilateral arrangements could use one or other local currency (either the rouble or the local currency of the partner) in clearing settlements. For example, under the Soviet–Indian trade agreement all payments for goods delivered to India from the Soviet Union

---

[21] This differential arrangement was applied in other areas and it is unlikely that military equipment was exempted. Brezinski, H., 'Economic relations between European and less-developed CMEA countries', *East European Economies: Slow Growth in the 1980s*, Selected Papers submitted to the Joint Economic Committee, Congress of the United States, vol. 2: Foreign Trade and International Finance, 28 Mar. 1986 (US Government Printing Office: Washington, DC, 1986).

[22] Described in chapter 10 in this volume.

[23] United Nations Conference on Trade and Development, Institute of Economics of the World Socialist System, *Innovations in the Practice of Trade and Economic Cooperation between the Socialist Countries of Eastern Europe and the Developing Countries* (United Nations: New York, 1970), pp. 8,10.

were made in Indian rupees into the account of the State Bank of the USSR at the Reserve Bank of India. Money held in this account was used to purchase Indian goods.

The prices of goods transferred were also fixed in the framework of these bilateral arrangements. According to President Gafaar Mohammed Numeiri of Sudan and General Sa'ad el-Din Shazli, a former Egyptian Chief of Staff, the prices of arms imported from the Soviet Union under bilateral clearing arrangements were established in roubles and then converted into local currencies.[24] These prices were 'fixed by the partners on the basis of world prices' but 'in determining the prices, the parties strive to eliminate the purely short-term and other accidental price fluctuations on the world market'.[25]

According to several accounts, payment schedules for bilateral trade under clearing arrangements were also adjusted according to the status of the particular recipient. Moshe Efrat refers to two categories of recipient. The first had a form of most-favoured-nation status and was offered a discount on the list price of equipment as well as being permitted to clear an account over a 20-year period at a rate of interest of 2.5 per cent per year. A second category of countries received no discount and was expected to clear an account over a 12-year period, also at an annual rate of interest of 2.5 per cent.[26]

This statement suggests that, as was noted above for intra-CMEA trade, the price index for arms sold to non-socialist countries was probably established with reference to available data on the market value of Western arms.[27] However, it is also known that these prices were adjusted according to the specific political and economic conditions prevailing at the time a deal was made. For example, Roger Pajak has suggested that Egypt was offered reductions of between 40 and 50 per cent on the official export price of Soviet arms during the 1960s.[28]

In 1975 the Egyptian Government presented to the United Nations an account of the value of equipment lost in the 1967, 1970 and 1973 wars against Israel along with the value of replacement.[29] According to these data Egypt received

[24] Quoted in Efrat, M., 'The economics of Soviet arms transfers to the Third World: a case study: Egypt', *Soviet Studies*, vol. 35, no. 4 (Oct. 1983), p. 440.

[25] United Nations Conference on Trade and Development, Institute of Economics of the World Socialist System (note 23), p. 9.

[26] Efrat, M., 'The defence burden in Egypt during the deepening of the Soviet involvement in 1962–73', University of London Ph.D thesis submitted May 1981, p. 35; Heikal, M., *Sphinx and Commissar: The Rise and Fall of Soviet Influence in the Arab World* (Collins: London, 1978), pp. 25–26, 32; and Mohrez Mahmoud El Hussini, *Soviet–Egyptian Relations 1945–85* (Macmillan: Basingstoke, 1987), pp. 96–97.

[27] Similarly, the prices of other commodities such as Egyptian cotton were adjusted from world market prices. Foley, T., 'The mighty transformation: Soviet aid and Arab liberation', *New World Review*, vol. 38, no. 4 (fall 1970), p. 35. In a large study of the economic aspects of Soviet–Egyptian military–technical cooperation Moshe Efrat concluded by examining a control sample of goods and commodities that there were relatively minor differences between the prices used in bilateral trade with the Soviet Union and with other industrialized countries. Efrat (note 26).

[28] Pajak, R., 'Soviet arms and Egypt', *Survival*, vol. 17, no. 4 (July–Aug. 1975), p. 165.

[29] United Nations, Permanent Sovereignty over National Resources in the Occupied Arab Territories, Report of the UN Secretary-General, UN document A/10290, 3 Nov. 1975, quoted in Efrat (note 24), p. 445.

**Table 4.5.** Major developing country debtors to the Soviet Union in 1990
Figures are in m. roubles. Figure in italics is a percentage.

| Country | Debt |
| --- | --- |
| Cuba | 15 490.6 |
| Mongolia | 9 542.7 |
| Viet Nam | 9 132.2 |
| India | 8 907.5 |
| Syria | 6 742.6 |
| Iraq | 3 795.6 |
| Afghanistan | 3 055.0 |
| Ethiopia | 2 860.5 |
| Algeria | 2 519.3 |
| North Korea | 2 234.1 |
| Ten largest developing country debtors as a % of total developing country debt to the Soviet Union | *81.0* |

*Source: Izvestiya*, 1 Mar. 1990.

equipment worth $10.2 billion in the period 1967–73. A very high proportion of this would have been from the Soviet Union. Comparing these data with other public statements by Egyptian officials—for example, the head of the Economic Committee in the Egyptian Parliament and the Deputy Prime Minister for Economic Affairs—Moshe Efrat estimates that Egypt received discounts of roughly 33 per cent before 1967 and around 50 per cent after 1967.

As with the rules governing multilateral clearing within the CMEA, for bilateral clearing arrangements the objective of both sides was that trade should be balanced on an annual basis. In practice this was not achieved. In 1991 the Soviet Union released data showing the scale of the debts owed by various countries (see table 4.5). All the countries on the list were recipients of Soviet arms.

Russia has subsequently tried to recover these debts but the process has been complicated by both political and technical problems. In some cases, for example, that of Syria, Russian efforts to address the issue of debt have been made more difficult by the general deterioration in bilateral political relations. In other cases, such as that of India, the bilateral political relationship has remained strong but there have been technical problems in calculating the debt.

Even in 1997 neither the rouble nor the rupee is fully and freely convertible at market rates—that is, it is not possible to buy rupees outside India or roubles outside Russia. The discussion in 1992 revolved around what would be a reasonable rate at which to convert roubles to rupees.[30] One element in the discussion was the respective value of roubles and rupees against the US dollar.

An agreement was reached during the visit of President Yeltsin to India in January 1993. In India the government was criticized for accepting an exchange

---

[30] *Aviation Week & Space Technology*, 25 July 1994, p. 58.

rate which favoured Russia as the basis for converting India's debt. Roughly two-thirds of the debt was converted at a rate of 19.92 roubles to the rupee. This part of the debt was to be repaid over a 12-year period at an annual rate of interest of 2.4 per cent. The remaining third of the debt was to be converted at a rate of 31.57 roubles to the rupee but repayable over a 45-year period with no interest charged.[31] At the time the intergovernmental agreement was reached the debt was valued at between $9.3 billion and $11.6 billion including the interest payments.[32]

Under the agreement reached India will pay Russia $800–$900 billion each year between 1994 and 2006 to clear the largest part of the debt. This money is paid to Russia's account at the Central Bank of India and is available for the purchase of Indian goods or to finance joint projects in India.[33]

## Soviet arms transfers paid for in hard currency

During the cold war there were countries which seem to have conducted their arms trade with the Soviet Union almost entirely on a hard currency basis. Oleg Baklanov, Secretary of the Communist Party Central Committee, estimated in 1990 that in a normal year about one-third of Soviet arms transfers were made in hard currency.[34] It is likely that this applied mostly to the countries with large oil revenues such as Angola, Iraq and Libya.

The Soviet Union apparently did not receive payment in advance from these countries. Iraq was said to have 'an unusually good repayment record. With hard currency earnings from oil exports, Iraq was better able than any other Soviet client to meet its repayment obligations to Moscow'.[35] However, in 1990 it was revealed that Iraq was among the countries that owed large debts to the Soviet Union.

In some cases countries which had bilateral clearing arrangements permitting use of local currency to finance arms imports also occasionally conducted arms trade with the Soviet Union on a hard currency basis. This seems to have been particularly true for Arab countries that made financing arrangements which involved third parties. For example, imports by Egypt were part-financed using grants provided by other Arab countries. During the 1973 October War between

[31] While both the rouble and the rupee have lost value against the US dollar in recent years, the depreciation in the rouble has been much faster. Indian critics argued that the rapid decline in the value of the rouble was predictable at the time the agreement with Russia was reached and should have been taken into account in deciding an exchange rate. *Financial Times*, 29 Jan. 1993, p. 3; *Far Eastern Economic Review*, 18 Feb. 1996, p. 18; and 'Focus on technology transfer, new weapons', *The Hindu*, 22 July 1993.

[32] *Asia–Pacific Defence Reporter*, Feb.–Mar. 1993, p. 22; *Hindustan Times*, 22 July 1993; *Segodnya*, 25 Oct. 1994 (in Russian) in FBIS-SOV-94-207, 26 Oct. 1994, pp. 11–12; and *Defense News*, 9–15 Jan. 1995, p. 25. Not all of this debt was incurred through arms purchases. The Soviet Union supplied India with large quantities of energy, heavy industrial goods and both raw and semi-processed materials.

[33] According to Alexander Belikov, Deputy Head of the Asia Department, Russian Ministry of Foreign Economic Relations, quoted by Interfax, 2 Aug. 1995 (in English) in FBIS-SOV-95-149, 3 Aug. 1995, p. 8; and *Financial Times*, 24 May 1996.

[34] Information provided in an interview between Baklanov and Milton Leitenberg, 12 Nov. 1990.

[35] Pajak, R., *Soviet Arms Aid in the Middle East* (Center for Strategic and International Studies, Georgetown University: Washington, DC, 1976), p. 30.

Israel and a coalition of Arab states Libya is believed to have provided $500 million to Egypt and Syria to pay for 70 MiG-21 fighter aircraft of different versions.[36] These agreements, reached outside the framework of normal trade channels, reflected the immediate requirement of Egypt for rapid delivery of equipment.

In some cases it appears that hard currency payments were not made directly but integrated into financial arrangements involving several countries. For example, in some sources it is claimed that Libya transferred to the Soviet Union the right to the proceeds from the sale of 70 000–80 000 barrels of crude oil per day, part of which was to cover the costs of Libyan arms imports and part of which was to cover the cost of assistance to Syria.[37] This oil generated revenue when it was sold on the world market by brokers.

In some cases the Soviet Union was prepared to defer or relieve debts. According to some sources debt rescheduling (often involving a degree of debt relief) was a regular occurrence.[38] However, there were cases of relief not being available. In one rather specific case, after the decision by Egypt to break its ties with the Soviet Union in 1974, Soviet leaders refused to reschedule Egypt's debts.[39]

## IV. The impact of domestic reform on foreign trade

The defence industrial sector has been deeply affected by the changes which followed the end of the Soviet Union. Political and economic reforms have changed the relationship between the state and manufacturing industry. Price and currency reforms have changed the terms of trade.[40]

Within the state socialist system the needs of the military were given special priority. Consequently, according to a view expressed by the Soviet General Staff in the early 1960s, 'the country's entire economy is constantly subordinated to military planning, in particular, to the requirements for mass production of modern weapons'.[41] This approach was a product of the Stalinist world view compounded by the experiences of World War II and the cold war. Across time as the threat of a central confrontation receded and pressures for investment in civilian economic development grew the impact of this way of thinking was attenuated. Nevertheless, the organizational structures established to meet

[36] Glassman, J. D., *Arms for the Arabs: The Soviet Union and the War in the Middle East* (Johns Hopkins University Press: Baltimore, Md., 1975), p. 146; and Pajak (note 35), p. 38.

[37] To add to the complexity, these deals were apparently brokered by a Finnish trading company operating on the international oil market. *The Times*, 15 Feb. 1978. See also Pajak, R., 'Arms and oil: the Soviet–Libyan arms supply relationship', *Middle East Review*, vol. 13, no. 2 (winter 1980/81), pp. 51–56.

[38] *New York Times*, 5 Sep. 1967, pp. 1, 24.

[39] *New York Times*, 2 May 1975. Moreover, after the break between Egypt and the Soviet Union a cooling of relations between Egypt and Libya meant that Egypt no longer received as much external financial assistance.

[40] See chapter 8 in this volume.

[41] Sokolovsky, V. D. (Marshal), *Military Strategy: Soviet Doctrine and Concepts*, translated by R. L. Garthoff (Praeger: New York, 1963 edn).

what was perceived to be an overriding security requirement also created what Michel Checinski has called 'structural causes' for Soviet arms exports.[42]

The Soviet economy used the concept of price in a way which was different from its use in a market economy. This was particularly true where military production was concerned. Checinski noted that the decision to retain a massive arms production capacity could only be transformed into operational reality if three questions were addressed: what numbers of which weapons were to be produced; over what time-scale production plans were to be fulfilled; and how the bottlenecks in production and distribution that were ever-present in Soviet industry could be overcome.[43] Soviet economic planning gave high priority to addressing these problems and price policy was one important element in this planning system. As in a market economy, prices were seen as an instrument to achieve efficient distribution of goods and services. However, efficiency was measured against a narrow definition of military security and not against wider social and economic considerations.

As a result of this set of priorities, under the state socialist system neither costs nor prices were established through bargaining in a market but were established centrally by administrative decision. Numerical requirements were turned into rouble-denominated quotas by applying centrally maintained price lists to the number of any given item to be acquired. These quotas were translated into micro-decisions through national planning agencies which would distribute production between state-owned enterprises.[44] The enterprises could receive instant payment in local currency against certification that a specific quota obligation had been met using the price schedules determined by the planning authorities.

While the needs of the Soviet armed forces were the dominant factor in planning, at different times the existence of foreign suppliers and foreign markets was probably helpful both from a production perspective (to fill gaps in any given production line) and also in price setting. In Soviet foreign trade different price lists were used as the basis for negotiations with foreign buyers. However, the final price in any given transaction appears to have been set in negotiations and could vary for the same weapon system on a case-by-case basis. In this way foreign sales may have given some indications about the accuracy of domestic price lists. Goods produced for export were integrated into the overall defence order alongside goods produced for the Soviet armed forces. In practice the

[42] Checinski, M., 'Structural causes of Soviet arms exports', *Osteuropa Wirtschaft*, vol. iv, no. 3 (Mar. 1977) (in English).

[43] Checinski (note 42), p. 174.

[44] In practice the process was probably more complex in that managers at particularly important enterprises could and did lobby central authorities to gain preferences in either production quotas or unit prices. For example, Arthur Alexander describes how on at least 2 occasions chief designers at the Yakovlev and Tupolev design bureaux overturned decisions taken by the planning apparatus by making direct appeals to Stalin and Khrushchev, respectively. Alexander, A. J., 'Decision making in Soviet procurement', eds D. J. Murray and P. R. Viotti, *The Defense Policies of Nations: A Comparative Study* (Johns Hopkins University Press: Baltimore, Md., 1982) pp. 161, 175–76. According to Alexander this was not unusual behaviour, although the impact of these lobbying efforts remains controversial between analysts.

application of this system meant that the price charged to a foreign buyer for any item was not coupled to the price at which the state trading company acquired it from the manufacturer. In these circumstances the trading companies may have been able to generate significant profits for the state by exploiting differentials in price that existed between foreign and domestic trade.

There are some suggestions that the prices of Soviet goods sold in foreign markets were not always low. Soviet intelligence and planning authorities collected information about weapon prices in the United States and elsewhere and used this as a guideline to establish foreign trade price lists. However, revenues from foreign sales were never passed directly to manufacturers who instead had access to a hard-currency allocation provided to them by the relevant sectoral ministry as a privilege. The sums involved were described by one Soviet designer as 'miserable'.[45]

In 1992 two important economic policy decisions were taken which should have had a major impact on Russian trading practices. First, the government decided to remove some internal price controls, thereby changing the costs of production for defence manufacturers and the relative advantage of exporting manufactured products. Second, it was decided that all foreign trade negotiations would be conducted on the basis of prices quoted in hard currency.

At the same time in some of its features the Russian defence sector differs from the wider economy. Its domestic prices remain fully regulated, and the prices in foreign trade are heavily influenced, by the state authorities. State authorities still manage the revenues from export sales. Before 1994 the MFER was responsible for distributing hard currency proceeds from arms sales. In 1994 this function was taken over by Rosvooruzhenie.[46] This continued state control over the distribution of proceeds from arms sales has led to arguments between government and industry about whether the money received has been distributed fairly and honestly. In 1993 the MFER was criticized by industrialists, in particular by the League of Assistance to Defence Enterprises and its chairman, Alexander Shulanov.[47] In an August 1994 interview then Rosvooruzhenie General Director Viktor Samoylov described the payment system in operation. According to Samoylov, Rosvooruzhenie retained between 1.5 and 3 per cent of the purchase price for itself. Around 10 per cent of the purchase price was used to cover costs of insurance, transport and related services. The remaining money was distributed to the manufacturers. However, Samoylov added that there was no clear method for determining the distribution of funds between the design bureau that created a system, the plant which manufactured it and the plants which made components that went into the system.[48]

[45] Bogdanov, O., 'Antonov Design Bureau and its activities in the new environment'. Unpublished paper, Apr. 1993.

[46] ITAR-TASS, 26 Jan. 1994 (in English) in FBIS-SOV-94-018, 27 Jan. 1994, pp. 22–23.

[47] *Kommersant Daily*, 16 Apr. 1993 (in Russian) in FBIS-SOV-93-075, 21 Apr. 1993, p. 35; and *East Defence & Aerospace Update*, May 1993, p. 2. The criticism of the MFER was echoed to some extent by the then State Committee on Defence Industries.

[48] Moscow Russian Television, 20 Aug. 1994 (in Russian) in FBIS-SOV-94-183, 21 Sep. 1994, pp. 17–20.

In spite of these explanations Rosvooruzhenie was heavily criticized by industry for a variety of reasons. Some complained that it retained too high a share of payments as compensation for its own service and complained of the lack of control and transparency in its accounting practices; some made accusations of outright corruption.[49]

## The relationship between price and cost

In general the pricing methods used by the Soviet Union benefited the manufacturing industry. In 1990 the State Committee on Statistics (Goskomstat) compared Soviet trade assuming world market prices with trade at official prices in order to estimate the impact of abolishing price controls on the terms of trade.[50] The outcome suggested that the price controls which operated in the energy and raw material sector kept prices of these inputs well below their true market value and in this way represented a large subsidy to producers of manufactured goods.

Defence manufacturers are heavy consumers of, for example, energy and non-ferrous metals. Domestic prices of these inputs in Russia were increased but not decontrolled and did not reach world market prices for some key items.[51] Nevertheless, the prices charged to manufacturers have risen significantly in recent years.[52]

Under the conditions in 1992–93 some Russian defence manufacturers also took advantage of the relative absence of enforceable state regulations to sell stockpiles of raw and semi-processed materials, which had been bought at internal, regulated prices, on foreign markets.[53] This was usually accomplished

[49] After an investigation of Rosvooruzhenie in Nov. 1994 Samoylov was sacked. Press reports of the decision listed irregularities in the handling of payments as one of the reasons. *International Defense Review*, July 1995, pp. 55–56.

[50] Tarr, D. G., 'The terms-of-trade effect of moving to world prices on countries of the former Soviet Union', *Journal of Comparative Economics*, vol. 18, no. 1 (Feb. 1994).

[51] Price controls from 2 sources remain on key inputs. Some are imposed on producers who are designated as having a monopoly in a given area. Others are imposed (usually but not always) by the Ministry of Economics. In spite of controls, prices have usually been increased in line with overall inflation in wholesale prices. Webster, L. W., Franz, J., Artimiev, I. and Wackman, H., *Newly Privatized Russian Enterprises*, World Bank Technical Paper no. 241 (World Bank: Washington, DC, 1994), p. 23.

[52] The impact of cost increases is offset to some extent by the fact that defence manufacturing enterprises have been allowed favourable conditions regarding value-added tax, favourable corporate tax rates (ranging from reductions of 50% in tax to complete tax exemption) and access to credit on favourable terms. From 1 Jan. 1996 military equipment and armaments were among the categories of Russian goods relieved of export tariffs. Enterprises regarded as particularly important to the defence industrial base are also eligible for direct funds from the federal budget for plant reconstruction, buying new equipment, developing manufacturing techniques and developing new materials. Interfax, 1 Sep. 1995 (in English) in FBIS-SOV-95-171, 5 Sep. 1995, p. 25.

[53] The impact of price liberalization on industrial enterprises (not specifically defence enterprises) is described in Moody, S. S., 'Decapitalizing Russian capitalism', *Orbis*, vol. 40, no. 1 (winter 1996). For sectoral discussions, see Evangelista, M., 'From each according to its abilities: competing theoretical approaches to the post-Soviet energy sector', ed. C. A. Wallander, *The Sources of Russian Foreign Policy after the Cold War* (Westview Press: Boulder, Colo., 1996); and Haglund, D. G. and MacFarlane, S. N., *The Former Soviet Union in International Minerals Markets: The Resurrection of 'Strategic Minerals' Policy?*, Occasional Paper no. 47 (Centre for International Relations, Queen's University: Kingston, Ontario, June 1994).

through dealers located in neighbouring countries (the Baltic states being particularly prominent).[54]

As the Goskomstat simulation predicted, price reforms seem to have had a severe impact on Russian manufacturing industry in general and some have expressed concern that Russia may face 'deindustrialization' as manufacturing has become an economically irrational activity.[55] The defence sector—which remains the most closely controlled element of the state sector in Russia—has probably been affected more directly than any other group of enterprises.

Whereas state procurement plays a limited (and steadily declining) role in setting prices in the Russian economy in general, in the defence sector equipment prices are still heavily regulated. In 1992 and 1993 the Ministry of Defence prepared a draft Law on the Defence Order and the Status of Plants which Fulfil It, which was to have been completed by May 1993. Under this law the relations between the Ministry of Defence and the manufacturers would have been regulated by state contracts. While there is a definition of state contracts in the Law of the Russian Federation On Deliveries of Products and Goods for the State, this does not apply to the Ministry of Defence. Under the draft law, different standard contracts for scientific research on and development and purchase of military equipment were being developed by the Ministry of Defence. These contracts would include the work schedule, a protocol of agreement about prices, a protocol of agreement about the dispensation of funds and compensation for default, and a protocol of agreement about modifying the contract price.[56]

If it had been adopted, this practice of using contracts to regulate procurement would have forced the Ministry of Defence to accept the implications of changes in the cost of production. However, the draft never became law and in practice procurement discussions with industry still refer to a central index of prices.[57] Another dimension of the proposal to move to a contract-based procurement system was that prime contractors would have become solely responsible for managing relations with subcontractors and suppliers of other inputs. In practice these relationships are still managed to some extent by state organizations—notably the Ministry of Defence Industry—on behalf of manufacturers.[58] An exception to this may be those subcontracting relationships that exist between Russian enterprises and enterprises located in other members of the CIS. In interviews with Russian enterprise managers partners in other CIS

[54] Kolpakov, S. and Drugov, Y., 'Effects of industry demilitarization and radical economic reform in Russia on the branches providing materials for military production'. Unpublished manuscript, Apr. 1996.

[55] For a general discussion, see Hanson, P., 'The future of Russian economic reform', *Survival*, vol. 36, no. 3 (autumn 1994).

[56] Vlasov, V. I., 'The supply of arms and military equipment for the Russian armed forces: tendencies in development of Russian defense industries'. Unpublished paper, Apr. 1993.

[57] This was partly because of the difficulties of negotiating with industry against a background of massive inflation. However, the discussions also became part of a wider discussion about the division of responsibility between government agencies in the management of the Russian defence industry. This discussion principally involved the Ministry of Defence, the State Committee on Defence Industries (later the Ministry of Defence Industry) and the State Committee for Property Management.

[58] In Mar. 1997 the Russian Government abolished the Ministry of Defence Industry with implications that are not yet clear.

states were often named as the worst offenders as regards late payment or payment default.[59]

## Russian weapon prices in foreign trade

Manufacturers of defence equipment therefore find themselves 'squeezed' between the need to pay increased prices for inputs and the inability to pass on these costs in full to their only domestic customer, the government. This means that in Russia the prices applied in domestic trade are still established on a different basis from the prices applied in foreign trade.

For a brief period in 1992–93 Russian manufacturers and trading organizations believed that they could deal in arms in the same way as other goods and services. However, efforts to negotiate contracts with foreign governments without state assistance usually failed and it is now understood that neither government nor industry can conduct large-scale arms exports successfully unless they cooperate.[60]

Statements by Rosvooruzhenie suggest that price negotiations take into account both what is known about Western pricing policies and information from Russian enterprises about their cost base after the partial liberalization of input prices described above.[61]

Since 1994 Russia has moved towards a system in which negotiations with foreign governments are undertaken by mixed teams of government officials, including representatives from several ministries, and representatives of industry. The negotiations move in stages. First, a decision is reached about the types of system which may be desired by the buyer and whether or not these will be released for sale by the Russian side. After the release of the systems requested by the buyer has been approved, questions of quantities and prices are addressed. In these discussions the needs and views of Russian industry now receive a much more prominent place than was the case in the Soviet period. After a broad framework of quantities and prices has been agreed between the Russian Government and the foreign buyer, enterprises discuss with the Russian Government who will produce which items.

Available evidence suggests that compared with the Soviet period more recent arms exports have gradually increased the share of currency in overall payment. Under the 1991 agreement with China to supply Su-27 fighter aircraft as much as 70 per cent of the value of the deal was to be covered by transfers of Chinese consumer goods to Russia.[62] After 1992 Russian negotiators appear to have reversed the balance so that 70 per cent or more of the value of contracts with China are paid in hard currency.

[59] Webster *et al.* (note 51), p. 17.

[60] According to Rosvooruzhenie only one of the enterprises permitted to conduct independent foreign trade activity—aircraft manufacturer MiG-MAPO—has chosen to do so. Tarasova, O., [Rosvooruzhenie calls for unity], *Segodnya*, 1 Nov. 1996.

[61] 'Russian defence exports: the insider's view', *Military Technology*, Sep. 1996, pp. 65–67; and *Nezavisimaya Gazeta*, 28 June 1996 (in Russian) in FBIS-SOV-96-126, 28 June 1996, pp. 26–27.

[62] For details, see chapter 11 in this volume.

**Table 4.6.** Financial aspects of the 1994 Malaysian MiG-29 agreement

|  | Malaysian $m. | US $m. |
|---|---|---|
| Total cost of aircraft | 1 516.35 | 590.02 |
| Cost of training package | 3.40 | 1.32 |
| Avionics retrofit | 238.22 | 92.69 |
| Simulator | 114.27 | 44.46 |
| Infrastructure support | 142.60 | 55.49 |
| **Total** | **2 014.84** | **783.98** |

*Note:* Converted at the exchange rate existing at the time the agreement was signed.

*Source: Asian Military Review*, Aug.–Sep. 1993, p. 16.

One of the test cases through which the procedures for negotiating arms contracts were developed was the agreement with Malaysia over the transfer of Russian fighter aircraft.

### The case of MiG-29 fighter aircraft supplied to Malaysia

In June 1994 Russia and Malaysia signed an agreement on the transfer of 18 MiG-29 fighter aircraft. This case has provided fairly detailed information about the economic and financial aspects of a bilateral arms transfer.

The agreement included the supply of 16 MiG-29M multi-role fighters and two MiG-29UM trainer aircraft. However, the trainer aircraft were to be equipped with all systems needed to make them fully combat-capable. Table 4.6 summarizes the financial details of the agreement. Russia agreed to supply the armament for the aircraft under a separate agreement. The figures in the table below therefore exclude R-27 medium-range air-to-air missiles, R-73 short-range air-to-air missiles and internal twin-barrel 30-mm calibre guns.

While the aggregate value of this agreement was over $780 million excluding the primary armament for the aircraft, which would in itself have a significant value, this does not translate into equivalent revenue for Russia because of the way in which the agreement was structured.

First, two elements of the overall package were to be supplied by third parties. The training package was to be implemented by a team of Indian pilots, technicians and engineers who were already operating the MiG-29 in Indian Air Force service. The avionics retrofit was to be conducted by British company GEC Marconi which supplied the aircraft with new tactical navigation and attack systems, a new identification/friend or foe (IFF) system and new ultra-high frequency (UHF) and very high frequency (VHF) telecommunications.

Second, the Russian parts of the agreement (together worth $690 million) were not all to be financed through currency payments. Around 60 per cent of the value of the contract was to be in hard currency while the remaining 40 per cent was to be provided in goods such as palm oil and textiles.[63] The structure of the offset element of the package was itself complex. The entire value of the contract was to be provided to Russia immediately. The Malaysian Government was to borrow this money and a series of different lending options were considered including borrowing from banks in Singapore, from a consortium of European banks or from the Russian Central Bank. Under the agreement Russia would meet its offset obligations in two ways. Those Russian enterprises involved in the programme would guarantee to buy goods in Malaysia up to a value of $150 million which would be credited to Russia's offset account.[64] In addition, the Russian Government would provide certain services to Malaysia which would also be credited to Russia's offset account. In one joint initiative, Russian technicians would be assigned to the Aerospace Tech Systems Corporation. This company, registered in Malaysia, is expected to provide repair and maintenance for the MiG-29 aircraft beyond the warranty period under which they are maintained by Russian personnel under the original agreement. In a second initiative, Russian engineers and technicians provide courses at the University Sains Malaysia north of Kuala Lumpur.[65]

In this case the full value of the equipment and services provided under the Russian-controlled elements of the agreement was transferred in cash. Under these conditions exports certainly contribute directly to revenues.

## V. Conclusions

In the introduction to this chapter two questions were posed. First, were arms exports profitable to the Soviet (and now Russian) economy? Second, did arms exports yield hard currency and, if so, how much? The information available suggests tentative answers to both questions.

During the Soviet period arms exports seem to have brought significant economic benefits if allowance is made for the peculiarities of the overall economic and industrial system in which they were located. Since the production system was developed primarily to meet Soviet military requirements—and would have existed regardless of decisions to export or not to export—the costs of production for export were treated as marginal costs. However, there is evidence that in a large number of its bilateral arms relationships the Soviet Union acquired either currency or goods that were needed and would otherwise have been difficult to obtain.

The amount of hard currency derived from arms sales during the Soviet period remains impossible to quantify in spite of the recent release of information about the earlier period by Russian authorities. The data which have been released still appear to refer to the estimated value of exports and so do not

[63] *Asian Recorder*, 2–8 July 1994, p. 24070; *Aviation Week & Space Technology*, 8 Aug. 1994, p. 28; and 'MiG-29 planes to be delivered to Malaysia', ITAR-TASS (in English), 17 Aug. 1994 in FBIS-SOV-94-160, 18 Aug. 1994, p. 11.

[64] *Asian Recorder* (note 63).

[65] *Defense News*, 10–16 Mar. 1997, p. 16.

allow for special factors in Soviet trade such as clearing arrangements in multilateral and bilateral trade, non-cash payment (i.e., barter) and non-payment or default.

The information available suggests that Soviet arms exports were far more profitable to the central state authorities than they were to the manufacturing enterprises. However, this appears to be changing in line with the overall process of economic and political reform. As government and industry develop and implement procedures that enable them to work together there is evidence that enterprises (in particular those that can sell large, complex systems) will prefer exports over sales to the Russian Ministry of Defence.

Paradoxically, this fact is not necessarily good for Russian economic prospects. Some of the factors which assist Russia in exporting arms have a negative impact on other economic areas. First, achieving profits through the distortions created by maintaining price controls means that an effective subsidy is paid to manufacturers by other parts of Russia's economy. Second, the measures taken to give financial relief to manufacturers in the form of special tax exemptions and centrally financed funds and subsidies is a barrier to the development of a more simple and more enforceable system of financial regulation.[66] Third, because foreign trade is often denominated in dollars Russian producers who depend on exports prefer a weak rouble which has a corresponding impact on the costs of imports.

The main barrier to successful exports from a company perspective is the reality of the post-cold war arms market in which foreign contracts are relatively few and difficult to win against fierce competition.

At the same time equipment has also been transferred to CIS states as grant military assistance as part of the attempt by Russia to develop cooperation in defence and security matters.

---

[66] This is not unique to Russia, although the need for a more effective system of regulation is probably greater in Russia.

# 5. The influence of external factors on Russia's arms export policy

*Sergey Kortunov*

## I. Introduction

In the former Soviet Union, with its administrative–command economy, there existed an effective system of state control over arms exports by the executive authorities.[1] Moreover, the USSR supplied weapons primarily for political and ideological reasons—often using concessional economic arrangements that were, in effect, a form of military assistance—as part of its overall competition with the United States in particular. Under these conditions, the arms export policy of the Soviet Union can be seen as a subordinate element of an overall security concept shaped by the cold war.

Domestically, the policy-making system was designed to prevent damage to the military and economic potential as well as to the political and defence interests of the USSR. It balanced a number of institutional actors, each of which contributed with its specific expertise. The system was based on a number of decrees and other government acts which took into consideration the USSR's international commitments. However, the system of policy making was characterized by a complete state monopoly as well as strict secrecy in decision making. Neither the broad criteria according to which decisions were taken nor the decisions themselves were subject to public discussion.

The fact that policy was made and implemented in a command economy also contributed to the specific administrative form that the process assumed.

With the disintegration of the USSR the international and domestic framework in which Soviet arms export policy operated practically ceased to exist. Politically, the collapse of the Soviet Union led to a 'de-ideologization' of all aspects of decision making in post-Soviet Russia. At the same time, Russia has initiated a transition from an administrative–command economy to a market economy. At the international level it has sought to escape its economic isolation and bring about extensive integration with the global economy. Taken together, these factors have inevitably affected the arms trade and its supervision by the executive authorities.

Domestically, a bitter struggle has started and is still continuing in Russia between various institutions, each of which would like to increase its control over security policy, including policy in the area of military–technical cooperation with foreign states, the right to trade in arms and the right to control this trade. Until this struggle is resolved it will not be possible to create a national

[1] This system is described in chapter 3 in this volume.

consensus on these matters. In the meantime the continuous changes in the Russian system for military–technical cooperation, which sometimes cause understandable astonishment in foreign as well as domestic political and public circles, are likely to continue. However, this situation has its objective causes and reflects the transitional nature of the present historical period in Russia. It will pass.

This chapter discusses the impact of the changes in the international and domestic environment on some of the broad issues of military–technical policy and arms exports.[2]

## II. Competition in the international market and state protectionism

The new Russian national system of military–technical cooperation and export controls has been built by the executive authorities in an extremely complicated international environment. This external context has affected Russian domestic processes both directly and indirectly.

After the cold war the central structural features of the international arms market were more fluid and difficult to understand and predict than had been the case under previous conditions. What had seemed to be fixed elements in the market—such as the arms transfer relationships within major alliances or with friendly states in the developing world—could no longer be relied upon. At the same time, forms of cooperation which proved to be impossible during the cold war—notably multilateral discussions of arms export policy and control—now seemed to have some prospects for success.

The immediate consequence for Russia of the changed conditions was a sharp fall in the volume of arms sales in the international market in the period 1990–93. Russian industry failed to make use of its fairly powerful export potential. There are several reasons for this but one of the most important was that Russia was unprepared to adapt to one new characteristic of the market—competition among the main arms exporters.

With the end of the East–West confrontation, the demand for weapons in the Euro-Atlantic area suddenly fell, dealing a heavy blow to the interests of arms producers in major Western countries. The governments of the principal arms-exporting countries (including Russia) were put under pressure by domestic arms manufacturers to pursue a policy of increased protectionism with regard to orders placed with their domestic defence industry. Governments also came under pressure to participate more actively in promoting the products of the arms industry in international markets. Top government officials in France, the UK and the USA—up to and including the heads of government—became active in marketing arms. The efforts in the autumn of 1992 by US President George Bush and Secretary of State James Baker to promote the sale to Taiwan

---

[2] A more detailed description of the changes in regulations and administrative procedures is presented in chapter 6 in this volume.

of 150 F-16 fighter aircraft and of French President François Mitterrand to sell Taiwan 60 Mirage fighter aircraft attracted particularly widespread attention.

To a certain degree this trend may have been intensified by the 1990 Treaty on Conventional Armed Forces in Europe (CFE Treaty), which established ceilings for specified conventional armaments in the national armies of state parties. The indirect impact of the treaty and the improved European security environment may slow down the future purchases of military equipment by defence departments of state parties.

The end of the cold war was also probably a contributing factor in heightening tensions and increasing the temptation of some states to pursue national ambitions which had been kept in check by the global military and political confrontation between the Soviet Union and the United States. Perhaps as a result of uncertainty about their national security in the future international system, there was a marked increase in demand for new arms in particular in countries of South-East Asia. Iraq might be cited as a country which probably could not have pursued its recent policies of national aggrandizement during the cold war.

The 1991 Persian Gulf War also highlighted the need for the most modern weapons among developing countries, presenting these countries (including many former Soviet clients) with the problem of how to re-equip their armed forces.

This combination of domestic crisis and the emergence of what seemed to be new international opportunities led arms producers in all industrialized countries that manufactured military equipment to place greater emphasis on exports within their business strategies.

As a result of the above factors, what had been a rather predictable and tightly managed international arms trade system was replaced by a market in which the major arms exporters—France, Russia, the UK and the USA—now pursued an unprecedented competition for new and old clients no longer restrained by the political or ideological considerations taken into account during the cold war. Safeguarding defence industrial interests became the dominant interest within the military–industrial complexes in these countries.

It is common knowledge that the modern market requires certified products, reliable after-sale service and supply of spare parts. This was always a bottleneck of the Soviet arms trade. For decades, this trade was built on barter agreements with developing or underdeveloped countries prepared to accept minimal after-sale service. Good weapons landed up in not very skilful hands that were not always ready to use them properly. This is illustrated by the experience of the earlier conflicts in the Middle East and the Persian Gulf War.

Apart from producing excellent military equipment it was necessary to ensure its reliable functioning and skilfully organize sales on the world market—something Russian manufacturers had never learned to do, nor could they promptly find their way in this market. For example, the market situation for sophisticated high-technology products is at present fairly difficult, while the market for spare parts for outdated technology in a world saturated with Soviet-made

MiG-21 and MiG-23 aircraft looks very different. In many countries this equipment is still in operation and requires service, repairs and modernization. Russian manufacturers have apparently failed to appreciate the potential of this vast and promising market, which is gradually becoming filled with Western products even when Soviet military technology is involved.

## Multilateral discussions of arms transfer control

During the cold war it proved practically impossible to organize multilateral discussions of arms transfer control. Although the United States and the Soviet Union held bilateral discussions in the late 1970s, these discussions broke up without any results.[3] For most of the cold war, the only multilateral export control regime was the Coordinating Committee for Multilateral Export Controls (COCOM), which was a mechanism used by Western powers to undermine Soviet military–technical capabilities. After the end of the cold war COCOM began to undergo a transformation that eventually led to its dissolution in March 1994.

After the invasion of Kuwait by Iraq in August 1990 there was unprecedented cooperation between the five permanent members of the UN Security Council (the P5) in agreeing a series of measures to prevent Iraq from recreating its prewar arsenals. A series of Security Council resolutions laid down far-reaching arms control measures to be applied to Iraq. In the aftermath, the idea of cooperative actions by the P5 to moderate regional arms programmes in the Middle East appeared to offer some hope of success. An initiative on Middle East arms control was presented in May 1991 by President Bush which included the suggestion that the P5 should take special responsibility in the area of arms transfers.[4] The Bush initiative set the agenda for a process which led to meetings of the P5 in Paris in July 1991 and in London in October 1991.[5] At the October meeting the P5 agreed among themselves a series of Guidelines for Conventional Arms Transfers.[6]

The P5 met once more, at the level of senior officials, in Washington in May 1992. By this stage it was already clear that there was little further progress in the discussion of conventional arms transfers. The public documents and statements after the May meeting related almost entirely to the issue of proliferation of nuclear, biological and chemical weapons along with missile delivery systems. It was not possible at this stage for the P5 to reach agreement about giving one another advance notification of arms agreements. It was tentatively agreed that the representatives of the P5 would meet again in late 1992 in Moscow. However, in October 1992 the Chinese Government suspended its

---

[3] Four rounds of the Conventional Arms Transfer Talks (CATT) were held between Dec. 1977 and Dec. 1978. Spear, J., *Carter and Arms Sales: Implementing the Carter Administration's Arms Transfer Restraint Policy* (Macmillan: London, 1995), chapter 7.

[4] Middle East Arms Control Initiative, White House Fact Sheet, 29 May 1991.

[5] Anthony, I. *et al.*, 'The trade in major conventional weapons', *SIPRI Yearbook 1992: World Armaments and Disarmament* (Oxford University Press: Oxford, 1992), pp. 292–94.

[6] These guidelines are reproduced as appendix 1 in this volume.

participation in the talks when the United States announced the sale of F-16 fighter aircraft to Taiwan.[7]

The talks of the major arms exporters underlined that they were not interested in establishing any mechanism for controlling and limiting the conventional arms trade at this stage. By late 1991 the desire to maintain flexibility and make deals without any international controls began to outweigh the desire to work out a mutually acceptable code of behaviour and detailed policies defining responsible behaviour in the sphere of arms trade. Once the USA found out that the practice of prior notification of arms supplies would not be accepted (as is known, it attempted to impose this procedure upon other negotiating parties primarily in order to improve its own chances against other exporters) it lost virtually all interest in the talks.

As a result, the guidelines, worked out with such difficulty in 1991, were sacrificed, not in any long-term political interests but for more immediate financial gains.

France was perhaps relieved to see China leave the talks, particularly because the suspension of the talks could formally be attributed to the USA. A few months later, the contract between France and Taiwan on the supply of 60 Mirage-2000-5 fighter aircraft was announced, a deal which would itself have complicated the P5 discussions. After mediation by Russia, China subsequently agreed to resume the work of the P5 in a new format. However, by this time France had changed its position and (contrary to the previous understanding reached between it, Russia, the UK and the USA) used various pretexts to avoid resuming the discussion.

Considering the place and the role of the military–industrial complex in the Russian economy (and previously in the economy of the USSR) the end of the cold war had harsher consequences for Russia than for any other major arms supplier. In 1992 alone, Russia's arms production dropped by over 60 per cent as compared to the levels recorded for 1990.[8] The drastic fall in military purchases by the state led to work being stopped in whole factories. Huge numbers of employees in the defence industry, which had always enjoyed a privileged status in Russia and hence constituted a stable social base of the former regime, lost this position and turned instead into a potential source of social instability.

The Russian Government could not disregard this fact and it was one of the reasons which forced Russia to follow the example of the principal Western governments and adopt a policy of state protectionism in respect to its domestic arms manufacturers. In spite of its domestic problems, Russia did not suspend its participation in the P5 discussions. However, the Russian position regarding the need for multilateral controls did undergo a change. It is enough to recall the evolution of the views expressed and the statements released by the Russian Foreign Minister, Andrey Kozyrev, on this question after 1991. In January

[7] Anthony, I. *et al.*, 'Arms production and arms trade', *SIPRI Yearbook 1993: World Armaments and Disarmament* (Oxford University Press: Oxford, 1993), pp. 460–61.

[8] Glukikh, V., 'Reform and stabilization of the defence industry', *Conversion*, vol. 1, no. 3 (1994), pp. 17–18.

1992, in an article intended to underline the differences between the new Russian and the former Soviet foreign policy, Kozyrev observed that Russia 'will share with and help those who are in real need and use the resources obtained not for building up their military and police forces, but for the socioeconomic development of the countries'.[9] After meeting Secretary of State Baker in Moscow in March 1992, Kozyrev expressed his support for efforts to reduce the arms trade in the Middle East, a region he described as 'saturated with weapons'.[10] By 1993, when the P5 talks had been suspended, the Russian position increasingly tended to reflect the need to increase export sales. Although in 1994 Kozyrev still favoured cooperation in defining the principles that should govern exports of military technology, he qualified this with the observation that partnerships 'cannot negate a firm, even aggressive, policy of defending one's own national interests'.[11]

Studies of the practices adopted by major Western arms-exporting countries in ensuring effective control over sales in free market conditions have been important for the development of a Russian control system. In this connection it was essential that the emergence of a national export control system in 1992–93 be accompanied by the establishment of contacts and a widening dialogue between Russia and Western countries in this field. This dialogue, if it is continued, should eventually bring about a harmonization of export control systems for conventional weapons. The Wassenaar Arrangement on Export Controls for Conventional Arms and Dual-Use Goods and Technologies, the newly emerging international mechanism for export control which is replacing the disbanded COCOM, is bound to play an important role in continuing this work.[12]

At the same time, it is worth noting that control of conventional arms supplies is one of the few areas that lack a developed international legal framework, although the Wassenaar Arrangement now provides a mechanism for international talks and consultations.

The main international obligations of Russia in the field of conventional arms control follow from the 1991 Guidelines for Conventional Arms Transfers. These offer some criteria for deciding whether or not a specific delivery should be permitted. The P5 countries, including Russia, undertook to avoid arms deliveries which could prolong or exacerbate an existing armed conflict, increase tension in a region, introduce a destabilizing military potential, violate an embargo or other internationally agreed restrictions, be used for other pur-

---

[9] Kozyrev, A., 'Transformed Russia in a new world', *Izvestiya*, 2 Jan. 1992 (in Russian) in Foreign Broadcast Information Service, *Daily Report—Central Eurasia* (hereafter FBIS-SOV), FBIS-SOV-92-001, 2 Jan. 1992, pp. 77–81.

[10] Moscow Mayak Radio Network, 'Kozyrev, Baker comment during news conference', 12 Mar. 1992 (in Russian) in FBIS-SOV-92-050, 13 Mar. 1992, pp. 1–2.

[11] Kozyrev, A., 'No sensible choice but a true US–Russia partnership', *International Herald Tribune*, 19–20 Mar. 1994, p. 6.

[12] On the Wassenaar Arrangement, see Anthony, I. and Stock, T., 'Multilateral export control measures', *SIPRI Yearbook 1996: Armaments, Disarmament and International Security* (Oxford University Press: Oxford, 1996), pp. 542–45; and Anthony, I., Eckstein, S. and Zanders, J. P., 'Multilateral military-related export control measures', *SIPRI Yearbook 1997: Armaments, Disarmament and International Security* (Oxford University Press: Oxford, 1997), pp. 345–48.

poses than providing for the legitimate defence of the receiving state, support international terrorism or seriously undermine the economy of the importer.

Decisions on arms supplies from Russia are also made with reference to decisions by international organizations. The UN has banned deliveries to certain countries and requested greater transparency in armaments. The countries currently subject to mandatory UN arms embargo include Iraq, Libya and Yugoslavia (Serbia and Montenegro). The Organization for Security and Co-operation in Europe (OSCE) has also established the principles that should govern arms transfers, although these are politically rather than legally binding measures.[13] It is the responsibility of the Russian Ministry of Foreign Affairs to monitor the observance of Russia's international obligations in this sphere.

These decisions by international organizations provide some guidelines for arms transfer policy. However, Russia has not yet found a stable and consistent national arms transfer policy.

## III. Main issues in domestic discussions of arms transfer control

There are three central issues which have been under intense discussion in Russia since 1992. The first of these is the overall national security policy concept which should guide Russian decisions, including those on arms transfers. The second is the proper relationship between government and industry, and the third is the division of responsibility for different issues between state agencies and authorities and the relationship of these areas of government to one another.

Russia has yet to determine its national interests and is still in the process of finding a national identity. The absence of a national security concept to replace that which guided decisions in the Soviet Union leads to a specific dilemma in the field of arms transfers. Russia needs to seek and win international markets for arms in order to help an industry in crisis. At the same time unlimited exports of sophisticated weapons and high technologies could damage national security.

In some cases there will be no conflict between the desire for commercial benefits from exports and the need for security. However, in cases where this is not so clear Russia has been searching for a way to strike a balance between a state arms trade policy in which political and military aspects are predominant and fulfilling the economic or commercial interests of the country. In another set of cases it could be useful to export regardless of commercial considerations. Is the priority to earn money from arms sales or are there circumstances where, for example, attracting this or that country into the orbit of one's political influence justifies the use of military assistance?

No country has fully resolved these questions. However, in seeking the balance between political and commercial interests Western countries have established legal and administrative means for evaluating the alternative options. Russia has just started to develop such means.

[13] Reproduced in appendix 2 in this volume.

In the former Soviet Union priority was given to political considerations in arms transfer decision making, often to the detriment of commercial considerations. For decades the Soviet military–industrial complex received guaranteed payments from the government for arms manufactured for export. A significant portion of this military equipment was either sold at concessional rates to foreign countries or, on occasion, given away. The exceptions were probably the supplies to Iraq and Libya, which produced more or less stable revenues— although even here the debts owed by these countries were increasing at the time of the dissolution of the Soviet Union.

This pattern of business was inconsistent with the new efforts, which began in the late Soviet period, to find a new balance between the role of the state and the role of other actors within the economy.

A period of what might be called 'market romanticism' started around 1990. There was a belief that economic conditions would improve if the state withdrew entirely from economic decision making. Although the Soviet Union was participating in the discussions of multilateral arms export control, this period saw the first attempts to put military–technical cooperation on a pure business footing, giving maximum freedom to manufacturers. During the period of President Mikhail Gorbachev's administration, priority was already being given to the 'sell to anybody who pays' principle. The only exceptions were countries under UN sanctions or those obviously hostile to the USSR. All foreign transactions were intended to be conducted on the basis of hard currency payment.[14]

It soon turned out that in the field of arms transfers the slogan 'cash during the year of delivery' was unrealistic in a situation of crisis in the defence industry—which was not ready to face international competition and had no experience of how to conduct itself in a market environment. Hard-currency income from arms sales actually dropped dramatically. The idea of relying on hard-currency earnings from arms sales to finance conversion of the defence industry to civilian production did not prove workable. At the same time, the state monopoly on the arms trade was weakened.

The effort to find rules which balance the principle of a free-market economy with strict control over exports by the executive authorities is also new to Russia. While Western countries do not have a single model or approach to the issue of ownership and control over arms industries, they have found mechanisms for allocating responsibility which are more efficient than those currently in existence in Russia. This allows them to combine an efficient policy of promoting their weapons on international markets without sacrificing state oversight and control. Russia has just embarked upon the path of searching for such a balance.

Increasing re-examination of the relationship between the state and industry was bound to make an impact on the Russian national system of arms exports

[14] Kortunov, S., 'Russian aerospace exports', ed. R. Forsberg, *The Arms Production Dilemma: Contraction and Restraint in the World Combat Aircraft Industry*, Center for Science and International Affairs Studies in International Security no. 7 (Massachusetts Institute of Technology Press: Cambridge, Mass., 1994), p. 93.

that was emerging after 1992. Issues of privatization, for example, were the subject of fierce struggles internally in Russia (including struggles both within and between various groups in the government). The defence industry and state control over the arms trade were not exempt. A very wide spectrum of views was expressed, including views advocating a return to a state monopoly and others advocating a very liberal regime that would grant extensive rights to manufacturers (although still maintaining a state system of export licensing).

The various agencies and groups which participated in this struggle over policy were not always themselves completely of the same mind. For example, although the State Committee on Defence Industries (Goskomoboronprom) generally advocated more freedom of action for defence factories, this view was not always shared by the factories themselves, some of which preferred a direct role for the state.

These struggles over national security and defence industrial policy have been conducted with fluctuating success for different points of view. This is the main cause of the never-ending series of reorganizations of the national export control system. As any given point of view gained the upper hand within the executive branch, this would be reflected in decisions and decrees.

Russian manufacturers were not discouraged by the setbacks in the world arms market. During 1993–94 they continued to fight for their independence (partly through the State Committee on Defence Industries). Objectively, this was to result in the weakening of the state monopoly in this sphere.

In May 1994 a special decision of the Russian Government approved an ordinance on certification of companies for the right to export arms and *matériel*, as well as work and services.[15] Enterprises which developed and produced arms and *matériel*, once they were certified and registered as participants with foreign economic activities in the field of military–technical cooperation, were allowed to look for foreign customers in countries with which such cooperation was not forbidden. They were also allowed to demonstrate arms and hand over, during the course of negotiations, tactical and technical specifications of arms and *matériel* approved for export, to convey duly agreed approximate prices, to do the marketing, to sign contracts and, on the basis of duly obtained licences, to export independently arms, *matériel*, work and services produced in excess of the government defence orders.

## IV. The national control system in transition

The 'ebb and flow' of different interests can be traced through the development of the Russian national export control system.[16]

---

[15] Decision of the Government of the Russian Federation on granting the enterprises of the Russian Federation the right to participate in military–technical cooperation with foreign countries, no. 479, 6 May 1994, reproduced in appendix 3 in this volume as document 10.
[16] A detailed description of the administrative steps taken to create an export control system is presented in chapter 6 in this volume.

In 1991, at the end of the Soviet period, two state-owned foreign trade companies, Oboronexport and Spetsvneshtekhnika, which had been subsidiaries of
the MFER, were hived off, acquired the status of independent legal entities and
were licensed to trade in arms. At the same time, unhappy with what they saw
as the record of inefficiency of these structures when they were within the
ministry, certain military aviation enterprises also began to seek independent
foreign-trade rights. The State Committee on Defence Industries began to claim
that it now represented the interests of producers. Finally, a range of new or
potential actors began to emerge claiming that they could manage arms exports
within the private sector, without assistance from state agencies, experts 'or
professionals.

Direct contact between defence enterprises and foreign partners was allowed
as early as 1991 to increase the effectiveness and profitability of arms sales and
make enterprises less dependent on intermediate state agencies. Directors of
arms enterprises proclaimed their readiness to find buyers for their products and
conduct negotiations by themselves. Once put on a commercial basis, it was
argued, military–technical cooperation would provide equal opportunities for
manufacturers and trading companies to secure financial gains directly, in contrast to the previous practice when all the earnings went to the state budget
before being passed on to the defence industry.

At the same time, the reality of the market place was a drastic reduction in the
volume of state orders, which left defence enterprises facing a crisis—especially after the failure of an ill-prepared government conversion programme.
They attempted to sell the *matériel* they produced by any available means.
Factory managers were backed by thousands of employees whose jobs were in
danger. At the same time there was a perception that the collapse of the defence
industry would lead to the loss of unique technological processes and design
work.

These conditions inevitably led to the weakening of foreign trade controls in
1991. The main manifestation of this, along with the emergence of new sovereign states and the absence of interstate border controls in certain sectors of the
post-Soviet space, was initially almost uncontrolled export of non-ferrous
metals and valuable strategic raw materials. Companies which had only been
licensed to look for potential buyers—a marketing exercise—for Russian products often attempted to sell these as well. While this was mostly occurring in
the non-ferrous metal and raw materials sector, in a number of cases it also
happened with defence goods.

It quickly became apparent that neglect of a more systematic trade policy was
leading to chaos—in fact to the disappearance of potential markets. The same
arms were being offered to customers by different sellers who would engage in
a price war that led to the items being offered below the cost price. The absence
of coordination between the manufacturers in certain markets posed a danger to
the entire future of military–technical cooperation with those countries.

In 1992–93 the balance in state policy began to swing towards a more regulated system. On 22 February 1992, the Russian President signed a decree 'On

types of products (work, services) and industrial wastes that cannot be sold openly' which introduced obligatory licensing for sales of arms and *matériel* as well as of other special items. In order to ensure a unified state policy in the field of military–technical cooperation, the Interdepartmental Commission on Military–Technical Cooperation between Russia and Foreign States (Komitet voyenno-tekhnicheskogo sotrudnichestva, KVTS) was appointed by presidential decree 507 on 12 May 1992.[17] A regulation of the Council of Ministers of 28 January 1993 established a list of military products (work and services) that could be imported into or exported from Russia subject to a licence and defined licensing procedures.

At the same time, Russian salesmen and arms manufacturers were arguing that in order to advance into new non-traditional markets and to participate successfully in the international market place they needed new types of assistance from the state. For instance, hard-currency credit was urgently required to assist in developing technologies that could be competitive internationally and to support the capital investment programmes of manufacturers capable of producing modern technology.

The government was unable to provide such credit but in November 1993 a new state agency, Rosvooruzhenie, was created. Among its tasks was taking charge of investing private and government funds in the Russian military–industrial complex to develop, on contract basis, weapons that would be in great demand in the world market.

During 1994 further modifications to the decision-making process were made in an effort to improve coordination between the ministries and agencies that had an interest in military–technical cooperation. For example, in the autumn of 1994, the president created the post of special assistant to the president on military–technical cooperation—a post held by Boris Kuzyk. In December 1994 the president also set up a State Committee on Military–Technical Policy (Gosudarstvenny komitet po voyenno-tekhnicheskoy politike, GKVTP) under his authority. According to the decree which established it, no. 2251, 'On the State Committee of the Russian Federation on military–technical policy', this committee had wide-ranging authority over military–technical cooperation, the future direction of policy on military technology development, the state defence order, modernization of armaments and conversion. It also brought together representatives of different interested agencies from ministries, industry and the armed forces in an effort to develop a unified policy on these issues. These changes underlined another unanswered question in establishing Russia's national export control system: Would the system be more effective if it were the responsibility of the presidential administration or coordinated between government ministries by a committee, or should this responsibility be given to a new agency established specifically for the purposes of export control?

---

[17] 'On military–technical cooperation of the Russian Federation with foreign states', Presidential Decree no. 507, 12 May 1992.

To summarize, since 1992 a framework of export controls has started to take shape in Russia which pays due regard both to the country's military–economic and to its military–political interests. This framework is being improved with maximum regard to international experience in the field of export controls and to the experience in the market-place of the major arms exporters.

In order to establish a legal framework of military–technical cooperation, a bill on military–technical cooperation between Russia and foreign countries was prepared at the Russian Government's request and sent to the parliament for discussion. This has not yet been passed into law. Another document, 'The concept of military and technical cooperation between Russia and foreign countries', has also been drafted but never formally approved or accepted, and a decree of the Russian Government of May 1994 intended to reflect the new power structure in the field of military–technical cooperation between Russia and foreign countries has never been implemented.

## V. A future agenda for Russian export control

Russia is putting in place an export control system not as a favour to Western countries but to protect its own interests. Russia has inherited an unstable periphery and would be an early victim of widespread conflict, perhaps a major victim if conflict involved the use either of non-conventional weapons or of advanced conventional weapons.[18] However, Russia has not yet managed to put in place an export control system which balances different economic, political and military interests or which defines and balances the roles and functions of different state agencies and administrative units.

The creation of the GKVTP could have been a big step forward in consolidating the strict state monopoly in the field of arms trade and military–technical cooperation. Internationally, a greater role for centralized executive authorities is the main tendency in the development of national systems for arms export control today. It is increasingly common for export control to be regarded as a separate government function requiring its own specialized agency rather than being the domain of any ministry or department. There are reasons to believe that this tendency is at an early stage and will be further strengthened in future through the Wassenaar Arrangement as the current problems facing this body are resolved.

Unfortunately, it is possible that an earlier mistake is being repeated under the present organization of Russian authorities. The promotion of Russian arms in world markets and decisions about export control are being entrusted to the same authority. The development of exclusive competence for the GKVTP could, in practice, have led to a weakening of political control over the arms trade by the Ministry of Foreign Affairs and Federal External Intelligence Service.

[18] Kortunov, S., 'Russian–American cooperation on counterproliferation', *The Monitor*, vol. 1, no. 4 (fall 1995), pp. 1–9.

Once the tasks of trade promotion and the implementation of licensing decisions were reallocated, policy-making bodies could focus their attention on meeting the fundamental challenge of coordinating export policy with national security policy. It would be necessary to ascertain the balance between and coordinate the following industrial policy processes: reductions in the state orders for arms and *matériel*, conversion, diversification and transformation of the forms of ownership. A market mechanism for implementing conversion and diversification programmes in defence production still needs to be developed.

In Russia policy as regards conventional arms is separate from policy on dual-use goods.[19] At present there are no clear links between Russian policy with regard to the various international export control regimes such as the Wassenaar Arrangement and the Missile Technology Control Regime (MTCR). Controls over information classified as national secrets and control over exports of goods and services which are or can be used in manufacturing various types of arms and *matériel* also function separately.

In the near future it may become logical to link up and eventually merge government control mechanisms for military–technical cooperation and the transfer of information, manufacturing technology, dual-use products and materials.

Additional measures of government regulation are required to cover the issue of scientific knowledge and intellectual property—which are so far not subject to export control in Russia. The government needs specific powers to oversee four specific areas: (*a*) the transfer of technologies and scientific and technical information in high-priority fields of science and technology, which are crucial for sustaining the scientific and technical potential of the country and guaranteeing its defence; (*b*) the power to control transfers of intellectual property developed by the state in order to defend the state's interests, including the investment of this kind of intellectual property in joint enterprises in Russia and abroad, obtaining foreign patents and selling licences and know-how; (*c*) the registration of international scientific and technical exchanges where results of scientific and technical activities are being transferred; and (*d*) as an enforcement mechanism, the power to make selective checks and inspections of international scientific and technical exchanges in order to identify those who violate existing legislation and to bring them to justice. Instructions to intelligence agencies would also be required in this regard.

The classifying and declassifying of information in the fields of defence, the economy, science and technology should be linked with the processes of controlling exports of information in these spheres and regulated within a single frame of reference. On the basis of the existing regulations and the new powers referred to above, a centralized policy-making authority could therefore have the following additional functions: (*a*) working out and inventorying scientific and technical activities developed during the Soviet period and standard regula-

[19] For a discussion, see Kortunov, S., 'National export control system in Russia', *Comparative Strategy*, vol. 13 (1994), pp. 231–38.

tions to establish ownership rights and the rights to use and have access to results from these activities; (b) creating legislation to protect rights over intellectual property and manage the information resources of the Russian Federation, including scientific and technical results obtained while carrying out government defence orders, which are federal property; (c) developing legislation to deal with questions of infrastructure and state support for research and development; (d) organizing, managing and running Russia's international cooperation in the field of military and dual-purpose technologies; (e) controlling transfers of technology and scientific and technical information in those high-priority fields which are crucial for sustaining the scientific and technical potential of the country and guaranteeing its defence; (f) control over state intellectual property in order to defend the property interests of the owner, including the investment of intellectual property in joint enterprises in Russia and abroad, obtaining foreign patents and selling licences and know-how; (g) building an information infrastructure for the conversion, transformation into joint-stock companies and privatization of state enterprises; (h) registration of international scientific and technical exchanges where the scientific and technical results obtained while carrying out government defence orders are being transferred; and (i) carrying out selective checks and inspections of international scientific and technical exchanges in order to identify those who violate the existing legislation and to bring them to justice.

Implementing these proposals would facilitate a comprehensive solution to the problems facing the military–industrial complex. This solution would also involve the participation of industry. Further, it would help sustain the scientific and technical potential of industry at an adequate level while still ensuring a state control not only over individual programmes, but also over a most important and still manageable part of Russian society.

# 6. The process of policy making and licensing for conventional arms transfers

*Peter Litavrin*

## I. Introduction

The background to Russian policy making is the fact that Russian arms transfers fell year by year between 1987 and 1994. The result was the loss of the dominant role that Soviet arms exporters had played on the world market, reflecting the serious economic and political problems of the country. Russian policy has been to try to halt and then reverse this tendency. The Russian leadership has made several attempts to support the military–industrial complex. However, the main emphasis has been on organizational and structural changes in the sphere of military–technical cooperation.

Since 1992 Russian policy on arms transfers and military–technical cooperation has been subject to constant changes and reforms. There is no sign that the process of change has ended. At the end of 1996 additional changes were being made to the process of decision making for arms transfers. In negotiating and implementing arms transfers and conducting military–technical cooperation there are also regular changes in approach. After 1993 Rosvooruzhenie, the newly established state company for the export and import of arms, became the major supplier of Russian arms. Nevertheless, since 1994 several decrees of the President of the Russian Federation have allowed other producers of conventional arms to enter the international market.

After the collapse of the previous Soviet system that governed arms transfers the new one has been emerging in an atmosphere of crisis, difficulty and constant reorganization. This is hardly surprising. The Russian Federation itself is in a state of profound change and the political strife around the military–industrial complex and such potentially beneficial related areas as arms transfers is understandable.

## II. The organization of arms transfers in Russia

The most important elements of arms transfer policy are controlled by the president of the Russian Federation.

The president adopts decrees governing this area and decides all of the sensitive and most important issues. As of late 1997 there is no law on Russian military–technical cooperation with foreign countries as the draft law is yet to be adopted by the State Duma. The president decides questions which establish new precedents in state policy. He decides on the list of weapons proposed for export for the first time. He also makes decisions on the establishment of military–technical cooperation with foreign countries. Under Russian rules even a shipment of a couple of revolvers to a foreign country sometimes demands an internal decision on the establishment of military–technical cooperation with that country.

If problems occur with any particular decision it may be brought up in the Security Council, under the chairmanship of the president, and it can decide what to do and how to handle that case.

The Government of the Russian Federation is the main body that executes military–technical cooperation with foreign countries on a day-to-day basis. It decides what kind of military production may be exported on what terms and on what conditions. It gives permission for the export and import of ordnance, military equipment and services as well as approving lists of states to which the export of particular types of arms is prohibited. By the end of October 1996 the key agencies involved in this process were the MFER, the Ministry of Defence, the Ministry of Foreign Affairs, the Ministry of Defence Industry, the Federal External Intelligence Service and the Federal Security Service.[1]

The State Committee on Military–Technical Policy (Gosudarstvenny komitet po voyenno-tekhnicheskoy politike, GKVTP) was established by Presidential Decree no. 2251 in December 1994 and was supervised by the president himself.[2] It prepared conceptual documents on military–technical cooperation with foreign countries, issued licences and resolved the practical questions of arms trade. Decisions were made on the basis of documents presented by a subsidiary organ, the Interdepartmental Coordinating Council for Military–Technical Policy (Koordinatsionny mezhvedomstvenny sovet po voyenno-tekhnicheskoy politike, KMSVTP), which was headed by First Deputy Prime Minister Oleg Soskovets until his dismissal by the president. The KMSVTP replaced the previous Interdepartmental Commission on Military–Technical Cooperation between the Russian Federation and Foreign States (Komitet voyenno-tekhnickeskogo sotrudnichestva, KVTS). Its main task was to coordinate and exercise control over the different bodies involved with transfers of arms and related technologies and services: for example, it coordinated the preparation of lists of controlled items subject to export licensing and lists of countries to which exports of each type of technology are allowed. In preparing such lists each agency represented on the KMSVTP offered views and information based on its own competence.[3]

---

[1] On 17 Mar. 1997, President Yeltsin signed decree no. 249 which reorganized the Russian Government. Among the changes were the abolition of the Ministry of Defence Industry. Its responsibilities were placed in the Department for the Reformation of Industry and Defence Industry Conversion. Directive of the Russian Federation no. 484-r, 8 Apr. 1997, *Rossiyskaya Gazeta*, 22 Apr. 1997, p. 4 (in Russian) in Foreign Broadcast Information Service, *Daily Report–Central Eurasia* (hereafter FBIS-SOV), FBIS-SOV-97-106, 22 Apr. 1997.

[2] 'On the State Committee of the Russian Federation on Military–Technical Policy', Decree no. 2251, 30 Dec. 1994, *Sobranie zakonodatelstva Rossiyskoy Federatsii* [Collection of legislative acts of the Russian Federation], no. 10 (1995), article 865; and Statute of the State Committee of the Russian Federation on Military–Technical policy, *Sobranie zakonodatelstva Rossiyskoy Federatsii*, no. 41 (1995).

[3] Transcript of Press Conference with Vice-Premier Yakov Urinson at Russian Federation Government House, Moscow, 25 Aug. 1997, provided by Federal News Service, Inc., Washington, DC.

In August 1996 the GKVTP, the body responsible for coordinating Russian arms exports, was disbanded and licensing authority was transferred to the MFER.[4] The GKVTP (an agency that was directly accountable to the president) was unable to consolidate its position within the system and lacked the resources and influence to play a leading role in the decision-making process. The other agencies preserved their status after this change, although further changes cannot be ruled out. One effect of these changes was to increase the role of the Ministry of Defence and the Ministry of Defence Industry in giving advice and information about aspects of export licensing decisions and this will have the indirect effect of reducing the importance of Rosvooruzhenie in the process.

In 1997 further changes in the policy-making process occurred. During the first eight months of the year several decrees were issued revising aspects of military–technical cooperation. In July a decree was issued charging the Chairman of the Government of the Russian Federation (Viktor Chernomyrdin) with supervision of military–technical cooperation.[5] In August 1997 two more followed, of which the first defined the roles of the different economic actors in military–technical cooperation.[6] These actors are the enterprises granted the right to conduct exports, together with state authorities. By this decree the Rosvooruzhenie state trading company was transformed into a wholly government-owned unitary enterprise. The decree also created two additional government-owned unitary enterprises—Promexport and Rossiyskiye Tekhnologii.

Rosvooruzhenie is charged with managing complex export contracts requiring the coordination of many enterprises. Promexport is charged with managing follow-on support (spare parts, components and service support) as well as disposal of surplus equipment from the Russian armed forces. Rossiyskiye Tekhnologii is charged with managing exports of intellectual property (licences and know-how) connected with controlled items. The main functions of Promexport were the subject of a separate decree,[7] and the new status and tasks of Rosvooruzhenie were elaborated and a supervisory commission established to oversee the activities of the company in another decree again.[8]

The basis for policy is that every arms sale from Russia is permitted provided that the exporter has a licence. As noted above, these licences are now issued by the MFER.

---

[4] 'On the structure of the federal organs of executive power', Presidential Decree no. 1177, 14 Aug. 1997, reproduced in *Rossiyskaya Gazeta*, 16 Aug. 1997 (in Russian) in FBIS-SOV-96-162, 20 Aug. 1996.
[5] 'On measures to improve the system of management of military–technical cooperation with foreign states', Presidential Decree no. 792, 28 July 1997, reproduced in *Rossiyskaya Gazeta*, 2 Aug. 1997, p. 6 (in Russian) in FBIS-SOV-97-232, 20 Aug 1997, and in appendix 3 in this volume as document 25.
[6] 'On measures to strengthen state control of foreign trade activity in the field of military–technical cooperation of the Russian Federation with foreign states', Presidential Decree no. 907, 20 Aug. 1997, reproduced in appendix 3 in this volume as document 26.
[7] 'On the Federal State Unitary Enterprise Promexport', Presidential Decree no. 908, 20 Aug. 1997, reproduced in appendix 3 in this volume as document 27.
[8] 'On the Federal State Unitary Enterprise Rosvooruzhenie', Presidential Decree no. 910, 20 Aug. 1997, reproduced in appendix 3 in this volume as document 28.

Decisions about what to sell and to whom involve several steps. However, the actions of government (which has a control function) and the exporting agent (which is responsible for negotiations with a customer) are now separate.

Any organization applying for a licence must present the following documents in order to have its request considered: (*a*) a permit from the Russian Government authorizing the applicant to engage in arms trade; (*b*) a licence application; (*c*) the original of an end-user certificate which must be issued by a responsible organization in the recipient country; (*d*) a signed export or import agreement; and (*e*) permission given by a responsible organization of the recipient country which names the partner firm in Russia that will conduct the export–import operation. This partner firm must be registered in Russia and authorized to conduct export–import operations.

During the evaluation of the export request, different aspects related to the sale, including political and military factors, the current state of bilateral relations with the future recipient, security interests and of course commercial conditions, are studied. If the deal is particularly sensitive a special decree by the president is needed. All the key ministries and agencies mentioned above are involved in the process of preparing a governmental or presidential decision.

After presidential or governmental approval has been granted and provided that the necessary information has been provided by the exporter, the MFER issues a licence. Once this has been done spare parts, components and auxiliary equipment may be exported without new, special permission from the government provided that they are required for the arms already transferred.

## III. The changing relations between government and industry

One important feature of the present situation in Russia regarding arms exports is the ongoing struggle on the part of producers to receive permission from the government to conduct military–technical cooperation themselves. This permission is required before a producer can seek foreign partners, conduct negotiations, sign contracts or organize exhibitions of military equipment either in Russia or abroad. Until recently only Rosvooruzhenie, the Moscow Aviation Production Organization (MAPO) and a few other producers had this permission, although all were still required to follow the licensing process before a transfer could be made.[9] It is important to note that only producers of arms may apply for such permission and not trading companies which are only engaged in import–export transactions. To receive permission the applicant must pass a rigorous certification procedure so that the government is confident that they can meet the necessary standards. There were many applicants for permits but it was not until 1994–95 that new agents were certified. In 1996 seven new enter-

---

[9] Decision of the Government of the Russian Federation on granting the enterprises of the Russian Federation the right to participate in military–technical cooperation with foreign countries, no. 479, 6 May 1994. Reproduced in appendix 3 of this volume as document 10.

prises were granted this right provided that they only traded the arms they produced themselves.[10] Not all the producers which received permission to export were making major complex systems. Large state companies like Rosvo-oruzhenie are not flexible and dynamic enough to discover and fill all the niches that exist in the international market. The activity of MAPO, producer of the MiG-29, is an example of such a success.

The state approach is not to allow too many potential producers to get access to foreign markets because of the possibility that existing rules and procedures for arms exports will be violated. This concern on behalf of the state is understandable since many gross violations in the past have come to light.

While often referred to, the leaking of state secrets and advanced technologies and the 'brain drain' are only part of this problem. Sporadic, unauthorized contacts by Russian arms producers with foreign counterparts in the period 1992–94 sometimes led to direct damage to the military potential of Russia. Some sophisticated military systems or unique technologies were transferred for nearly nothing. Lack of coordination between Russian exporters also resulted in a harmful competition among themselves that weakened the position of Russia on the world market.

On the other hand it is not possible to re-establish the old system under which the state was the only producer and exporter of arms. According to Russian terminology the policy of arms transfers is still a state monopoly. In reality this means that the government strictly controls the process of sales and licensing. The Russian state is no longer, as was the case, the absolute and sole owner of weapons produced in Nizhniy Novgorod or by the Sukhoi plants. Moreover, the recent steps taken in the field of conventional arms transfers demonstrate the tendency to permit new producers to enter the world market.

The realities of the current situation facing the defence industries in Russia dictate that the voice of arms producers should be decisive among the various agents engaged in arms transfers. Unless the interests of manufacturing industry are taken into consideration and unless all those involved in producing arms have their share of profits alongside those who are marketing, advertising and trading them, the overall state of Russian arms production and exports cannot turn radically for the better. Equally, trading from existing stockpiles will not change the general situation of decline and destruction in the industrial sphere.

According to estimates of specialists from the then State Committee on Defence Industries (Goskomoboronprom, which became the Ministry of Defence Industry) the volume of production of military equipment in Russia dropped by 17 per cent in 1995 compared with 1994. On the basis of these and earlier figures some predict that in five years the Russian defence industry will

[10] Instructions of the Government of the Russian Federation, no. 202-i, 19 Feb. 1996; Instructions of the Government of the Russian Federation, no. 203-i, 19 Feb. 1996; Instructions of the Government of the Russian Federation, no. 204-i, 19 Feb. 1996; Instructions of the Government of the Russian Federation, no. 205-i, 19 Feb. 1996; Instructions of the Government of the Russian Federation, no. 206-i, 19 Feb. 1996; Instructions of the Government of the Russian Federation, no. 207-i, 19 Feb. 1996; and Instructions of the Government of the Russian Federation, no. 208-i, 19 Feb. 1996. Reproduced in appendix 3 of this volume as documents 16–22.

be completely ruined.[11] The only way to reverse this tendency is urgent invest-ment—at least 3–4 billion dollars—in the military economy. By dramatizing the situation the 'captains of defence industry' seek government support for this investment for at least the near future.

The demise of many defence enterprises and companies is a natural process not only in Russia but also in Europe and the United States. The difference is that in the West the acceleration of this process in the 1990s was a result of competition under conditions of reduced procurement expenditure. The strongest could reasonably be expected to survive. In Russia the situation looks very different. Lack of financing has meant that essentially competitive indus-tries are also on the verge of collapse. Sometimes a factory or enterprise is unable to bridge the period before a new product that is in demand becomes available on the market.

Reflecting the need to take account of factors such as these, the role of indus-try is growing within the government decision-making process. The gradual increase in the number of enterprises able to conduct foreign trade activities and the decision to give the MFER enhanced responsibilities are examples.[12]

## IV. New principles underlying decision-making procedures

In the 1960s and throughout the 1980s, Soviet arms shipments were strongly ideologically motivated. This situation started to change in 1990–91 when an attempt was made to lay down a more commercial principle as the basis for a new Soviet arms trade policy. The USSR declared that it was ready to sell ordnance to any country that could pay in the hard currency it so badly needed. Military grants to allies in the developing world were reduced to a minimum.

Unfortunately for the ideologists of this 'new thinking', neither Western states nor many others lined up to buy cheap Soviet weapons. At the same time, the markets in former clients like the East–Central European countries, Cuba, North Korea and Syria were almost entirely lost. Military–technical cooperation with Iraq, Libya and the former Yugoslavia was curbed for political reasons: mandatory UN arms embargoes were in place.

As a result of this short-sighted policy the USSR lost traditional (albeit not profitable) markets but instead of gaining beneficial relations with new cus-tomers—which would have been a good substitute for the old clientele—it gained nothing. This outcome was logical as the new approach had been based on a wrong assumption—that the Soviet Union could easily find its place in the crowded world arms market.

[11] See chapter 8 in this volume.

[12] As part of the reorganization of the Russian Government in Mar. 1997 (see note 1), the Ministry of Foreign Economic Relations became the Ministry of Foreign Economic Relations and Trade. The minister responsible, Oleg Davydov, kept his title as the Minister for Foreign Economic Relations but lost his position as deputy prime minister. In future the ministry, now responsible for both foreign and internal trade issues, will come under the supervision of the Deputy Prime Minister and Minister for Economics, Yuriy Urinson. *Business Law Review*, Interfax (in English), 25 Mar. 1997, in FBIS-SOV-97-062, 25 Mar. 1997.

What was not taken into consideration was that selling weapons—not just giving them away—demands certain features which could not be met by Soviet producers and exporters such as a developed system of marketing and an after-sales system for supporting the equipment throughout its lifetime with spare parts and maintenance. These arrangements are demanded by rich clients such as those of the Persian Gulf. In addition, Western countries have not hurried to become customers of Moscow for political and military reasons.

The outcome of these changes was evident even before the collapse of the Soviet Union. Soviet arms shipments abroad had already fallen drastically due to the wrong decisions.

In the late Gorbachev era the shift of decision making from the CPSU Central Committee bureaucracy to the 'merchants' from the MFER and governmental experts could not prevent arms exports shrinking at the moment when the country was falling apart and its economy was entering a profound crisis.

In Russia in 1992–93 the difficulties in the defence industry were made worse by the collapse of the organizational structure of the arms trade. The administrative inter-agency coordinating structure was largely maintained, including the Ministry of Defence, the MFER and the Federal Security Service—although now without the Central Committee of the CPSU—but the outside situation had changed. Moreover, the former system was designed to take into account the interests of the agencies concerned and to prevent the leaking of state secrets. It was never intended to be effective mainly from a commercial point of view.

After the decisive role of the Communist Party in this process evaporated all the players involved started a tug of war aimed at increasing their influence. The clashes of interest between different agencies sometimes blocked prospective deals: for example, the sale of T-80 tanks abroad was prohibited at first even though this tank had won praise and awakened interest at the Abu Dhabi defence equipment exhibition. Once the export of the T-80 was permitted several opportunities had already been missed.

The Ministry of Defence and the defence industry enterprises often accused the MFER of inefficiency and red tape. At least in part this was true. On the other hand the Ministry of Defence refused for years to allow the export of systems that could have been extremely competitive internationally. In turn, the moves made by the defence industry in the international market in its eagerness to sell were so clumsy that manufacturers were sometimes their own worst enemies.

The tendency towards commercialization of arms transfers on the one hand and an idealized picture of international cooperation on the other, which characterized the last period of the Soviet Union, was inherited by Russia. In 1992–94 the Russian Federation, because of the serious economic situation, tried to sell arms not only for immediate payment in hard currency but also for barter and against long-term, low-interest credits. The incentive to do so was the attempt at economic survival on the part of a defence industrial base that could not be supported by the national procurement budget. In some cases, however, shipments of arms were made to reimburse debts owed by the former Soviet Union.

Russia did not achieve much success in its attempts to sell arms to Western states, although military–technical relations were established with France, Greece and Turkey.

A low profile was kept where former allies were concerned, for two main reasons. First, these states were not regarded as clients which would be able to bring Russia substantial new business. Transfers of arms to them could not bring money to the Russian treasury in the near term. Second, as Russia oriented its foreign policy largely to the West the Russian leadership came to see many former allies in the developing countries as, in the words of Foreign Minister Andrey Kozyrev, 'political hooligans'. Thus, close military–technical ties with these countries came to be regarded as giving support to extremism, terrorism and subversive activity even though most of these states—Syria and Viet Nam, for example—were not under any UN embargo or on a Western 'blacklist'. The Soviet Union had invested billions of dollars in these countries and it was only by preserving military–technical cooperation that Russia had any chance of gaining partial repayment of these debts. However, this argument was not taken into serious consideration until 1994. Several of these arguments also apply to the initial Russian inactivity as far as cooperation with the CIS was concerned in the period 1992–94.

As the hopes of the Russian leadership for closer partnership with the West in the military–technical sphere subsided and attempts to restore Russian arms trade policy started to emerge, the 'romantic period' came to an end.

The key elements of the new policy were stricter governmental control over the arms transfer process, attempts to re-establish closer military–technical relations with former partners—such as the East–Central European states, Cuba, India and the CIS countries—and an active search for new non-traditional markets in East and South-East Asia, Latin America and Southern Africa.

With a certain improvement in the economic situation in Russia and the first signs of an increase in arms sales in 1994–95, politico-military factors started to play a greater role in the decision-making process. It became absolutely clear that policy neglect in the field of military–technical cooperation with countries like Belarus, Kazakhstan and Ukraine and former developing country allies was counter-productive. Active state financial and political support was increasingly recognized as an essential element in export success. By refusing to transfer arms and spare parts at reduced prices or on a barter basis, Russia lost important benefits: political influence, the confidence of partners, opportunities for future cooperation and the chance to recover debts. Monopolists like Rosvooruzhenie, working purely on a commercial basis, were not interested in taking these matters into account.

## Future prospects

Russian moves on the world market have become more determined, consistent and oriented towards success in the market-place. It was acknowledged that decisions had to be taken based first of all on the interests of the producers.

Government Decree no. 479 of 6 May 1994 gave more rights to the arms producers[13] and, at the same time, the Russian state started to give them direct support.

This closer coordination of state and industry led to the realization of profitable deals with Malaysia and the United Arab Emirates in 1994. In these two cases only direct government intervention and guarantees secured deals that the Russian companies could easily have lost.

In 1996 Russia still lacked a strong centre for coordinating activities in the field of arms exports. There are still a number of agencies trying to increase their influence and gain a place in the sun. Until recently the Federal Security Service has also been one of the key players in this area. The Ministry of Defence, which was once a central agent in the arms export control system, has lost its position as a result of Presidential Decree no. 1008 of 5 October 1995[14]—although it has not lost the capacity to block any individual deal for reasons of national and military security. During 1993–97 the Ministry of Foreign Affairs has been much preoccupied with issues of compliance with the international obligations of the Russian Federation. It has not been actively engaged in supporting national arms exports. In 1996 the Ministry of Foreign Affairs started to take a more active role, partly as a result of the entry of Russia into the multilateral Wassenaar Arrangement.[15]

Russia's ability to re-emerge as a major arms seller is largely dependent on whether it can reorient itself towards new markets, mainly in those countries whose financial solvency is good.

The development of trade with the financially and politically stable countries in the Middle East and the Asia–Pacific region is therefore of vital importance to Russia. Russian arms exporters are also trying to break back into the markets of East–Central European countries which are faced with the task of modernizing existing weapons of Soviet origin. It is possible that this will be done by installing Western technology.

At present, China and India are the largest consumers of Russian services both in the field of ordnance and in the licensed production of military equipment, including design for use by military units.

Russian arms exporters also attach great importance to improving their after-sales servicing of military equipment to bring it to the same level offered by competitors on the world arms market. In conditions where potential buyers have limited finances—particularly with developing countries—the modernization of existing equipment rather than its replacement with new production is also given preference.

---

[13] Decision no. 479, 6 May 1994 (note 9); and Regulations on the certification and registering of enterprises for the right to export armaments, military equipment and military-purpose work and services, approved by Decision no. 479, 6 May 1994. Reproduced in appendix 3 of this volume as documents 10 and 11.

[14] 'On military–technical cooperation of the Russian Federation with foreign countries', Presidential Decree no. 1008, 5 Oct. 1995. Reproduced in appendix 3 of this volume as document 1.

[15] See chapter 5 in this volume.

The situation in Russian arms export policy is improving but very slowly. The significance of the fact that Russia ranked relatively high among arms exporters in 1995 and 1996[16] should not be exaggerated. The successes achieved were partly a result of mobilizing the last resources of the military–industrial complex developed in the past. To a great extent the slow pace of change can be explained by the chronic political and social instability in the country. As arms exports are one of the few things that bring money into the Russian treasury, the struggle of political groups around them is inevitable.

Further changes can be expected in the political priorities and administrative arrangements in Russian arms export decision making. However, two things are obvious. First, state control over this matter is going to become stricter. Second, the number of Russian arms producers with access to the world market will increase. These developments are an absolute necessity if Russia is to restore its position as a major exporter of weapons while at the same time being a responsible supplier.

---

[16] Anthony, I., Wezeman, P. D. and Wezeman, S. T., 'The trade in major conventional weapons', *SIPRI Yearbook 1996: Armaments, Disarmament and International Security* (Oxford University Press: Oxford, 1996), pp. 463, 465; and Anthony, I., Wezeman, P. D. and Wezeman, S. T., 'The trade in major conventional weapons', *SIPRI Yearbook 1997: Armaments, Disarmament and International Security* (Oxford University Press: Oxford, 1997), pp. 267–68.

# 7. The role of the Ministry of Defence in the export of conventional weapons

*Yuriy Kirshin*

## I. Introduction

Weapons and combat equipment from the major arms-producing countries have been dispersed across all continents and to all states through the international arms trade. The developing countries are incapable of independent arms production. However, their rich natural deposits—especially of oil and gas— enable some of them to spend large sums of money, some of it on military hardware. Historically, the majority of states that became independent in the period between 1960 and 1980 preferred to buy their weapons from the Soviet Union rather than from their former colonial rulers. As described in chapter 3, the export agencies which participated directly in this trade during the Soviet period were the Central Engineering Directorate (Glavnoye inzhenernoye upravleniye, GIU), the Central Technical Directorate (Glavnoye tekhnicheskoye upravleniye, GTU), and the Central Directorate of Collaboration and Coopera- tion (Glavnoye upravleniye po sotrudnichestvu i kooperatsii, GUSK).

With the disintegration of the Soviet Union and the elimination of the WTO, these agencies underwent certain changes. The GIU and GTU both began to lose their prestige and influence. This was one reason for a sharp fall in the vol- ume of military–technical cooperation. In this context, a decree of the Russian President in November 1993 instituted the state company Rosvooruzhenie, which was based on the foreign economic associations Oboronexport (the successor to the GIU) and Spetsvneshtekhnika (which was created from the former GTU and GUSK).[1]

Rosvooruzhenie was formed with a view to bolstering military–technical cooperation with foreign states and also to ensure that the state maintained a monopoly over the export and import of combat *matériel*. It is accountable to the President of the Russian Federation and works under the direct supervision of the Russian Government. It is an executive rather than a decision-making agency. Proceeding from its position as a state monopoly, it handles the export of weapons and combat equipment on the basis of decisions taken by the president or the government. It organizes its export activity in close contact and cooperation with the Ministry of Defence and the State Committee on Defence Industries (later the Ministry of Defence Industry, now abolished) which

---

[1] 'Weapons in clean hands', *Moskovskie Novosti*, no. 5 (30 Jan.–6 Feb. 1994), p. 14 (in Russian) in Foreign Broadcast Information Service, *Daily Report–Central Eurasia* (hereafter FBIS-SOV), FBIS-SOV- 94-024, 4 Feb. 1994, p. 24; and Oslikovsky, S., 'On the way to increasing the effectiveness of military– technical cooperation', *Military Parade*, Nov.–Dec. 1994, p. 13.

coordinates and controls agencies making up the defence industry and other organizations involved in producing military hardware.

Military–technical cooperation embraces a broad spectrum of activities, including many which only the Ministry of Defence is competent to perform. For example, potential buyer states are likely to require field demonstrations of equipment and information about how it should be operated and maintained. Training of foreign servicemen is undertaken by the Main International Cooperation Department of the Ministry of Defence. Before 1992 most foreign officers were trained free of charge; tuition fees have since been introduced.[2]

In order to ensure competent engineering support for exports and, in particular, the effective transfer and maintenance of weapon systems, the Ministry of Defence has sent a number of officers to Rosvooruzhenie on secondment. These individuals, all of whom are uniformed officers, hold some high positions, including those of deputy to the general director, chiefs of administration, and heads of departments and groups. This enables Rosvooruzhenie and the Ministry of Defence to coordinate their activities with regard to exports, presenting equipment at exhibitions and demonstrating weapons and combat equipment to potential customers at testing grounds. Some officers of the Ministry of Defence are posted overseas to act as representatives of Rosvooruzhenie. These representatives are tasked with the search for weapons markets.

The main responsibility for market research rests with Rosvooruzhenie. However, the role of the Ministry of Defence in this sphere has changed in recent times. It has a variety of specialist organizations for military–technical cooperation which carry out different aspects of research into the weapon market, and there are groups of military specialists who analyse the situation in foreign countries. The ministry is receiving increasing numbers of requests from different states which may lead to the purchase of weapons and combat equipment from Russia. Most of these are addressed directly to the defence minister. Many countries prefer to import weapons confidentially, without involving a broad circle of persons in the process and to tackle the issues of arms purchases, and military–technical cooperation as a whole, only through contacts with the military—specifically with the Ministry of Defence. However, in Russia the Ministry of Defence has no right to perform this function. Its role in the arms export decision-making process is more limited, restricted to certain checking functions and participation in the drawing up of government documents and decisions.

Market research is carried out by Rosvooruzhenie in close cooperation with the Ministry of Defence Industry and enterprises of the military–industrial complex along with the Ministry of Defence and the Ministry of Foreign Affairs. Market research is not only carried out with the purpose of deriving maximum economic advantage. It is also important to understand the market

[2] Bogdanchikov, V., 'Training of military personnel in Russia for foreign armies', *Military Parade*, July–Aug. 1997, pp. 64–67.

with a view to providing for Russia's security—preventing weapons from spreading to 'flashpoints' and preventing the export of banned armaments.

Rosvooruzhenie, which was formed in 1993, has not been able to solve all problems of military–technical cooperation. This is because it does not have a complete monopoly on exports. The right to export arms has been granted to an agency of the Ministry of Defence Industry (Promexport) and licences to export have also been obtained by a number of research and production associations and even individual defence enterprises. Such an amorphous export system has begun to take shape because the state has proved incapable of financing the state defence order, including its export part. All economic entities have therefore rushed into the foreign weapon market to try to earn the revenue they need for their survival.

This absence of coordinated effort on the supply side in Russia has also caused an unhealthy rivalry among Russian participants in military–technical cooperation. This has led to a drop in prices for Russian-made weapons, caused disorientation among foreign partners and led to a loss of potential sales, undermining Russia's prestige as a weapon exporter.

Representatives of the Ministry of Defence take part in the teams of Russian officials that conduct military procurement negotiations, defining the quantitative aspects of any deal. The ministry also offers support by demonstrating weapons and combat equipment and assisting with the physical deployment of the weapons ordered to the buyer states. However, representatives of the Ministry of Defence do not take part in the assessment of prices or commercial aspects of the deal as these are considered commercial secrets.

If it has been stipulated in a relevant government decision, the Ministry of Defence can, upon agreement with Rosvooruzhenie, provide technical assistance in the transport, loading and protection of exported weapons in transit. The costs of this are reimbursed by the weapon seller and by Rosvooruzhenie.

The contract for the purchase of weapons and combat equipment may also contain clauses on further deployment and preparation for service of weapons in the buyer country and on the training of national military personnel. These elements are assessed and included in the general price of the transaction. The Ministry of Defence may help with these elements of the transfer and the costs incurred by the ministry are reimbursed by Rosvooruzhenie. The training of national military personnel from the purchaser country is carried out at educational establishments of the Ministry of Defence, and its expenditures are reimbursed by the arms seller provided this has been stipulated in the contract for the export deal. In the event of a separate contract being concluded for the training of military personnel between the purchaser country and the Ministry of Defence of the Russian Federation, the costs of the latter are offset by the buyer state according to the terms of the contract. Oversight of the economic aspects of training foreign servicemen in Russian military educational establishments is the responsibility of the government. Part of the income obtained by the Ministry of Defence is channelled to the state budget, and the other part is retained to meet the needs of the ministry.

The right to negotiate and sign contracts and to fix rates for the training of foreign servicemen has been granted to the Main Administration for International Military Cooperation (Glavnoye upravleniye mezhdunarodnogo voyennogo sotrudnichestva, GUMVS) under the General Staff of the Armed Forces of the Russian Federation. GUMVS also has the right to strike individual deals related to sending military specialists to assist with the deployment of weapon systems and maintaining them in a combat-ready state in a foreign country. Rosvooruzhenie can also conclude such contracts. In this case, military specialists can be sent to the importing country as part of a mixed team including both individuals from the Ministry of Defence and civilian specialists from ministries or enterprises of the military–industrial complex.

## II. The role of the Ministry of Defence in arms export control

In Russia a national system of export control has been established under the overall control of the president. Although the specific form of the national export control system has changed regularly, the Ministry of Defence is one of the agencies designated to play an important role within this control system.

The objective of the Russian authorities is to produce an export control system which avoids the possibility of damaging the political, economic or military interests of the Russian Federation. As part of this export control system it is necessary to determine which materials, goods, services and technologies should be subject to control. Proceeding from requests from buyer states, arms trade organizations authorized to carry out military–technical cooperation must submit their proposals to the government. In deciding whether or not to approve an export, the views of the Ministry of Defence on certain aspects of the proposal must be taken into account.[3]

### Ministry of Defence advice on export requests

In 1992 an Export Control Committee (Komitet eksportnogo kontrolya Ministerstva oborony, KEKMO) was established as part of the Ministry of Defence.[4] Its tasks include examination and approval of requests to export specific types of weapons to specific countries. In cooperation with the Ministry of Foreign Affairs, the Ministry of Defence has prepared both an equipment list and an evaluation of the strategic situation in potential recipient countries for this purpose.[5] The purpose of the equipment list is to decide on the defensive or offensive nature of the weapons to be exported. The evaluation also assesses whether the purpose of the weapons that are being requested is offensive or defensive.

[3] The licensing process is described in chapter 6 in this volume.

[4] Felgengauer, P., 'Russia's arms sales lobbies', *Perspective*, vol. 5, no. 1 (Sep.–Oct. 1994), pp. 1–8.

[5] Interview with Col-Gen. Vladimir Zhurbenko, *Krasnaya Zvezda*, 27 Aug. 1994, p. 4 (in Russian) in FBIS-SOV-94-169, 31 Aug. 1994, pp. 30–35.

The main criterion guiding the work of KEKMO is to prevent any threat to Russia's security in the military–technical field. It examines possible exports on a case-by-case basis and can give approval, issue a refusal or impose more restrictions on deliveries. The list includes not only weapons but also technologies and know-how. KEKMO is also part of the process of granting licences or permits to enterprises and production associations that give them the right to trade in equipment, spare parts, tools, accessories and other military items, and dual-use equipment.

KEKMO is headed by the First Deputy Defence Minister. It includes representatives of the Chief of the Armament Administration of the Ministry of Defence, the Main Administration for International Military Cooperation, the Main and Central Administrations of the Ministry of Defence and the General Staff of the Armed Forces of the Russian Federation. Also included are representatives of the commanders of each of the fighting services and their main headquarters, representatives of the Federal Security Service, and Rosvooruzhenie. It meets as required, depending on the urgency of the issues to be examined, but at a minimum once a month. Members are advised of the matters to be considered at each session in good time. They are expected to study them thoroughly and prepare draft conclusions which are examined in the working sessions. After discussions of the various draft conclusions, decisions are taken collectively. If controversy arises, discussion is suspended until the particular issue has been re-examined by experts. In the event of complicated matters being examined, experts or other interested persons may be invited to contribute their expertise to the working sessions. Decisions taken are formalized in minutes signed by the chairman, his deputies and the secretary.

A decision from KEKMO is obligatory for all military–technical cooperation requests. Without such a decision, members of the military–industrial complex may not proceed with an export. KEKMO decisions are executed by the Armaments Administration of the Ministry of Defence and the staff of the First Deputy Defence Minister.

Periodically, and as the need arises, the chairman of KEKMO or, on his instructions, one of his deputies (for example, the chief of the Armament Administration of the Ministry of Defence or the Chief of the GUMVS) reports on the work accomplished to the government. If in an individual case a particular controversy arises—related, for instance, to the possible transfer of an advanced weapon system or the participation of defence enterprises, design bureaux and other organizations in military–technical cooperation with a foreign partner—the case may be submitted to the Interdepartmental Coordinating Council for Military–Technical Policy (Koordinatsionny mezhvedomstvenny sovet po voyenno-tekhnicheskoy politike, KMSVTP) for consideration.

## III. Arms exports and Russia's security

The export of weapons and combat equipment is the most lucrative aspect of the process of manufacturing armaments in Russia. It can also accelerate the pace of technological development, affecting the types of weapon that can be deployed in the Russian armed forces.

Export of weapons and combat equipment speeds up scientific and technological progress, which allows more advanced systems to be produced. Wars and military conflicts also stimulate progress in the military–technical field. For example, during World War II dozens of new weapon systems were produced.

Buyer countries also strive to acquire the most modern weapon systems. This has both positive and negative aspects. On the positive side, a purchaser state may receive defensive weapons which maintain a balance with those in the possession of a likely aggressor country. On the negative side, the country that has acquired such weapons does not always have the skill to use and operate them effectively. This compels the purchasing country to employ advisers and specialists, for example, to train its personnel or manage logistic and spare parts-related issues more or less constantly.

Taking these facts into consideration, Russian weapon export organizations are often called upon to give appropriate recommendations to buyer states.

Weapon systems are not in themselves necessarily offensive or defensive. One and the same weapon may be defined as offensive or defensive depending on the circumstances. If a weapon is in the hands of an aggressor, it may be referred to as offensive, and if it is in the hands of a victim or possible victim of aggression, it may be regarded as defensive. Nevertheless, there are some purely defensive weapon types and weapon complexes. These are diverse types of air defence weapons and anti-tank weapons. Unless an aggressor uses aviation and tanks, these weapons will not be used either.

In selecting and assessing the character of exported weapons, the buyer country is first considered and the weapon is checked with the list of weapons banned for export. Arms exports are prohibited by Russia to extremist states, countries with terrorist activity, and countries which are forbidden to buy weapons by decisions of the UN. Nuclear, biological and chemical weapons and other weapons of mass destruction are also banned for export.

The contracts that govern a specific arms transfer also necessarily stipulate a number of binding limitations on the purchaser country in terms of the use it can make of weapons supplied. Contracts also state the penalties if the purchaser violates relevant provisions of the contract. These can include the prohibition of either follow-on exports or modernization.

A thorough process of selection of purchaser countries is intended to prevent the destabilization of the situation in a particular region and the accumulation of weapons at the borders of states likely to fall victim to aggression, as well as to avert threats to the security of Russia.

In the export of weapons, an assessment is made as to whether the specific export is consistent with the sufficiency of armaments in a given state. Before endorsement of an arms export proposal from a trade organization, an assessment is made by the Ministry of Defence and the Ministry of Foreign Affairs. If this assessment reveals that weapons have accumulated in excess of the reasonable requirements of the purchaser country for repelling aggression, the ministries and the government do not allow the export.

Thus, the Ministry of Foreign Affairs and the Ministry of Defence play the main role in the political and strategic assessment of the consequences of export deliveries. Particular attention is given to safeguarding Russia's security.

The economic and commercial aspects of an export deal are the prerogative of trade organizations—primarily Rosvooruzhenie—which has the right to fix price policy and define the rules for trading through intermediaries.

The clauses of a contract which do not contain information that is a commercial secret are discussed with all interested ministries and agencies, including the Ministry of Foreign Affairs, Ministry of Defence, at one time the Ministry of Defence Industry but now the relevant department of the Ministry of the Economy.

The existence of a state monopoly over arms exports provides a balanced approach to the sale of weapons. It helps to stave off any attempt to export weapons and combat equipment at any price to any state indiscriminately. However, at the same time the state is interested in deriving the maximum profits from legitimate export deliveries to help cover the costs of weapons manufacture and research on and development of new weapon systems in particular. One of the most important considerations at the present stage of Russia's democratic development is to maintain employment at defence enterprises.

# 8. Russian defence firms and the external market

*Elena Denezhkina*[*]

## I. Introduction

In the Soviet period practically all the economic activities of enterprises were centralized. They were the preserve of sectoral ministries and departments, to which enterprises were strictly subordinate.[1] If this was the case for domestic state orders, it was even truer for relations with foreign clients, which were the exclusive prerogative of central government. More generally, the effectiveness of defence enterprises was not judged in terms of their earning money but rather in terms of how efficiently they were able to meet deadlines and fulfil targets established by the central planning authorities. Performance was the key to future financing and development but it was not measured according to market principles.

It is true that under the previous system military–scientific production complexes played a significant role in the state defence procurement system. In parallel with these military–scientific complexes, enterprises also coordinated the manufacturing development of new products, participated in working out production plans and pursued quality assurance on behalf of the state. However, interviews with defence industry managers confirm that the old procurement system was highly centralized and formal and lacked certain forms of systematic information exchange. For example, the failure to identify existing or overlapping capacities led to the development of duplicate facilities to meet different state orders. There was little emphasis on identifying areas of potential horizontal cooperation between enterprises.

After the dramatic changes which occurred in Russia after 1989–90, both these basic conditions changed. Domestically, there has been a major shift away from the principle that the government alone should set the priorities for arms production. Within the Russian industrial sector in general there are now enterprise-level initiatives regarding the development of production profiles, market research and the search for industrial partners, customers and suppliers. The sector which produces arms and military equipment has also seen the emergence of enterprise-level initiatives. At the same time, both the manufacturing sector and trading companies that specialize in selling products that they

---

[1] See chapter 3 in this volume.

[*] This chapter draws on the results of a research project funded by the British Economic and Social Research Council on the economic transformation of the Russian defence sector, for which interviews with defence industry managers were carried out by the author in St Petersburg, Nizhniy Novgorod and Sverdlovsk Oblast in 1994–95.

do not themselves manufacture have acquired some rights to operate in international markets.

There has been a very significant structural shift within the Russian defence industry brought about by privatization, changed company status, and the gradual formation of financial–industrial groups and transnational companies. Overall, there has been a major shift in terms of the discretion available to enterprise directors and a corresponding decrease in central control. It is true that the process of privatization (or the preparations for privatization) have not been as far-reaching in the defence sector as in other areas of economic life. However, some of the new patterns of behaviour can also be observed in enterprises still wholly owned by the state. The most important factor has been the apparent inability of the central authorities to exercise coordination or regulation in a consistent or effective manner within those enterprises that they still own. While the old instruments of state control have been transformed or eliminated, no new system has emerged to replace them.

The directors and chief designers who were interviewed for the research on which this chapter is based were careful to emphasize that the old system, while rigid and bureaucratic, was characterized by clear-cut laws and rules regarding the relations between different levels in the hierarchy. The dismantling of the system took away from most defence enterprises the pool of knowledge and experience of external economic activities. They lacked information about both supply and demand in the international and domestic markets. To this was added a new lack of information and clarity regarding the government's own procurement strategy. For defence enterprises, therefore, survival depended not only on the technical parameters of the firm, but also on the ability of directors and specialists to adapt and reorganize to meet the challenge of these changed circumstances.

No clear picture has emerged of this element of Russia's industrial base. There remain major gaps in our understanding resulting from the lack of clarity as regards central issues such as ownership patterns and the financial mechanisms and operations whereby defence enterprises manage to survive.

It could be concluded that, far from being directed by the state, the strategic management of defence enterprises has in effect gone underground. At present the reality of defence industry management can only be described effectively by case-study methods, as quantitative measurement is likely to be unreliable and is unlikely to develop sufficiently complex categories of data. However, since the operations of enterprises are frequently convoluted and often seem to be specific to a particular case, no single instance can serve as a basis for generalization.

This chapter offers a tentative assessment of how some enterprises have reacted to the new conditions with special attention to the external dimensions of their economic activities. This includes both industrial cooperation and foreign sales. However, it is first necessary to describe briefly the domestic conditions in which the new Russian managers have had to construct their business strategies.

## II. Confusion in state defence procurement

As explained in chapter 3, military production for domestic and foreign customers was integrated into a single state order during the Soviet period. Within
that overall state order, the needs of the Soviet armed forces were the most
important element.

According to interviews in a series of defence enterprises in St Petersburg,
Nizhniy Novgorod and Sverdlovsk Oblast, as well as in federal and regional
policy-making bodies and financial institutions, at present orders from the
Russian Government are no longer seen as desirable by industry, on account of
chronic late payment. Moreover, whereas with commercial orders the only limit
on what may be paid to employees is the size of the order itself, with government orders there is a stipulation that no one may earn more than six times the
minimum wage—which is itself hardly attractive. The state has thus increasingly come to be seen as a short-sighted and untrustworthy partner in business.
Worse still, enterprises which work on state orders have become notorious for
causing delays in the supply chain because of their financial paralysis due to
non-payment by the government. As a result, enterprises which are not working
for the government have become unwilling to deal with those that are. This is
one factor which helps to explain the slow pace of integration of defence enterprises with other sectors of the economy (which have already left the system of
dependence on state orders, with its associated problems) as well as some of the
difficulties faced by defence enterprises attempting to enter the open market.

**The domestic procurement process**

One of the peculiarities of the procedure for issuing state procurement orders
for defence is that these are set out before the start of the calendar year, and
therefore before the finance to pay for equipment ordered has been appropriated. The customers of the Russian defence industry include not only the
Ministry of Defence but also the Ministry of Atomic Energy, the Ministry of the
Interior, the Federal Security Service, the Federal Border Troops and the mer-
chant fleet. The power of procurement is now placed with these ministries
rather than with the State Committee on Defence Industries (later the Ministry
of Defence Industry and the Ministry of the Economy), which assists with the
identification of suppliers and coordination of horizontal ties between enterprises.[2] Agreements on annual equipment requirements are concluded with the
leading enterprises concerned and the enterprises, for their part, now conclude
agreements with suppliers of components, raw materials and energy. What is
supposed to happen next is for the combined state defence order to be confirmed in the state budget and the necessary funds appropriated.

---

[2] 'Industrial restructuring and defence conversion in Russia', TACIS [Technical Assistance to the
Commonwealth of Independent States] and the Commission of the European Communities Delegation in
Moscow, Moscow, May 1995, p. 6.

**Table 8.1.** Official Russian defence budgets, outlays and deficits/surpluses, 1992–95
Figures are in current m. roubles.

| Year | Defence budget | Defence outlay | Deficit/surplus |
|------|---------------|----------------|-----------------|
| 1992 | 384 | 855 | 471 |
| 1993 | 8 327 | 7 210 | – 1 117 |
| 1994 | 40 626 | 28 028 | – 12 598 |
| 1995 | 59 379 | 47 600 | – 11 779 |

*Source:* International Institute for Strategic Studies, *The Military Balance 1996–1997* (Oxford University Press: Oxford, 1996), p. 108.

This system has not been revised in line with the realities of a more stringent financial climate. In the years 1992 and 1993, state defence orders were not reduced in line with reductions in the defence procurement budget, although this was explicable by the fact that the budget was not agreed until months after the initial deadline. At this stage it was clear that issues of military manpower and spending on equipment were secondary to macroeconomic objectives such as control of state expenditure.[3] In 1994 the budget was not agreed until August for similar reasons and in 1995 the budget was agreed in March but subsequently revised twice towards the end of the year.[4]

To add to the problems of industry, not only are some agreements not funded but the money which is allocated to defence within the framework of the national budget is not always paid in full. According to the data in table 8.1, the defence budget has not been paid in full since 1992.

As a result of these failures to pay out budgeted funds to meet commitments, there is a large mismatch between orders issued, finances assigned and defence production. More is being produced than has been ordered but not necessarily according to any logical set of priorities. As the money runs out, some production which would have been deemed necessary is not taking place, whereas much that would perhaps have been regarded as unnecessary is produced. Customer–supplier relationships throughout the defence industry have become characterized by mutual debt and paralysis.

The reverse problem also exists: plants considered strategically important are maintained without sufficient orders to utilize more than a small proportion of available capacity. The manager of the Votkinsk Mechanical Plant in Udmurtia, producer of long-range ballistic missiles, noted in an interview, 'I am supposed to maintain the entire production process for the sake of the two missiles that are ordered from us'.[5]

---

[3] Bergstrand, B.-G. *et al.*, 'World military expenditure', *SIPRI Yearbook 1994* (Oxford University Press: Oxford, 1994), pp. 421–31.

[4] George, P. *et al.*, 'Military expenditure', *SIPRI Yearbook 1996: Word Armaments and Disarmament* (Oxford University Press: Oxford, 1996), pp. 333–38.

[5] General Director Viktor Tolmachev, interviewed on Moscow NTV, 1 Nov. 1996 in Foreign Broadcast Information Service, *Daily Report–Central Eurasia* (hereafter FBIS-SOV), FBIS-SOV-96-222, 1 Nov. 1996.

**Table 8.2.** The level of state debt to the defence industry, 1993–96
Figures are in 1996 b. roubles.

| Date | Debt |
|------|------|
| 24 Dec. 1993 | 700 |
| 16 Sep. 1994 | 2 100 |
| 1 Jan. 1995 | 15 000 |
| 10 Jan. 1996 | 9 000 |
| 15 May 1996 | 6 300 |

*Sources:* Moscow Mayak Radio Network, 24 Dec. 1993 in FBIS-SOV-93-247, 28 Dec. 1993, p. 11; *Military Technology*, Feb. 1995, pp. 68–69; and *New Europe*, 25 Feb.–2 Mar. 1996, p. 8.

The scale of the problem is reflected in data of the State Committee on Statistics (Goskomstat) for the year 1992. According to these data, the total value of defence production to meet state orders in 1992 was 30 per cent higher than the amount approved in the budget. Significant discrepancies exist for all the years up to 1996. One consequence of this practice of ordering goods for which no funds are approved in the budget has been a growth in government debt to industry. Table 8.2 summarizes some of the publicly available information about levels of indebtedness arising out of late payments. At worst these figures suggest a mounting stockpile of redundant armaments and military technology produced at a substantial loss by defence enterprises, which must therefore have an even greater incentive than otherwise to seek external customers for their products.

In the period before the 1996 Russian presidential election, reducing the level of government debt to the defence industry was one issue which received some priority. According to Deputy Minister for Defence Industry Yuriy Glybin, the Central Bank of Russia was instructed to transfer 5000 billion roubles to repay debts incurred through under-funding of the state defence order in 1996. According to an article published in November 1996 using data provided by the state trading company Rosvooruzhenie, the debt owed by the Ministry of Finance to the defence industry for work already contracted for amounted to $600 million which, compared with the situation in May, represented a reduction of around $1 billion.[6] If correct, these data would suggest that the government debt to industry was reduced significantly in 1996. However, other statements by Russian officials give a contradictory impression. For example, in testimony before the Duma, Deputy Minister for Defence Industry Yuriy Starodub said that in the first 10 months of 1996 the government had covered only 29.4 per cent of the value of contracts awarded.[7] If correct, this would mean that the problem of payments has not been solved but only temporarily alleviated and can be expected to re-emerge in 1997.

---

[6] Tarasova, O., [Rosvooruzhenie calls for unity], *Segodnya,* 1 Nov. 1996 (in Russian).
[7] Interfax, 26 Nov. 1996 (in English) in FBIS-SOV-96-229, 26 Nov. 1996.

## Micro- and macroeconomic strategies and the defence industry

The main source of contradiction between the economic policies of government and the economic strategies of defence enterprises lies in the apparent arbitrariness of the former and the (ultimate) subordination of the latter. There is still little clarity or cooperation in relations between government and defence enterprises and the long-term strategy remains somewhat opaque. This combination of a lack of clear policy from above with a mass of enterprise-level survival strategies from below has not helped to resolve the situation, whether through conversion, entry onto the world market or outright closure.

The contradiction shows itself most sharply in the case of scientific institutes involved in defence, in the 70 or so one-factory towns and in those sectors in which the manufacturing cycle involves a whole supply chain of different enterprises and organizations.

According to N. I. Kvasha, former Director of TsKB Lazurit, a leading designer of submarines):

the majority of defence firms are put in the same difficult position—on their knees, with no prioritization whatsoever. We don't have any idea how many of our products will be in demand in five or 10 years' time, and what's more nobody else does. Do we actually need an army or a fleet? So far there is no programme for the future and we are paralysed by uncertainty. We're forbidden to do anything and we have no investment capital for changing our product profile. My colleagues and I need at least some clarity from the state and then we can decide our development strategy.[8]

The fact that Kvasha spoke of being forbidden to take certain steps underlines the fact that there are many areas in which the Russian Government continues to monitor and regulate the defence industry. This was confirmed in the Law on Defence adopted on 1 June 1996. By that law the Ministry of Defence is tasked with ensuring that state-owned enterprises in fields such as communications and transport are able to meet the needs of national security.[9] There are also many areas in which government policies are either not yet formed or contradictory.

First, there is a general lack of agreement or cohesion on what should be the appropriate macroeconomic policy for the defence industry between the Ministry of Defence, Ministry of the Economy, Ministry for the Defence Industry and Ministry of Finance. There are also some differences between these ministries and the Duma (particularly where the financing of state orders is concerned).

Second, budget constraints prevent the armed forces from achieving the level of modernization needed. This discourages potential foreign buyers of Russian armaments, since they do not have the reassurance that systems have been bought by the country's own armed forces.

[8] Author's interview with N. Kvasha, then Director of TsKB Lazurit, Nizhniy Novgorod, Sep. 1995.

[9] *Defense News*, 24–30 June 1996, pp. 4, 74. However, First Deputy Minister of Defence Andrey Kokoshin stressed in parliamentary hearings that the Ministry of Defence would prefer to carry out this task through increased allocations to enterprises—paying them to retain certain capacities—rather than by administrative decision. Interfax, 25 July 1996 in FBIS-SOV-96-145, 26 July 1996, p. 11.

Third, the economic situation—characterized by rapid increases in prices, high taxation and high costs in terms of energy, transport, raw materials and components—works against successful export whether of military or of civilian hardware. The high costs of production reduce the competitiveness of Russian products on world markets, despite the low wage levels in Russia.

These high input costs mean that even where arms can be sold their profitability is low and they do not generate revenues that can be used for investment. According to Viktor Glukhikh, former Chairman of the State Committee on Defence Industries (Goskomoboronprom), speaking at a press conference in January 1996, some armaments only make 6 per cent profit when exported. His successor Zinoviy Pak (Glukhikh was dismissed in January 1996) gave the figure of a 35 per cent loss on defence sales, whether domestic or for export, on account of high production costs.[10] This could only mean that items were being sold at prices below the cost of production to maintain some cash flow and/or win market shares. This does not of course rule out the possibility that some enterprises may earn profits through arms sales and, in any case, Russian profit margins should be treated with caution as there is little incentive to report a profit and the distribution of production costs is not entirely transparent.

Finally, there are substantial areas of contradiction between the main pieces of legislation governing the area, notably the Law on State Procurement, the Basic Propositions of the Military Doctrine, the Law on Conversion, presidential decrees on privatization of defence industry enterprises and the Law on the Formation of Financial–Industrial Groups.[11]

## III. The position of the defence industry

There are many competing definitions of the Russian defence industry complex and many conflicting statements about its size. For the purposes of this paper, the definition applied is that used by the former State Committee on Defence Industries. According to Viktor Glukhikh, the sector consists of 1800 enterprises and organizations and officially employs 3 million persons. Of these 500 000 are in scientific institutes. According to these data, the defence industry accounts for only 4 per cent of the nation's industrial potential but as much as 65 per cent of its scientific potential.[12]

Table 8.3 shows the fall in defence industry production between 1991 and 1994/95 (over 60 per cent) according to data from the information and statistics directorate of the State Committee on Defence Industries (published in January 1996). It can be seen that according to these data there has been a fall in the

[10] *Kommersant Daily,* 9 Feb. 1996.
[11] Denezhkina, E., 'Problems of conversion and the military–industrial complex of St Petersburg', eds P. Opitz and W. Pfaffenberger, *Adjustment Processes in Russian Defence Enterprises Within the Framework of Conversion and Transition,* Beiträge zur Konversionsforschung no. 2 (Literatur Verlag: Münster, 1994) (in English).
[12] Press conference by V. Glukhikh, Moscow, Jan. 1996.

**Table 8.3.** Index of production volumes in the Russian defence industry as classified by the Ministry of Defence Industry, 1991–96

| Year | Total output | Civilian output | Military output |
|------|------|------|------|
| 1991 | 100.0 | 100.0 | 100.0 |
| 1992 | 80.4 | 99.6 | 49.5 |
| 1993 | 64.6 | 85.6 | 32.5 |
| 1994 | 39.2 | 52.6 | 19.9 |
| 1995 | 31.2 | 41.3 | 16.6 |
| 1996 (est.) | 22.9 | 27.1 | 14.3 |

*Source: Krasnaya Zvezda,* 3 Aug. 1996, p. 3 (in Russian) in FBIS-SOV-96-152, 6 Aug. 1996, pp. 13–14.

volume of both civil and military production. However, the scale of the reduction has been greater in the military sector.

How far do the figures reflect the real position? They reflect the change in the volume of production, not sales, profit or capacity utilization. However, they do accord with the anecdotal evidence from interviews and case studies, and it seems reasonable to use them as a rough guide to the changing pattern of production.

The largest falls in production, according to data up to November 1995, have been in the communications industry (40.8 per cent), armaments (24.9 per cent), radio (24.2 per cent) and electronics (24.2 per cent). Shipbuilding was the only sector which recorded an increase (141.4 per cent) from the start of 1995. This relatively strong performance is reflected in the higher salaries paid in the shipbuilding sector, where, according to data from October 1995, the average salary is 594 000 roubles, compared with a defence industry average of 401 000 roubles.[13] According to a number of senior managers in shipbuilding firms in St Petersburg and Nizhniy Novgorod who were interviewed in August and September 1995, this increase in activity reflects the fact a number of vessels which had been delayed for between three and five years were launched in 1995.

Data released by the Ministry of Defence Industry in August 1996 suggested that in 1996 the fall in the volume of production in the defence sector was continuing. Three sectors—electronics, shipbuilding and aviation—recorded particularly steep reductions.[14]

The overall fall in production reflects not only cutbacks in military procurement at home but also the failure of the Russian defence industry to make inroads into new markets abroad.

The general process of restructuring the pattern of ownership of manufacturing industry, which has been an important element of economic reform, has also affected the defence industry. Central control over decision making has been

[13] *Kommersant Daily*, 26 Dec. 1995, p. 2 (in Russian) in FBIS-SOV-95-248, 27 Dec. 1995, pp. 21–22.
[14] Vitaliy Vitebskiy, Head of Economics Department, Ministry of Defence Industry, interviewed in *New Europe*, 1–7 Sep. 1996, p. 12.

diluted. According to Viktor Glukhikh, 36 per cent of defence production is still carried out in wholly owned state enterprises, 34 per cent in joint-stock companies which are partly state-owned and 30 per cent in joint-stock companies without any state involvement. Defence science is still primarily undertaken in state-owned establishments and primarily commissioned by state-owned industrial organizations. Only 14–16 per cent of defence scientific work is carried out for private joint-stock companies, as opposed to 70 per cent for state-owned defence enterprises. The remaining 15 per cent of work by defence science establishments is commissioned by civilian users.

Responsibility for developing a new relationship between government and industry has been contested by several different agencies. For example, under the Law on Privatization of State-owned and Municipal Enterprises of 3 July 1991, the State Committee for the Management of State Property (Gosudarstvenny komitet po upravleniu imushchestva, GKI) was authorized to initiate the privatization process in enterprises.[15] However, whether the authority of the GKI should extend to defence enterprises was a controversial question from the outset. In November 1993 a presidential decree 'On particular aspects and extra measures regarding privatization and state regulation of the operations of enterprises and organizations of the defence sectors of industry' provided an official basis for classifying defence enterprises into 'core military suppliers' and 'civilian enterprises which produce some military products'.[16] The former were exempted from privatization until at least the end of 1995 while the latter (over 70 per cent of the enterprises comprising the defence sector) would have been eligible for privatization.[17]

There have been cases of overlapping or contradictory competence for this aspect of policy. Moreover, a new factor has to be taken into account: the wishes of the major enterprises themselves, some of which prefer to be privatized even though they are core elements of the defence sector. Some enterprises heavily engaged in arms production have been privatized without the consent of the State Committee on Defence Industries or the Ministry of Defence. In 1995 it transpired that the main firms involved in the production of Sukhoi fighters, KA-50 helicopters and even some strategic weapons had been privatized by auction. Fears that this would compromise government policy and the security of armaments programmes led to 30 privatizations being reversed by presidential decree no. 541 of 13 April 1996.[18]

---

[15] Busza, E., 'Strategies for privatization: the options', ed. M. McFaul, *Can the Russian Military–Industrial Complex be Privatized? Evaluating the Experiment in Employee Ownership at the Saratov Aviation Plant*, Report of the Russian Defence Conversion Project (Center for International Security and Arms Control: Stanford, Calif., May 1993).

[16] Denezhkina, E., 'Is there a future for Russia's defence industry? Conversion in the context of current economic reforms', Lectures and Contributions to East European Studies at the Swedish National Defence Research Establishment, no. 7, Stockholm, 30 Aug. 1994.

[17] For a detailed discussion, see Denezhkina (note 11), pp. 49–66.

[18] *Military Parade*, July–Aug. 1996, p. 116; and *Segodnya*, 23 July 1996, p. 1 (in Russian) in FBIS-SOV-96-143, 24 July 1996, p. 31.

## IV. External economic operations of Russian defence firms

Analysis of the workings of a series of defence enterprises shows that there are four main types of external economic activity: (a) through the government and the various federal agencies involved in the management of Russian arms exports; (b) wholly or partly independently by the enterprise after the ownership of a firm has been changed to that of a joint venture or a financial–industrial group; (c) through receipt of international grants by tender, or through foreign direct investment; and (d) illegally, for example, by establishing 'phantom firms' or through illegal financial operations, smuggling, or illegal trade via third countries.

Although each of these types of foreign economic activity is characterized by a distinct set of administrative procedures and financial mechanisms, it is not always possible to distinguish them from each other in practice. Any specific transaction may have elements of more than one of these types of operation.

### Access to foreign markets via the government

One of the main instruments for accessing foreign markets has proved to be trade exhibitions, where defence firms display their products but where Rosvooruzhenie is responsible for all contracts made with foreign clients.[19]

The process of export licensing is described in chapter 6 of this volume; the documents which form the basis for the process are reproduced in appendix 3. On the basis of this description and documentation, it might appear that there are no deficiencies in the system of controlling and coordinating organs that has been built up in Russia. However, from an industrial perspective several directors of defence firms stated in interviews that the system is characterized by a lack of clarity and by shortcomings which hold back the entry of Russian defence industry into the world arms market.

First, while it is true that there is a high degree of centralization of decision making, the way in which the various roles and functions are dispersed among a whole range of federal organs diminishes the responsibility of any single agency for the final result.

The defence manufacturers themselves do not believe that the post-Soviet system has solved the problem of equitable distribution of the receipts from foreign sales. For example, the sale in 1995 of S-300 (SA-10) air defence systems via Rosvooruzhenie failed to deliver the expected profit to the factory which produced them. For several months after the items were delivered, the factory received no payment from the Ministry of Finance. Under current conditions in Russia, long delays before payment is made to producers also mean that inflation reduces the real value of the money received.

The tax aspects of international sales are also unsatisfactory from the point of view of arms producers. Another major problem for producers is the practical

---

[19] *Krasnaya Zvezda*, 7 Apr. 1995 (in Russian).

aspects of money transfers and subsequently the safe keeping of money received. The banks dealing directly with Rosvooruzhenie include Vneshtorgbank, Onaksimbank, Tokobank, Inkombank, Limbank and others. However, the fact that there are many delays and transfer problems in the Russian banking system gives producers an incentive to find alternatives. They may operate in cash, place balances in Western bank accounts or resort to barter trading. According to Vadim Yeremichev, Deputy Director of Rosvooruzhenie, 'barter is a necessary means of military–technical cooperation with countries which are unable to pay for armaments in hard currency'.[20]

Since arms exports remain low, Rosvooruzhenie is obliged by a presidential decree of 18 November 1993 to use its own resources to make direct investments in defence enterprises where these are necessary with a view to increasing international competitiveness.

Since 1994 a few defence enterprises have been given more independent room for action in foreign sales. Among the entities to have received permission to undertake independent initiatives are some major production associations including MiG-MAPO and Rosvertol.[21] However, for its part, Rosvooruzhenie has been sceptical about the idea of liberalizing access to the international market for defence enterprises and has frequently pointed out that past experiences have not been encouraging. According to the then Director-General of Rosvooruzhenie: 'The right of independent access to the market was granted to, for example, AO [joint-stock company] Kalashnikov, the republican firm Baikal, VO Proshenetorg and a number of others. However, in 1992–93 these enterprises, not having proper external economic experience, and not possessing the appropriate structures or suitably qualified staff failed, unfortunately, to use this right to realise any significant sales abroad'.[22]

Industrialists, however, would prefer to persist with the strategy of giving greater freedom to industry since this could bring success over the longer term. According to specialists at the Nizhniy Novgorod aerospace firm Sokol, which is involved in the sale of MiG-29 aircraft abroad, the reward gained from foreign contacts consists not only of profits, but also of the experience of strategy formulation and reorganization that go along with operating in these markets.[23]

Second, in industry there is a perceived lack of flexibility in the way the state organs deal with military exports. For example, it is claimed that in the United Arab Emirates (UAE)—a market in which Russian enterprises are very interested—it is expected that any arms contract will be accompanied by an offset arrangement that leads to an investment in the UAE by the seller of a sum not less than 60 per cent of the value of the contract. Compliance with offsets is

---

[20] *Military Parade*, July–Aug. 1995.
[21] See appendix 3 of this volume, documents 15–22.
[22] Alexander Kotelkin in *Military Parade*, May–June 1994.
[23] Author's interviews in Nizhniy Novgorod, Sep. 1995.

monitored by a group set up by the UAE Ministry of Defence in 1991.[24] For example, the French firm Aérospatiale, alongside its sale of Exocet anti-ship missiles, set up Tamko, a 49 per cent French-owned joint venture, to provide technical support to UAE civilian shipping. GIAT, supplier of Leclerc tanks, is building an air conditioner factory and a training facility in the UAE. Russian suppliers of similar equipment found the Russian Government unwilling or unable to help them compete in this type of market.

Third, since the end of the cold war there have been contradictions in Russian foreign policy which have in turn created contradictions in arms export policy. For example, the government has declared the development of relations with the states around the Persian Gulf to be a very high foreign policy priority. However, from the perspective of arms transfers, most of the Arab states in the subregion already have long and close associations with individual Western countries. In practice, Russia's aim of increasing arms transfers and military–technical cooperation with countries in this region is hindered by its ambiguous foreign policy approach regarding relations with Iran and Iraq—which most of the states on the Arabian peninsula regard as their primary security threats.

Another way for industry to enter the international arms market is through bilateral military–technical cooperation agreements, a number of which have been signed in recent years—notably with countries of the 'near abroad' and the former allies in East–Central Europe. After 1989 military–technical cooperation with countries in the region ended abruptly with the dissolution of the WTO and the CMEA. However, in 1995 there was some evidence that bilateral government-to-government agreements with at least some former allies might provide the umbrella under which industrial ties could be restored.[25]

Since the dissolution of the Soviet Union, defence industrial cooperation has been discussed in the framework of the CIS.[26] Moreover, the restoration of closer military–technical cooperation within the CIS is likely to accelerate.

The Russian Government has also concluded some agreements with West European countries regarding specific contracts. According to Alexander Kotelkin, Director-General of Rosvooruzhenie, agreements have been made with French, German and Italian firms, including one with a Franco-German joint venture to develop a new radio navigation system to be used on the version of the T-80 main battle tank offered to export customers.

[24] Redlich, A. and Miscavage, M., 'The business of offset: a practitioner's perspective', ed. S. Martin, *The Economics of Offsets: Defence Procurement and Countertrade* (Harwood: Amsterdam, 1996), pp. 390–91.
[25] In Mar. 1995 an agreement was made between the Russian and Bulgarian ministries of defence 'On coordination in the area of the development of armaments and military technology' and an analogous agreement was signed with Romania the following month. *Krasnaya Zvezda*, 3 Apr. 1995 (in Russian); *Krasnaya Zvezda*, 11 Mar. 1995 (in Russian); and *Krasnaya Zvezda*, 6 Apr. 1995 (in Russian). The restoration of these ties is discussed in more detail in chapter 10 of this volume.
[26] *Krasnaya Zvezda*, 24 Mar. 1995 (in Russian). Agreements on military–technical cooperation have been concluded with Armenia, Belarus, Georgia, Kazakhstan and Ukraine. Cooperation in the framework of the CIS is discussed in detail in chapter 9 of this volume.

## Establishment of joint ventures and financial–industrial groups

The establishment of joint ventures and financial–industrial groups is also seen as a means of improving access to export markets. Some of the new industrial units being created will be very large and several, in particular those in the aircraft industry, are making products which have a history of export success.[27]

In most cases the aim of joint ventures between Russian producers and foreign partners is to produce, sell and maintain civilian products developed through dual-use technology—that is, civilian technology which can have military applications. There have been many examples of joint ventures of this kind including those involving leading defence firms, with manufacturers of helicopters particularly prominent. In many cases the role of the Western partner is primarily seen as filling those functions in which Russian defence firms are weakest: marketing, sales and distribution.

Another important aspect from the perspective of the Russian producer may be the desire to leave part of the earnings in Western bank accounts and thereby avoid paying tax in Russia. This type of cross-border arrangement could even be described as amounting to a de facto financial–industrial group.

The development of financial–industrial groups in practice generally has little in common with the model of the financial–industrial group advocated by the League of Assistance to Defence Enterprises. Its President, Alexander Shulanov, describes financial–industrial groups as 'monopolistic conglomerates capable of producing both civil and military products. The integration of a wide range of activities and functions should permit Russian firms to rationalize, reorganize and enter world markets more effectively'.[28]

According to the Science and Technology Department of the former Ministry of Defence Industry, Russia will create around 30 military financial–industrial groups over the next few years.[29]


## Illegal methods

At present the mechanisms by which large corporations or groups conclude and meet export agreements are far from transparent and are often concealed altogether. This leads to the fourth type of export strategy mentioned above: the use of illegal methods. In most cases a specially founded joint venture creates one means of concealing the details of financial transactions. The top management of a firm usually handles all sales agreements directly, whether the agreement in question is with a bank clerk, a buyer or an intermediary. Where there is no transparency there is no way to measure whether profits derived

---

[27] E.g., the MiG-MAPO financial–industrial group (manufacturer of the MiG-29). Kogan, E., *Russian Defence Conversion and Arms Exportation*, PRIF Report no. 41 (Peace Research Institute Frankfurt: Frankfurt, Nov. 1995), pp. 11–23.

[28] Author's interview with Alexander Shulanov, Sep. 1995.

[29] *Defense News*, 8–14 July 1996, pp. 1, 27; and *Defense News*, 29 July–4 Aug. 1996, p. 11.

from a sale return to the firm or contribute towards the future development of the business.

The activities of arms-trading companies which sell items they do not produce have also attracted attention in frequent articles detailing the sale of arms from, for example, the inventories of armed forces units stationed in Russia, Belarus and Ukraine. These sales—which are illegal—are usually mediated through third countries and often seem to have been directed towards local conflicts and terrorist or paramilitary groups.

The scale of illegal sales is impossible to measure. However, anecdotal evidence suggests that arms shipments made without the necessary approvals and permits are not particularly rare. In one such case, an An-72 military transport aircraft was sent to Estonia containing weapons from the Smolensk army base. Such examples, often involving either the Baltic states or newly independent states in the Caucasus, are now routinely reported in the Russian press.[30]

The illegal trade in weapons is a two-way process and there are examples of foreign weapons being illegally imported into Russia.[31] The war in Chechnya has added to the demand for weapons.

Rosvooruzhenie was itself made the subject of criminal proceedings by the Procurator General of the Russian Federation, charged with concealing earnings for tax evasion purposes and carrying out illegal currency dealings.[32]

For some producers and trading companies the present regulations and tax system provide few incentives to make legal sales. A reduction in the costs of legitimate transactions could reduce the level of criminality. According to the commercial director of one defence firm, 'there is not, unfortunately, today, a sufficiently strong incentive for honest work'.[33]

## V. Examples of reorganization of Russian defence firms: the view from within

Among the best-known examples of Russian military technology in terms of actual or potential exports are the Sukhoi and MiG fighter aircraft, Kamov helicopters, conventional submarines (the Kilo Class but also miniature submarines of the Piranha type), surface warships and the S-300 series air defence system.

Rosvooruzhenie bears the main responsibility for managing the export of such products. But where do the actual producers stand in relation to the design and manufacture of military technology for export, and how have they met the challenge of modernization? A tentative answer to these questions is suggested by an examination of the main activities and attempts at reorganization of a number of leading Russian defence firms.

---

[30] ITAR-TASS, 11 Apr. 1996 in FBIS-SOV-96-071, 11 Apr. 1996, p. 22; and ITAR-TASS, 6 May 1996 in FBIS-SOV-96-088, 6 May 1996, p. 30.

[31] *Kommersant Daily*, 6 Apr. 1995 (in Russian).

[32] *St Petersburgskiye Vedomosti*, 28 Apr. 1995 (in Russian).

[33] Author's interviews, Yekaterinburg, Oct. 1995.

## AO Baltiyskiy Zavod

The AO Baltiyskiy Zavod shipyard in St Petersburg was founded in 1856 after the Crimean War to build warships and was to build over 100 before 1917, including most of the leading ships (such as the *Admiral Nakhimov* and the *Admiral Ushakov*) of the tsarist navy. Production spanned a wide range of different types and classes, from battleships to submarines. Baltiyskiy Zavod was one of the main centres of Soviet naval shipbuilding, producing nuclear-powered ice-breakers, research vessels and tankers as well as warships—notably missile cruisers such as the *Kirov*. One of the last ships of the Kirov Class, the *Pyotr Velikiy,* remained unfinished for several years owing to lack of money until it was finally launched in 1995. The proportion of the work at the shipyard covered by defence orders fell rapidly, from 80 per cent in 1988 to 7 per cent in 1993. This fall has (as at most defence enterprises) been largely 'spontaneous' rather than reflecting a deliberate or strategic programme of conversion.

Today Baltiyskiy Zavod finds itself in a relatively stable financial position, largely thanks to foreign orders. In 1993 the yard received an order from Germany for 12 chemical tankers, followed by a comparable order from Norway. According to company president O. B. Shulyavskiy, there has been a profound conflict between the economic interests of the enterprise and the commitment to fulfil the state orders: for example, the state failed to pay for the *Pyotr Velikiy* while the company faced bills from a whole range of suppliers, whose prices were rising all the time. This put the company in a critical position. Shulyavskiy noted: 'If the Ministry of Defence doesn't recognize the nature of the temporary economic situation—so that there is a major hiatus regarding defence orders—then the shipyard will reorganize itself so that defence orders will no longer be viable. There will no longer be sufficient skilled labour in the shipyard, or in the defence industry generally'.[34]

The client for the unfinished ice-breaker *Ural* is Murmansk Parakhodstvo, with finance provided by the Ministry of Transport and the Department of the Fleet. However, the ability of these state customers to pay is little better than that of their military equivalents.

It can be argued that the reorganization of the shipbuilding sector and its increasing orientation to the international market did not suddenly emerge, but had been developing for some time even before the dramatic changes since 1991. Technological and organizational obsolescence was frequently masked by the steady stream of state orders which had paid in full for the high costs of production. According to Shulyavskiy: 'We have been using techniques to build ships that are several times more expensive than those in Europe, to say nothing of South Korea or Japan. Now that there is a taxation regime and rising prices for energy and materials, our products are frequently uncompetitive. There is

---

[34] *St Peterburgskiye Vedomosti*, 16 Nov. 1994 (in Russian).

only one way out and that is investment, profitable sales abroad, and technical modernization'.[35]

This view is echoed by other shipbuilding firms. Thus, S. Karmanovskiy, Deputy Director of the AO Almaz, sees the best strategy for progress as being to work simultaneously in several different directions including fulfilling state defence orders (for example, building patrol hovercraft), designing commercial and scientific vessels for private clients at home or abroad, and buying licences to build and export shipbuilding components. According to Karmanovskiy, 'such a path of development will allow us to have varied sources of finance, continual market research at home and abroad, and the experience of international business cooperation'.[36]

## KB Malakhit

The KB Malakhit (Malachite Design Bureau) was established in Leningrad in 1948 to build submarines with energy sources independent of atmospheric oxygen. In 1952 this objective was revised by order of the government to mean building nuclear submarines. Within five years the *Leninskiy Komsomolets*— the first such submarine operated by the Soviet Union and designed by Malakhit—was built at Severodvinsk and launched into the Arctic Ocean.

In the late 1950s the *Volna*, the first Soviet submarine armed with ballistic missiles and with a non-nuclear propulsion system, was completed. Later the design bureau developed the high-speed cruise missile-carrying Papa Class submarine with a titanium hull as well as the Alpha Class submarine.

At recent exhibitions Malakhit has been showing the miniature submarines Piranha and Triton, the creations of chief designer Yuriy K. Mineyev. Weighing as little as 1.5 tonnes, the Triton has been adopted by the Russian Navy.[37]

The workforce at Malakhit contracted by more than one-third between January 1992 and June 1996. The organizational structure has been changed as have the economic mechanisms for dealing with customers and suppliers. A number of conversion programmes are in progress, including designs for vessels for geological survey and fish farming. As a *kazyonny* ('Treasury') enterprise, Malakhit is state-owned but with some degree of economic freedom. However, it is currently in a very weak financial state.[38]

A distinguishing characteristic of the work at Malakhit (as of other submarine design bureaux such as the TsKB Rubin in St Petersburg and the TsKB Lazurit in Nizhniy Novgorod) is that the designs demand inputs from skilled workers and technicians from almost all the engineering and scientific disciplines. Equally, when the submarines reach the phase of manufacture, they require

[35] *Konversiya, Politika i Vooruzhenie* 2 May 1994 (in Russian).

[36] Author's interviews with S. Karmanovskiy, Deputy Director of AO Almaz, Sep. 1994, Sep. 1995.

[37] Author's interviews with Yuriy Mineyev, Chief Designer, KB Malakhit, Sep. 1993, Oct. 1994.

[38] Author's interview with V. Barantsev, Deputy Director, KB Malakhit, Oct. 1995. *Kazyonny* enterprises are those exempted from the privatization programme by presidential decree.

cooperation from production enterprises spread across the territory not only of Russia but also of some other members of the CIS.

One response to the problem of managing these relationships under the new conditions is the setting up of a project-based management system, the temporary work collective. According to Deputy Director V. I. Barantsev, this marks a major departure in principle from earlier forms of work organization in the enterprise.

For each project a group of specialists is brought together, representing all the organizations involved—whether science, design or production. This group then hires the minimum number of staff required to carry out the job, coordinates the work and takes full responsibility for the project as a whole in terms of deadlines and quality. Payment is made according to results achieved by target 'milestones' which are subject to inspection. This innovation builds stronger links between the stages of the manufacturing cycle and between the partner organizations involved, as well as making the process as a whole more flexible and responsive. As Mineyev explains, 'we have almost ceased going to see civil servants in Moscow; instead we go direct to our partners, communicating horizontally, not vertically as before. We are now much more confident about being able to fulfil any order, including export orders'.[39]

As regards mechanisms for entering the international market, the Malakhit specialists interviewed are sceptical. In their view the selection of items displayed at international exhibitions and the fact that they are often presented abroad in an unprofessional way by government agencies—rather than by specialists from the design bureau itself—go a long way towards explaining the slow progress in international markets. Information about the scheduling and content of exhibitions is the property of federal departments. According to the enterprise representatives, they have been asked for technical documentation as late as two or three days before the exhibition. This is only available in Russian. The official government representatives who attend the exhibitions usually know little about what is on display.

At the same time, according to one senior manager of the design bureau, civil servants actively hinder the establishment of direct contacts between enterprise representatives and colleagues from other countries. For example, a project with US partners was discussed at a high level within government for more than six months without the necessary decisions being approved. In the eyes of industry, against the background of chronic non-payment for state orders, this kind of behaviour underlines that the government lacks any sense of responsibility either as a partner or as a client. At the same time, it insists on maintaining its role as a source of control.

As regards information about potential foreign clients or partners, the information department at Malakhit had put together a database containing the designs of products offered by potential competitors. However, they had very little data on potential foreign customers for either civilian or military products.

---

[39] Author's interviews with Yuriy Mineyev, Chief Designer, KB Malakhit, Sep. 1993, Oct. 1994.

Catalogues and directories of potential foreign cooperation partners, although essential in contemporary Russia, are in very short supply and expensive.

## TsKB Lazurit

Similar views on the problems of developing exports are voiced by the management of Russia's other two submarine design bureaux, TsKB Rubin (the Ruby Central Design Bureau) and TsKB Lazurit. TsKB Lazurit (the Azure Central Design Bureau) was founded in 1953 in Nizhniy Novgorod, specializing in ocean-going diesel-powered submarines and (later) in designing cruise missile-launching submarines. It also took part in the development of submarines with a titanium hull. From the 1960s it worked on various sub-systems including integrated search and rescue systems to respond to submarine accidents. Its submarines have pioneered the use of hydro-acoustic robot technology for navigation. Among its conversion projects are deep-sea resource exploitation applications, undersea restaurants, medical technologies, undersea oil exploration and transport. Together with a Canadian firm, Lazurit has developed a design for a deep-sea submarine, *Ocean Shuttle,* intended as a research vessel for use in the Atlantic, Arctic and Pacific oceans. The project has been seen as having major potential significance for global environmental research.

The Director of Lazurit, N. I. Kvasha, underlined that all the main commercial functions for trading internationally were concentrated in the central government. Moreover, Lazurit was obliged to abide by strict rules and licensing procedures in this regard. At the same time, there was a lack of clarity regarding any programme of strategic development for the military shipbuilding sector despite the need for external (including foreign) investment.[40] Within the design bureau a series of conversion programmes had been developed which were geared to the international market. These projects were intended to be the basis for discussions with Western investors aimed at putting together a workable business plan that was independent of state orders. To this end the design bureau prepared a marketing prospectus, parts of which were published in Russian and foreign journals. However, because of the status of the design bureau and its extensive use of dual-use technologies, all draft contracts first had to be agreed with the relevant Russian central authorities.[41]

The management at Lazurit has experienced the same lack of information and of financial support from its regional administration. It might have been expected that the regional consultancy firms set up in recent years would facilitate the reorganization of defence firms and the search for markets for high-technology products. Instead, according to the management of Lazurit and other defence enterprises in Nizhniy Novgorod, the interests of the defence sector have taken second place to the ambitions of regional leaders. According to Kvasha, entering foreign markets becomes extremely complicated when a firm

---

[40] Author's interview with N. Kvasha, Director of Lazurit, Sep. 1994.
[41] Author's interview with Yu. V. Postnov, Deputy Director of Lazurit, Aug. 1995.

has both local bureaucrats and central authorities to deal with. Under these conditions 'the drowning person must save himself by his own hands, and not count on anybody else'.[42]

## Kamov

The Kamov scientific–technical complex is one of the two leading helicopter firms in Russia. Founded in 1940 by order of the People's Commissar of the Aviation Industry, the firm set about building the first helicopter in the country. The Director was N. I. Kamov, his deputy M. L. Mil. In October 1941 the factory was evacuated to the Sverdlovsk Oblast (region) in the Urals and in 1943 it was disbanded. In 1946 OKB-2 was established at Sokolniki in the Moscow Oblast to design light helicopters, moving to Moscow in 1955. Here light helicopters such as the Ka-8 (1948), Ka-15 (1953), Ka-18 (1956), Ka-25 (1961) and Ka-26 (1965) and the Ka-22 heavy helicopter (1940) were developed. From the late 1970s onwards work was carried out on the design of combat helicopters in cooperation with a number of scientific research institutes and enterprises.

In contrast to submarine construction, where the basic product has proved to be difficult to adapt for non-defence purposes, helicopters can relatively easily find civil applications. Thus, according to S. V. Mikheyev, its chief designer, Kamov has been able to move quickly into the international and domestic civilian market. For example, Ka-27 helicopters and their derivatives have been adapted successfully for civil transport purposes, off-loading ships' cargoes (the Ka-32T) and accompanying convoys on Arctic shipping routes (the Ka-32C). In 1994 the firm concluded an agreement with the Moscow City Government to lease to the latter helicopters (the Ka-32-003) adapted for fire-fighting. The Ka-126, Ka-226 and Ka-62 helicopter types are all oriented specifically to the export market. At the same time work continues on orders for the Russian Ministry of Defence, for example, on the design of the Ka-31 VKRLF radio-locational helicopter.

Kamov was where the Ka-50 fighter helicopter, known worldwide through international exhibitions, was originally developed. Work on the Ka-50 started at Kamov in 1977 with the aim of replacing the Mi-24 attack helicopter in Soviet service. The new design was based on analyses by Kamov specialists of the vulnerability of helicopters in Afghanistan and Viet Nam as well as the availability of new materials and weapon systems.

However, Kamov has attracted few foreign orders for its helicopters. One reason for this is that the (since 1992) impoverished Ministry of Defence has not been able to replace its complement of helicopters with the Ka-50, but has instead continued to order the old Mi-24 attack helicopters. Seven Mi-24s were purchased in 1995, although two Ka-50s were said to be on order for 1996. The decision to purchase new models of the Mi-24 has come in for criticism, given that the model was planned to be 80 per cent phased out by 2005.[43]

[42] Author's interview with N. Kvasha, Director of Lazurit, Aug. 1995.
[43] *Krasnaya Zvezda*, 8 Feb. 1996 (in Russian).

The failure of the Ministry of Defence to buy newly developed defence products has the effect of making potential foreign buyers wary. As a result, much of the latest technology in the Russian defence industry is produced only on a one-off basis, which fails to provide an incentive to production enterprises in the sector which have traditionally been geared to long production runs.

## OKB Sukhogo (Sukhoi)

The OKB Sukhogo (Sukhoi Experimental Design Bureau) aviation complex was founded by P. O. Sukhoi, who was also its chief designer from 1939 to 1975. It was always geared to the international market via exhibitions and competitions, and its products were aimed to match specific Western aircraft. Thus the Su-24 bomber was meant as a direct competitor to the US F-111; the Su-27 fighter was ranged against the F-14, F-15, F-16 and F-18.

According to M. P. Simonov, chief designer from 1983, the development of the Su-27 was far from straightforward. During tests it did not meet the performance standard of the F-15, but it had already been put into batch production. The management, on their own initiative but with the approval of the then Deputy Minister for Aviation Industry, I. S. Silayev, had production of the aircraft halted (despite outstanding orders from the state) after 12 had been produced. A complete redesign then took place. The designers involved later said that all that then remained of the original design was the ejector seat and the tyres for the main wheels.[44] Thus an initiative 'from below' created an almost entirely new aircraft which was to break 34 world records in terms of performance and quality of weaponry, and which was to be the first of a whole new family of high-performance aircraft. The ability to modify the basic design became, as in Western aircraft, a selling-point. Sukhoi's designs allow for a wide variety of modifications to suit client preferences, permitting considerable flexibility in the world market.

The Su-24 bomber was exported to Iran, Iraq and Libya by the Soviet Union. Since 1992 the Su-27 has been exported to both China and Viet Nam.

Flexibility has also come to characterize the business organization at Sukhoi. For example, the single-seater Su-26 sports aircraft, the two-seater Su-29 and the new Su-31 have all been exported via a US distribution firm. Every year Sukhoi has been able to lower its costs and prices to stimulate demand (currently running at 20 aircraft per year at prices of $170 000–200 000 each). As production increases new suppliers (such as the machine construction factory at Dubno, near Moscow) have been engaged and the international reputation of the firm is further consolidated.

Success in export markets has helped to create interest at home. As a result, one of the most stable financial–industrial groups has been established, bringing together all the enterprises involved in the design and manufacture of Su-series military, training or sports aircraft around the Sukhoi bureau which provides the core of the group. Among the enterprises involved are the aerospace factories of Komsomolsk-na-Amure, Novosibirsk, Irkutsk and others. The financial part of the financial–industrial group is represented by the commercial bank Yalosbank which also has the function of attracting new participation by other banks.

Sukhoi demonstrates the extent to which a defence firm can carry through reorganization both of its business arrangements and of the technical content of its product and do so as a means of independently finding its niche in international markets. This is one example of a growing tendency within the Russian defence industry—increasing independence from the state in terms of finance, strategy and marketing, combined with increased development and use of military and dual-use technologies aimed at export markets.

### AO Irkutsk Aviation Production Association

Founded in the 1930s, the Irkutsk Aviation Production Association (IAPO) is a production facility making aircraft of different types, including Su-, MiG-, An-, Il- and Yak series. Its production processes are therefore geared towards multi-functionality and integration of large systems. Having all the necessary production equipment for the different types of aircraft, IAPO is capable of unit production of one-off models and flexible batch production of existing models with a wide range of modifications possible. For example, the facilities allow for switching of production between military and civil versions of Antonov and Ilyushin aircraft designs. According to the General Director of IAPO, Aleksey Fyodorov, the enterprise has all stages of the manufacturing cycle under its control and this provides the basis for flexible production. One example was the Su-27UB new-generation fighter (sometimes called the Su-30) which was test-flown in March 1985 and was in full production at IAPO by the middle of 1986.

Along with better-known classes of aircraft, IAPO has also been at the centre of the development of less well-known types, such as the Be-200 wing-in-ground effect or amphibious aircraft (to the design of TANTK Beriyev), the light four-seater Yak-112 and the Delta GALS-5.

In spite of this record of production, however, IAPO has no experience of exports or of sales to domestic buyers, including the Ministry of Defence. The facility has had 60 years of experience in aircraft manufacture and has a reputation for reliability. However, according to Fyodorov, 'for us, entering the market—including the export market—means a search for new paths of development'.[45]

## VI. Conclusions

It is clear from these examples of leading defence industry enterprises what an incentive they have to establish themselves in export markets. Some specialize in the export of military technology, while others (the majority) seek to export

[45] *Military Parade*, July–Aug. 1995.

civilian products embodying dual-use technology. The extent to which they are able to produce in a sufficiently flexible fashion to meet the demands of government or commercial clients depends very much on the extent to which they have successfully reshaped their organizational structures and procedures to encourage flexible operation.[46]

As far as the defence industry and arms exports as a whole are concerned, the picture is a contradictory one. On the one hand, there continues to be a lack of strategy or cohesiveness of any sort on the part of the government regarding the future of the Russian defence industry. It is neither being closed down nor supported. Under these conditions it has proved particularly difficult to rationalize and modernize. The same applies to its major potential clients, the armed forces. This lack of clarity also affects arms exports. While some support is given, on balance it may be said that government policy has worked to the detriment of exports (whether military or civilian) from Russian defence enterprises. This is particularly true of the financial regime within which defence enterprises are supposed to trade.

In this climate, where government influence is increasingly seen in negative terms by defence industry management, the initiative is passing increasingly to enterprises themselves to find a means of survival. The export of civilian technology is for them a high priority but, given that this frequently consists of production technologies that have both civilian and military applications, this may represent greater cause for concern than a more conventional arms export policy coordinated by the government.

[46] Denezhkina, E., 'Economic and managerial aspects of defence industry transformation', *Conversion in Machine-Building*, no. 2 (1996).

# 9. Military–technical cooperation between the CIS member states*

*Alexander A. Sergounin*

## I. Introduction

According to the Russian foreign policy concept adopted in January 1993, the CIS countries have first priority in Russian foreign policy. The aim is to establish fully fledged cooperation with the other CIS countries in economic, military, scientific and technological areas. However, this document points out that scientific and technical cooperation should be oriented to peaceful purposes and civilian use.[1] It includes no special provision on military–technical cooperation. This may be explained by two factors. First, at the time the Russian leadership was preoccupied by converting the defence industry. Second, Russia was cautious about stressing arms transfer policy and military cooperation within the CIS, anticipating a negative reaction from the West.

The Russian leadership very soon changed its mind. It was realized that conversion was impossible without proper funding—hence a new stress on an active arms export policy. At the same time, President Boris Yeltsin began a new policy aimed at further economic, political and military integration in the CIS.

Military integration had been among the first forms of integration—in the framework of the Tashkent Treaty on Collective Security of 15 May 1992 and the Principal Guidelines for the Evolution of CIS Integration adopted by the CIS heads of state in 1992.[2] By 1993 the former Soviet republics had made great strides in fostering economic, political, military, humanitarian and cultural ties through the CIS mechanism and bilateral relations. However, it had become clear that the integration process had to be based on a well-developed infrastructure and institutional network rather than on declarations and intentions. Without material and technical support, many of the agreements reached existed only on paper.

---

[1] 'Kontseptsiya vneshney politiki Rossiyskoy Federatsii' [The foreign policy concept of the Russian Federation], Special Issue of *Diplomaticheskiy Vestnik*, Jan. 1993, p. 6 (in Russian).

[2] For the text of the Tashkent Treaty, see *Izvestiya*, 16 May 1992, p. 3. The original signatories were Armenia, Kazakhstan, Kyrgyzstan, the Russian Federation, Tajikistan and Uzbekistan. By the spring of 1994 Azerbaijan, Belarus and Georgia had also joined. See also Samsonov, V., 'Political and military integration of CIS member states', *Military Parade*, Sep.–Oct. 1996, pp. 38–39.

* The research on which this chapter was based was supported by grants from the Copenhagen Peace Research Institute (1996) and INTAS (International Technical Assistance to the CIS Countries) (1995–96).

This chapter deals with military–technical cooperation within the CIS framework. Its evolution is described and a number of important issues are addressed. The legal bases for cooperation both in Russia and within the CIS and the process of decision making between Russia and CIS bodies are explained. Different forms of cooperation are identified and the impact of cooperation in the military–technical sphere on bilateral relations is examined.

First, it is necessary to make some observations about the information on which these analyses must be based. Recent events make it very difficult to compile a reliable database on this subject. The collective CIS agreements and bilateral agreements between states are available. However, published information on the establishment of financial–industrial groups is rare. Specific bilateral agreements on individual military–technical cooperation projects as well as detailed information on arms transfers are usually classified. Many reports in the media of the CIS countries and elsewhere are not confirmed by the official data available to the public. Study is also complicated by differences of opinion between experts about the credibility and methodology of official Russian and other CIS statistics and between assessments of arms transfers and defence industry development.

Section II examines the situation in which the defence industries of the newly independent states found themselves immediately after the dissolution of the USSR.

## II. The impact of the break-up of the Soviet Union

### The impact on the armed forces

The division of the military assets of the former USSR was a very important issue during the first stage of existence of the CIS, and a number of military–technical programmes, including arms and equipment transfers, were carried out. Most attention has been paid to agreements related to the nuclear inventory of the former Soviet Union, which are not discussed here. However, there was also very significant legal transfer of conventional military assets.

The scale of the transfer can be indicated by the quotas established at the Tashkent summit meeting of 15 May 1992 by Georgia and seven CIS states— Armenia, Azerbaijan, Belarus, Kazakhstan, Moldova, Russia and Ukraine—for items limited under the 1990 Treaty on Conventional Armed Forces in Europe (CFE Treaty). Kazakhstan also signed the agreement as a state, part of whose territory is covered by the treaty. The quotas are indicated in table 9.1.[3]

There were cases of armaments and military equipment being seized from Russian units deployed on the territory of the former union republics. For example, in 1992 military depots in Tbilisi and Akhaltsikh were seized by Georgians. While control of the depots has been returned to the Transcaucasus Military District (MD), the vehicles and equipment have not. According to the

---

[3] See also the text of the Tashkent Document in *SIPRI Yearbook 1993: World Armaments and Disarmament* (Oxford University Press: Oxford, 1993), pp. 671–77.

**Table 9.1.** The quotas for armaments under the Tashkent Agreement

| Type of armament | Armenia | Azer. | Belarus | Mold. | Georgia | Russia (in the area of employment) | Ukraine |
|---|---|---|---|---|---|---|---|
| Combat tanks | 220 | 220 | 1 800 | 210 | 220 | 6 400 | 4 080 |
| Incl. in regular units | 220 | 220 | 1 525 | 210 | 220 | 4 975 | 3 130 |
| Armoured combat vehicles | 220 | 220 | 2600 | 210 | 220 | 11 480 | 5 050 |
| Incl. in regular units | 220 | 220 | 2 175 | 210 | 220 | 10 525 | 4 350 |
| Of which APCS and combat vehicles with heavy weapons | 135 | 135 | 1 590 | 130 | 135 | 7 030 | 3 095 |
| Incl. combat vehicles with heavy weapons | 11 | 11 | 130 | 10 | 11 | 574 | 253 |
| Artillery | 285 | 285 | 1 615 | 250 | 285 | 6 415 | 4 040 |
| Incl. in regular units | 285 | 285 | 1 375 | 250 | 285 | 5 105 | 3 240 |
| Combat aircraft | 100 | 100 | 260 | 50 | 100 | 3 450 | 1 090 |
| Strike helicopters | 50 | 50 | 80 | 50 | 50 | 890 | 330 |

*Note:* Azer. = Azerbaijan; Mold. = Moldova.

*Source*: *Rossiyskiye Vesti*, 21 Dec. 1992, p. 2 (in Russian).

district command authorities, equipment worth more than 1 billion roubles (in 1993 prices) has been stolen in the Georgian capital alone.[4] In turn, the Georgian Government has accused Russia of making arms deliveries to Abkhazia (especially from the Russian military base in Gudauta). In Moldova, Russian forces are accused of transferring equipment of the 14th Army to the 'unconstitutional troops in Tiraspol'.[5]

In spite of these changes in distribution, the armed forces of the other CIS member states remain entirely equipped with weapons of Soviet origin.

## The impact on the defence industrial base

The collapse of the Soviet Union had many implications for the defence industrial base of Russia and other CIS member states.

First, the role of the defence sector in their economies has significantly decreased and the sector has found itself in deep crisis. The Soviet military–industrial complex employed roughly 7.5 million people in 2000 enterprises. The defence sector represented around 20 per cent of the total Soviet industrial

---

[4] Immediately after the dissolution of the Soviet Union there was particularly high tension between Russian and local populations in the Caucasus. In 1992 alone there were about 600 attacks against Russian military personnel in the 'near abroad', 80% of which took place in Azerbaijan and Georgia.

[5] *Rossiyskiye Vesti,* 21 Dec. 1992, p. 2 (in Russian).

labour force, 16 per cent of gross industrial output and 12 per cent of national industrial capital, and consumed 75 per cent of industrial R&D funds.[6]

Two years after the collapse of the Soviet Union, the number of people employed by the defence sector had shrunk by 3 million to 4.5 million. In 1992, the total output of the defence industry—including the nuclear industry—fell by 18 per cent with a further fall of 16 per cent in 1993. Military output fell by 38 per cent in 1992 and 30 per cent in 1993. During the first half of 1994, military output declined by 43 per cent and civilian by 40 per cent compared to the same period of 1993. As a result, in 1994 the output of weapons and other military hardware in Russia was less than 30 per cent of its 1990 level and the civilian share of total defence industry output had risen from 50 per cent to almost 80 per cent.[7] According to the Russian Ministry of Defence, by the end of 1993 around 70 per cent of defence plants were standing completely idle.[8] The situation has not improved since then. In the first half of 1996 military output fell by 26.8 per cent and the number of people employed in the defence sector declined by 13.6 per cent.[9]

Table 9.2 illustrates the economic decline of the Russian defence industries in 1992–96.

Second, the disruption of traditional economic ties had a severe impact on the new national economies of these states. The former Soviet Union consisted of a number of regions which differed greatly from each other in levels of economic development, specialization, raw materials, energy resources and climate. As parts of an integrated national economy these regions had a high degree of interdependence. In the early 1990s, about 23–30 per cent of the economic needs of any of the regions that subsequently became independent states were met by goods produced within what is now Russia.[10]

Third, the system of organizing economic management by sectoral branches created in the former USSR (and still applied today in many CIS countries) meant that the horizontal linkages between enterprises were not managed at the enterprise level. A medium-sized enterprise might have between 50 and 300 suppliers and customers in the total production–distribution chain.[11] This interdependence made inevitable the decline of production along the whole chain once relations with the centre were severed.

---

[6] Martel, W. and Hailes, T. (eds), *Russia's Democratic Moment? Defining US Policy to Promote Democratic Opportunities in Russia*, Air War College Studies in National Security, no. 2 (Air University: Montgomery, Ala., 1995), p. 187.

[7] Cooper, J., 'Transformation of the Russian defence industry', *Jane's Intelligence Review*, Oct. 1994, p. 445.

[8] Després, L., 'Financing the conversion of the military–industrial complex in Russia: problems of data', *Communist Economies and Economic Transformation*, vol. 7, no. 3 (1995), p. 334.

[9] *Krasnaya Zvezda*, 3 Aug. 1996, p. 3 (in Russian).

[10] Volosov, I., 'The Russian economy after three years of reform', *Peace and the Sciences*, Mar. 1995, p. 29.

[11] Krivokhizha, V., 'The reconstruction of the Russian military–industrial complex', *Peace and the Sciences*, Dec. 1994, p. 27.

**Table 9.2.** Indicators of economic decline in Russian defence industries, 1992–96

| | Total output | | | | Employment | | | | Salaries | |
|---|---|---|---|---|---|---|---|---|---|---|
| | 1992 as % of 1991 | 1993 as % of 1992 | 1994 as % of 1993 | 1996 as % of 1995 | 1992 as % of 1991 | 1993 as % of 1992 | 1994 as % of 1993 | 1996 as % of 1995 | 1992 as % of 1991 | 1993 as % of 1992 |
| Aircraft | 84 | 81 | 49 | .. | 91 | 90 | 85 | .. | 71 | 68 |
| Ammunition and special chemicals | 70 | 82 | 62 | 65.7 | 90 | 89 | 81 | .. | 71 | 63 |
| Armaments | 84 | 82 | 54 | .. | 93 | 91 | 85 | .. | 68 | 64 |
| Atomic industry | 100 | 103 | 77 | .. | 97 | 97 | 94 | .. | 114 | 119 |
| Communications equipment | 74 | 78 | 55 | 57.7 | 87 | 82 | 82 | .. | 56 | 51 |
| Electronics | 72 | 66 | 49 | 60.6 | 92 | 81 | 76 | .. | 54 | 44 |
| Radio technology | 84 | 93 | 55 | .. | 87 | 86 | 82 | .. | 53 | 53 |
| Shipbuilding | 89 | 88 | 76 | .. | 90 | 90 | 86 | .. | 77 | 87 |
| Space | 94 | 95 | 71 | .. | 89 | 89 | 82 | .. | 66 | 69 |
| Total defence complex | 82 | 84 | 65 | 73.2 | 91 | 88 | 84 | 86.4 | 69 | 67 |

*Sources:* Sköns, E. and Gonchar, Ks., 'Arms production', *SIPRI Yearbook 1995: Armaments, Disarmament and International Security* (Oxford University Press: Oxford, 1995), p. 473; and *Krasnaya Zvezda*, 3 Aug. 1996, p. 3 (in Russian).

Among the former Soviet republics, Russia has the most serious adjustment problems as it inherited three-quarters of all the defence plants (and roughly 90 per cent of the plants making finished products). The remaining 25 per cent were in Ukraine (14 per cent), Belarus (2 per cent) and other republics (8 per cent).[12] More than 70 per cent of workers engaged in weapon production were in Russia as against 17.5 per cent in Ukraine, 3.2 per cent in Belarus, 1.7 per cent in Kazakhstan, and between 1.4 and 0.1 per cent in the remaining republics.[13]

In Ukraine, 700 defence enterprises employing roughly 1.3 million people were inherited from the Soviet Union. This accounted for about one-third of Ukraine's GNP and 28 per cent of the industrial sector of the economy, and employed 18.6 per cent of all industrial employees by 1990. In 1992 Ukraine's defence production declined to only one-third of the level of 1991.[14]

The Ukrainian machine-building and metal-working industries made sub-assemblies that were sent to Russia rather than being used in local system integration. Plants in Ukraine produced half the total Soviet output of tanks, missiles, military optical products and radio communication systems. Ukraine also made half of all combat vehicles and the Nikolayev shipyards produced the majority of combat ships. Not only did these products include a high Russian content; they were made in enterprises concentrated in the east and south—a region with a high concentration of ethnic Russians that has generated the strongest opposition to Ukrainian nationalism and the strongest support for restoring ties with Russia.[15]

There were 196 industrial enterprises in Kazakhstan that were involved in military production. While some, mostly situated in the north, had thousands of employees, none had a complete circle of production or was able to produce finished products. With the collapse of the Soviet Union these enterprises were left without the defence contracts that formed the basis for their existence. This has produced disastrous consequences for enterprises such as the Petropavlovsk heavy-machine-building plant where 80 per cent of the output used to consist of military items that were to be incorporated in products that were made elsewhere in the Soviet Union.

Throughout Kazakhstan military orders have decreased by 82 per cent but the Petropavlovsk plant lost 100 per cent of its contracts. To make things worse, none of the military enterprises was capable of undertaking conversion without outside help. It is no wonder that thousands of their workers were living under conditions of heavy socio-economic and psychological strain in 1995.[16]

According to some Russian experts, the Kazakh Government bears a certain responsibility for creating this difficult situation. First, it has not established any

[12] Cooper, J., *The Soviet Defense Industry: Conversion and Economic Reform* (Council on Foreign Relations Press: New York, 1991), p. 21.

[13] *Strategic Digest* (New Delhi), Feb. 1994, p. 211.

[14] Kuzio, T., 'Ukraine's military industrial plan', *Jane's Intelligence Review*, Aug. 1994, p. 352.

[15] Kuzio (note 14).

[16] Kortunov, A., Kulchik, Yu. and Shoumikhin, A., 'Military structures in Kazakhstan: aims, parameters, and some implications for Russia', *Comparative Strategy*, vol. 14 (1995), pp. 301–309.

workable system of reciprocal payments with Russian clients and suppliers. Slow in introducing market reforms, the Kazakh Government has retained a pricing system for military equipment which does not compensate defence enterprises for the rapid increases in their costs due to inflation in the wider economy. Second, the Kazakh authorities have failed to extend any tax breaks to defence enterprises while prevailing interest rates have made all industrial production highly unprofitable. Under these conditions the prices of finished goods, including those produced at defence enterprises, have sky-rocketed. Third, since no law on conversion was adopted in Kazakhstan, managers in the defence industry are deprived of any legal protection should they make any important decisions about restructuring.[17]

## III. Incentives for military–technical cooperation

Russia and the other CIS countries all have reasons for promoting greater military–technical cooperation.

### Russia

Russia's vision of the role of the CIS in its new foreign economic policy is quite contradictory. The economic significance of the CIS for Russia is not obvious. The other CIS countries are insolvent and cannot be viewed as a potential source of investment: on the contrary, by 1996 they owed Russia $9 billion.[18] However, Russia has underlined that it is interested in developing production and technological relationships with the other CIS countries on the basis of cooperation and will not reserve the role of management exclusively for itself.

Russia has a number of specific interests in military–technical cooperation with the other CIS countries.

First, Russia needs access to certain types of strategic raw material that were traditionally obtained from the regions that now form the rest of the CIS—in particular, non-ferrous and rare-earth metals, cotton and foodstuffs. Russia also needs to ensure that some of the main transport routes through the territories of other CIS states function without interruption.[19]

Russian interest in chromium and silica from Kazakhstan and manganese and ferro-alloys from Georgia is partly determined by the opportunity to use these materials as part of a non-currency payment system. However, these resources also represent important inputs for the metallurgy complexes in the Urals and Western Siberia. Turkmenistan is rich in fossil fuels and at present is dependent upon Russia to transport these fuels to the market. Uzbekistan has enormous

[17] Kortunov *et al.* (note 16), p. 302.
[18] *Novoye Vremya*, no. 16 (1996), p. 16 (in Russian).
[19] *Kommersant Daily*, 3 Oct. 1995, p. 20 (in Russian).

gold deposits (one-quarter of the reserves of the former Soviet Union), is a leading cotton grower and is rich in natural gas.[20]

Second, Russia is interested in keeping and restoring a number of important technological chains and production facilities located outside Russia. It is advantageous for Russia to develop the space and aviation industries in cooperation with the other CIS countries. For example, production of Ilyushin aircraft is not possible without access to the output of the Tashkent Aircraft Plant and production of Antonov aircraft is not possible without the inputs of the Kiev Design Bureau. Russia depends on the Zaporozhe Aircraft Engine Plant in Ukraine to produce some civilian and military transport planes. However, in the longer term the need for this cooperation may diminish in most cases since only Belarus and Ukraine have significant scientific potential.

Third, the other CIS member states with their total dependence on Soviet equipment are a logical market for Russian weapons. The analytical memorandum of the Ministry for Cooperation with CIS Member States (Minsotrudnichestvo) of 22 September 1995 states that the other CIS countries are almost the only markets for Russia's finished products, and especially its machine-building output.[21]

Since the other CIS countries are insolvent, Russia usually transfers arms on a grant basis. While this is a burden for the Russian federal government, it helps to keep the Russian defence industry afloat. In addition, Aman Tuleyev, the new Russian Minister for Cooperation with CIS Member States, put it, Russia hopes to redeem part of the debts owed by other CIS member states with property, assets and shares in companies in debtor countries.[22]

Fourth, from a political and strategic point of view military–technical cooperation with the other CIS member states is important for Russia as an instrument of strategic control over the 'near abroad'. It has become especially important in view of the debate over NATO enlargement. Russia is not only trying to attract its neighbours into a new security arrangement but is also trying to ensure that the national armies of the other CIS member states are supplied with arms, infrastructure and an officer corps that are compatible with those of Russia and dependent on Russia.

The other CIS member states—which inherited equipment from the Soviet armed forces as well as some production capacities—offered arms for sale, thus competing with Russia. Ukraine established a new centre under the office of the President to coordinate arms exports and pursued many of the same clients as Russia—namely China, India and Iran.[23] Ukraine has sold artillery shells to Pakistan and helicopters to Algeria and upgraded Libyan warships.[24] Ukraine and Pakistan have reached agreement on the transfer of 330 T-80UD tanks.[25]

[20] Olcott, M. B., 'Sovereignty and the "near abroad"', *Orbis*, summer 1995, p. 355.
[21] *Kommersant Daily*, 3 Oct. 1995, p. 20 (in Russian).
[22] Russian television programme Vremya: interview with Aman Tuleyev, Minister for Cooperation with the CIS Member States, 9 Sep. 1996.
[23] Kuzio (note 14), p. 352.
[24] *Izvestiya*, 16 Apr. 1996, p. 3 (in Russian).
[25] *Krasnaya Zvezda,* 11 July 1996, p. 3 (in Russian).

In 1994 Moldova delivered 12 MiG-29 fighter aircraft to South Yemeni sepa-ratists during the civil war in that country.[26] Belarus has transferred 21 BMP-1 armoured vehicles to Bulgaria which were then re-exported to Angola, as well as transferring T-72 tanks to North Korea.[27] Kazakhstan has exported Su-25MK fighters to some Middle Eastern countries.[28] Apart from the competition that these sales offer to Russian exporters, the quality of armaments and services offered by other CIS states has sometimes discredited Russian weapons. Russia is trying to develop a common CIS arms export policy and promote defence industrial cooperation partly to offset these negative developments.

*The use of arms transfers as a policy instrument: the Caucasus*

A very specific use of arms transfers by Russia as an instrument to achieve political and strategic objectives has been the supply of arms and ammunition to certain political groups in newly independent countries. For example, Russia supported either the government in power or the opposition in each of the three Transcaucasian republics (Armenia, Azerbaijan and Georgia), as well as in Moldova and Tajikistan.

During the conflict in Nagorno-Karabakh a large amount of firearms and ammunition from the 147th Motorized Division located at Akhalkalaki in Georgia was transferred into the hands of the Armenian *fedayeen*. The T-72 tanks from this division, which had Russian crews, launched successful offen-sive actions against Shusha, Lachin and other places. Military equipment belonging to both the 147th Motorized Division and Armenia damaged in the fighting was brought by train to the tank repair factory in Tbilisi.[29]

By March 1995, Russia and Armenia were ready for closer cooperation in the military field and the two presidents, Boris Yeltsin and Levon Ter-Petrosyan, signed the Russian–Armenian Treaty on Military Cooperation.[30] Also in March 1995 the two countries conducted joint exercises in the Armavir district, border-ing Turkey. The then Russian Defence Minister, Pavel Grachev, noted that the military and military–technical cooperation between Russia and Armenia could serve as a model for other CIS countries.[31]

After the division of Soviet military assets, Azerbaijan showed little interest in cooperation with Russia. Reports by Armenian officials (including the Armenian President) that illegal arms transfers from the Russian 4th Army to Azerbaijani forces were taking place were denied by Russia. In March 1992,

[26] Anthony, I., Wezeman, P. D. and Wezeman, S. T., 'The trade in major conventional weapons', *SIPRI Yearbook 1995: World Armaments and Disarmament* (Oxford University Press: Oxford, 1995), p. 495.

[27] United Nations, General and complete disarmament: transparency in armaments. UN Register of Conventional Arms: Report of the Secretary-General, 11 Oct. 1993, UN document A/48/344, p. 12.

[28] *Rossiyskaya Gazeta*, 17 Mar. 1992 (in Russian).

[29] Tbilisi 7 DGHE, 22–28 Sep. 1995, pp. 1–2 (in Georgian) in Foreign Broadcast Information Service, *Daily Report–Central Eurasia* (hereafter FBIS-SOV), FBIS-SOV-95-194, 6 Oct. 1995, p. 74.

[30] Armenia became the first CIS country to sign an agreement with Russia on the deployment of a Russian military base in the republic. A motorized rifle division and a squadron of all-weather multi-purpose MiG-23 fighter-interceptors will be deployed in Armenia. *Moscow News*, 31 Mar.–6 Apr. 1995, p. 4.

[31] *Krasnaya Zvezda*, 25 Mar. 1995, p. 1 (in Russian).

General Sufian Beppayev—then deputy commander of the Russian forces deployed in the Transcaucasus—stated that the creation of a national army in Azerbaijan would not be in Russia's interests.[32]

Russia supplied arms to opposition groups in Azerbaijan. For example, the Russian 104th Paratroop Division supplied military hardware (including tanks) to Suret Guseinov, a mafia member-cum-warlord operating in Gyandzha, who ousted President Ebulfez Elcibey. In 1994 Guseinov, then prime minister, organized a failed coup against President Geidar Aliev.[33]

The military rapprochement between Russia and Armenia affected the position of Azerbaijan. It tacitly supported the Chechen separatists, even allowing them to deploy assault aircraft and training bases and acting as a conduit for arms transfers from Islamic countries. Russia was irritated by this position and protested officially on a number of occasions.

Given Russian–Armenian defence cooperation, many observers anticipated that Azerbaijan would try to foster military ties with Turkey and Iran. Instead it approached Ukraine. In March 1995 President Aliev received a Ukrainian delegation headed by Vice-Premier and Defence Minister Valeriy Shmarov, who stressed that bilateral military and military–technical cooperation were promising areas to develop.[34]

This cooperation seems most likely to consist of Ukrainian arms transfers to Azerbaijan. In September 1993, in response to an official protest by the Armenian Foreign Ministry, the Ukrainian Foreign Ministry declared that Ukraine was repairing and returning Azerbaijani tanks rather than providing any new material. However, in its return to the UN Register of Conventional Arms, Ukraine lists 100 tanks and 10 combat aircraft transferred to Azerbaijan in 1993. Late in 1994 there were additional reports of new shipments of tanks from Ukraine to Azerbaijan.[35]

Immediately after Georgia became independent Russia used arms transfers to influence both Georgia's domestic and its foreign policies. Before the Tashkent Agreement, the Transcaucasus MD transferred to Georgia 70 T-72 tanks and 20 attack helicopters.[36] At an early stage of independence Georgia lacked an effective regular army. Two paramilitary units—the National Guard and the Mhedrioni, loyal to Tengiz Kitovani and Dzhaba Ioseliani, respectively—were the most significant non-Russian armed forces operating in Georgia. Russian military intelligence favoured the National Guard and sold arms from the inventory of the Transcaucasus MD, military maps and other documents to Kitovani. He also received assistance in the form of training.[37]

The Georgian President, Eduard Shevardnadze, initially resisted military cooperation with Russia because of Russian assistance to the Abkhazian rebels. The Russian military base in Abkhazia, the Bombora military airfield near

[32] TASS, 18 Mar. 1992 (in English) in FBIS-SOV-92-054, 19 Mar. 1992, p. 17.

[33] *Novoye Vremya*, no. 5 (Feb. 1996), p. 12 (in Russian).

[34] *Moscow News*, 31 Mar.–6 Apr. 1995, p. 4.

[35] Anthony *et al.* (note 26), p. 496.

[36] *Novoye Vremya*, no. 27 (July 1996), p. 19 (in Russian).

[37] *Novoye Vremya*, no. 27 (July 1996), p. 18 (in Russian).

Gudauta, reportedly played a key role in the defeat of Georgian armed forces in Abkhazia. As early as 14 August 1992 (the date when Georgian operations began in Abkhazia), Abkhazian separatists received up to 1000 assault rifles and machine-guns from the air defence unit stationed near Gudauta. In addition, Su-27 fighter aircraft, Su-25 close-support aircraft and Mi-24 attack helicopters based at Bombora airfield bombed Georgian Army positions in Sukhumi. Some of the pilots used in these operations were Abkhazian. The Bombora airfield also played an important role in delivering ammunition and supplies to the Abkhazians in preparation for the assault on Gagra on 3 October 1992. The airborne assault unit stationed here took a direct part in the assaults on Sukhumi in March, July and September 1993.[38]

When the status quo in Abkhazia had been restored, the Russian Government began pressing separatists to abandon their demand for independence and stay with Georgia. According to some accounts Russian military equipment was delivered to the Georgian armed forces through the Vaziani military airfield (including the comparatively modern Uragan multiple rocket-launcher). This equipment gave Georgian forces the possibility of attacking the Abkhazian separatists' command headquarters in Gudauta (30 km from the front line) for the first time.[39]

Russia also served as a mediator in the talks between Abkhazia and Georgia and deployed 3000 peacekeepers in a security zone separating the forces. This created favourable conditions for a gradual rapprochement between Russia and the government in Tbilisi, which was consolidated after Russia backed Shevardnadze during an abortive attempt to regain power by ousted president Zviad Gamsakhurdia in 1993.

Georgia turned to its giant northern neighbour for economic and military help after two years of trying to go it alone. Shevardnadze stated, 'we realize more and more that the temporary coolness in relations between [Georgia and Russia] was a serious mistake which must be corrected'.[40]

## The other CIS countries

In turn, many of the other CIS member states are interested in military–technical cooperation with Russia.

According to leaders of the Progress machine-building design bureau in Zaporozhe, Ukraine (which builds aircraft engines) in the time of the Soviet Union this organization had production ties with 822 partners in the former Soviet republics of which 550 were in Russia. Ninety per cent of the materials used in production were received from Russia. Each engine crossed what is now the Russian–Ukrainian border between five and seven times during the process of its manufacture.[41] Under current conditions with high customs tariffs,

[38] See note 29.
[39] See note 29.
[40] *International Herald Tribune*, 2 Feb. 1994, p. 1.
[41] *Krasnaya Zvezda*, 20 July 1996, p. 4 (in Russian).

trade quotas and lack of cooperation between financial institutions, co-production with former partners is unprofitable.

Russians who worked in the factories that are now running at low capacity have suffered disproportionately from the decline, making local élites particularly eager to see the old inter-republic connections restored.

The coming to power of new élites connected with the local defence industry was one factor that gave momentum to renewed military–technical cooperation. For example, Leonid Kuchma (first Prime Minister and now President of Ukraine) was formerly the Director of the Southern Machine Construction Plant in Dnepropetrovsk—the largest rocket and missile production plant in the world.

None of the newly independent states other than Russia has the necessary prerequisites to develop a wide range of weapon systems and to compete internationally. Ukraine is unable to sustain the powerful missile production and military shipbuilding facilities which exist on its territory and Kazakhstan does not need its vast nuclear, missile and space test sites. These problems can only be solved on the basis of interstate cooperation programmes.

Repairing military hardware is another acute problem for the CIS armed forces. According to Colonel-General B. Y. Pyankov, First Deputy Chief of Staff for Coordinating Military Cooperation among CIS States, 'hardware goes out of commission, but repair plants are scattered across the territory of the former Union. Some states are able to repair only armoured vehicles, others only aircraft . . . Henceforth we will repair all hardware together'.[42]

To create national armies the other CIS countries need not only Russian arms and supplies but also well-trained officer corps. Many lack a system of military education and training. In addition, the ethnic composition of the officer corps is far from homogeneous. This sometimes leads to ethnic tensions in the army and the migration of Russian and other Slav officers from the Central Asian and Transcaucasian states.

According to some accounts, 98 per cent of Kazakhstan's officer corps are Russian or representatives of Slavic ethnic groups.[43] However, only Kazakhs are being promoted to the level of general in the new republic. A wall of mistrust is being erected between the army high command and the rest of the officer corps. At the same time the officer corps has problems attracting Kazakh nationals. There is one military school and one border guard school in Kazakhstan.

Around 2000 Russian officers are serving in the Turkmenistan border forces but neither money nor early promotions have succeeded in keeping them there. This is threatening a long delay in the formation of the army.[44]

Whether Russia or the other CIS member states are more interested in military–technical cooperation depends on the specific situation. Azerbaijan and Turkmenistan are reluctant to cooperate with Russia while Belarus and

---

[42] *Rossiyskiye Vesti*, 20 Sep. 1994, p. 3 (in Russian) in FBIS-SOV-94-184, 22 Sep. 1994, p. 1.

[43] Kortunov *et al.* (note 16), p. 306.

[44] *Rossiyskaya Gazeta*, 18 May 1995, p. 7 (in Russian).

Kazakhstan are willing to develop all forms of military–technical cooperation. Ukraine is eager for bilateral defence industry cooperation with Russia but fears the integration of its armed forces into the CIS. Accordingly, Russian strategy varies from country to country.

## IV. The legal basis and procedures for military–technical cooperation

The Russian draft military doctrine issued in November 1993 contains a special section on military–technical cooperation with foreign countries. It defines military–technical cooperation as including: (a) export and import of weapons and military hardware, military technologies and results of scientific and technical projects in the military sphere; (b) sending military advisers and specialists on official trips; (c) implementing commissioned and joint research and design projects to create new types of weapon and military hardware; (d) technical assistance in building military facilities and defence enterprises; and (e) other military–technical projects and services.[45]

The document describes the aims of cooperation in pragmatic terms: (a) to strengthen Russia's military–political position across the world; (b) to earn hard currency reserves for the state's needs, for the development of conversion and the defence industries, for the dismantling and salvaging of weapons, and for restructuring defence enterprises; (c) to maintain at the requisite level the export capabilities of the country as regards conventional weapons and hardware; (d) to develop the scientific–technical and experimental basis of defence industries, their research and design establishments and organizations; and (e) to provide social guarantees for the staff of enterprises, establishments and organizations which develop and produce weapons, military hardware and specialized equipment.

The doctrine states clearly that 'priority will go to the restoration and expansion on a mutually advantageous basis of co-production ties with other CIS countries'.[46]

Despite the significance of the CIS that this suggested, the Russian Government initially adopted little special legislation on arms transfers to the former Soviet republics. The other CIS countries were subject to general arms export–import regulations, although Russian regulations sometimes included special mention for the CIS member states. For example, under the Regulations on Military–Technical Cooperation of the Russian Federation with Foreign Countries (12 May 1992), all Russian executive agencies responsible for shaping arms transfer policy should coordinate their proposals on military–technical

---

[45] Quoted in 'Basic provisions of the military doctrine of the Russian Federation', *Jane's Intelligence Review*, Special Report, Jan. 1994, p. 12.
[46] See note 45.

cooperation with the CIS Joint Armed Forces Supreme Command.[47] On 24 July 1992 the Russian Government adopted regulations on licensing special assembly transfer for production of armaments and weapon systems in other CIS member states in order to foster cooperation between defence industries.[48] On 4 September 1995 the government issued new regulations which essentially liberalized arms and technology transfers between its defence plants and state organizations of other CIS countries. Fees for granting licences to plants and organizations cooperating with other CIS countries under intergovernmental programmes were also waived.[49]

The Ministry of Defence plays a special role in arms transfers to other CIS member states. For example, before October 1996 the principal government body responsible for licensing arms exports was the State Committee on Military–Technical Policy (Gosudarstvenny komitet po voyenno-tekhnicheskoy politike, GKVTP) and the state trading company Rosvooruzhenie had a leading role in negotiating, concluding and implementing agreements on arms and technology transfer.[50] However, during this period the Ministry of Defence transferred quantities of arms and ammunition to the armed forces of other CIS states on the basis of bilateral formal and informal agreements without the participation of either the GKVTP or Rosvooruzhenie.

In April 1995 the Russian Government adopted the Statute on the Procedure for Provision of Goods (Work, Services) Within the Scope of Production Cooperation and Specialization of Production between Enterprises and Sectors of the Russian Federation and other Members of the Commonwealth of Independent States.[51] It applies to Russian enterprises, associations, joint enterprises, financial–industrial groups and organizations, regardless of their form of ownership, which conclude contracts with enterprises and analogous structures in other CIS states which have adopted the standard documentation for orders for and customs certification of goods delivered under cooperation.

Provision of goods is understood to mean the delivery of raw materials, assemblies, parts, spares, intermediate products, semi-finished products, components and other goods necessary for technologically interrelated types of production and joint manufacture of finished products. The provision of services is

[47] 'Polozhenie o voyenno-tekhnicheskom sotrudnichestve Rossiyskoy Federatsii s zarubezhnymi stranami' [Regulations on military–technical cooperation of the Russian Federation with foreign countries], *Rossiyskaya Gazeta*, 16 May 1992 (in Russian).

[48] 'O poryadke litsenzirovaniya v Rossiyskoy Federatsii postavok spetsialnykh komplektuyushchikh izdeliey dlya proizvodstva voorizheniya i voyennoy tekhniki v ramkakh gosudarstv–uchastnikov SNG' [Regulations on licensing in the Russian Federation of special assembly transfer for production of armaments and weapon systems in the CIS member states], *Sobranie Aktov Prezidenta i Pravitelstva Rossiyskoy Federatsii* [Collection of legislative acts of the President and Government of the Russian Federation], no. 5 (1992), p. 247 (in Russian).

[49] 'Polozhenie o poryadke litsenzirovaniya v Rossiyskoy Federatsii eksporta i importa produktsii, rabot i uslug voyennogo naznacheniya' [Regulations on licensing in the Russian Federation of export and import of military products, works and services], *Kommersant Daily*, 10 Oct. 1995, pp. 68–69 (in Russian).

[50] 'Polozhenie o Gosudarstvennom komitete Rossiyskoy Federatsii po voyenno–tekhnicheskoy politike' [Statute on the State Committee on Military–Technical Policy], *Rossiyskaya Gazeta*, 10 Jan. 1995, p. 4 (in Russian).

[51] *Rossiyskaya Gazeta*, 5 May 1995, p. 14 (in Russian).

understood to be design and repair work, technical servicing and technology transfer.

The state entities identified in the statute as chiefly responsible for concluding and implementing agreements with CIS partners included the Ministry of Defence Industry (Minoboronprom, previously Goskomoboronprom, the State Committee on Defence Industries). Minoboronprom was the Russian representative on the Council on Defence Industries. Once an agreement has been concluded, the Ministry for Economic Cooperation with CIS Member States (Minsotrudnichestvo), the Ministry of the Economy, the Ministry for Foreign Economic Relations (MFER) and the State Customs Committee all perform certain specific technical functions in fulfilling agreements.

The Ministry for Economic Cooperation with CIS Member States and the Ministry of Economics are responsible for analysing the agreements concluded with a view to supporting and developing mechanisms to enhance production cooperation on the part of enterprises and sectors of the Russian Federation with their CIS counterparts.

Agreements usually include lists of enterprises and products and specify the volume of deliveries of the most important types of products or services. The lists of types of products and strategically important raw materials on which quotas are set are first submitted to the Ministry of Economics and the MFER by the Ministry of Defence Industry. Delivery from Russia of such products and materials must comply with separate legislation. Ensuring compliance and recording statistics of the trade are the responsibility of the State Customs Committee.

Accounting and payment between enterprises for the goods and services supplied under contracts are carried out through the Russian Central Bank or duly authorized Russian commercial banks.

To further CIS integration, President Yeltsin issued decree no. 940, 'Strategic policy of Russia towards CIS member states', dated 14 September 1995. The document itemizes the main tasks of Russia's policy towards its CIS partners as: (*a*) to ensure reliable stability in all its aspects, political, military, economic, humanitarian and legal; (*b*) to promote the establishment of the CIS states as politically and economically stable states pursuing a friendly policy towards Russia; (*c*) to consolidate Russia as the leading force in the formation of a new system of interstate political and economic relations in the post-Soviet space; and (*d*) to boost integration processes within the CIS.[52] It also states the intention to form 'a unified scientific and technological space' within the framework of the CIS and to implement agreements between the CIS member states in the defence sphere.

Within the framework of the document the Ministry for Economic Cooperation with CIS Member States, which is responsible for implementing Russia's economic and social policy on the CIS, obtained new powers. It coordinates the activity of federal agencies in the development of economic cooperation with

52 *Rossiyskaya Gazeta*, 23 Sep. 1995, p. 4 (in Russian) in FBIS-SOV-95-188, 28 Sep. 1995, pp. 19–22.

other CIS states and assists the foreign economic activity of Russian enterprises. However, the new institutional system does not eliminate overlapping competence or rivalry between executive agencies. Military–technical cooperation is still subject to a bureaucratic 'tug of war' between various executive and governmental bodies. The Ministry for Economic Cooperation with CIS Member States is seeking its own place in the administrative system. Apart from wasting time and resources this rivalry prevents the execution of genuinely promising co-production projects and arms transfer programmes. In addition, smugglers are using loopholes in the federal legislation and the general weakness of Russian state power to export armaments, technologies and strategic and raw materials illegally.

Military–technical cooperation with the other CIS countries is expensive for Russia since many programmes effectively represent military assistance. According to Colonel-General Vladimir Zhurbenko, Deputy Chief of the Russian General Staff, the other CIS states owe the Russian Ministry of Defence $6.7 million, mostly for training their officers at Russian academies. Nevertheless, the Russian Government decided to earmark funds in the 1997 federal budget to finance the training of up to 1000 cadets from other CIS countries.[53]

Military–technical cooperation is developing both through multilateral efforts at the CIS level and through bilateral relations.

## V. The CIS legal and institutional framework

It took a long time to establish a functioning CIS legal and organizational basis for joint military–technical policy.

In retrospect, the Kiev meeting of the CIS heads of state on 20 March 1992 proved to be a crucial moment for the development of CIS military integration in general and military–technical cooperation in particular.

The Agreement on the Powers of the CIS Supreme Defence Agencies of 20 March 1992 was the first document to touch on this problem. It established the CIS Council of Heads of State as a supreme defence agency. Among its tasks were: (*a*) to determine, together with the CIS Joint Armed Forces (JAF) Supreme Command, a coordinated programme of weapon manufacture and combat technology for the JAF, the volume of funding for the programme within the appropriations for defence and the maintenance of the JAF, and military contract handling priorities; (*b*) to establish the procedure for the standardization of weapons, combat technology and other *matériel* for the JAF, and corresponding logistic routines; (*c*) to determine defence R&D procedures to ensure, acting via member states' corresponding organizations, the provision of the JAF with weapons, combat technology and other *matériel* and services; and

---

[53] Open Media Research Institute, *OMRI Daily Digest*, no. 37, part I (21 Feb. 1996), URL <http://www.friends-partners.org/friends/news/omri/1996/02/960221I.htmlopt-tables-mac-english->. Hereafter, references to the *OMRI Daily Digest* refer to the Internet edition at this address.

(*d*) to produce war-oriented economic plans, *matériel* accumulation plans and reserve mobilization plans.[54]

The Council of Defence Ministers was set up to coordinate military developments within the CIS. In addition the JAF Supreme Command was formed to implement defence decisions of the CIS higher bodies.

It should be noted that this agreement was signed by only 7 of the 11 initial CIS members—Armenia, Belarus, Kazakhstan, Kyrgyzstan, the Russian Federation, Tajikistan and Uzbekistan. Four members were not ready for military integration in 1992.

At the same time Ukraine—which did not sign the document—proposed the Agreement on the Principles Governing the Provision of Arms, Military Equipment and Other Material Supplies for the Armed Forces of the Commonwealth Member States and the Organization of Research and Development Work, which was signed by the eight CIS countries on 20 March 1992. The representative of Moldova made the following entry: 'Moldova will decide matters set out herein only on a bilateral basis'.[55]

Nevertheless, at this early stage the heads of state acknowledged the need to preserve and extend partnership ties in the manufacture of military products, long-term economic relations and direct contacts. It was resolved that the development, production, delivery and procurement of weapons, munitions, technical production items and other *matériel* supplies to CIS states and their accumulation should be carried out in accordance with joint plans agreed between member states and paid for out of a common defence budget.

At the same time R&D, arms production and export–import regulations were to be the responsibility of the member states' governments. Arms, munitions and military–technical equipment should be repaired and manufactured at JAF depots. Under the agreement arms and munitions held in repair enterprises located in a country other than the owner may not be unilaterally reattached, reassigned or privatized.

Shipment of arms, munitions and other *matériel* to the JAF on the territory of member states should be effected by mutual agreement without hindrance or imposition of any duties. Member states must exercise the right of control over military cargoes being moved.

The agreement retained the institution of military representatives at industrial plants engaged in the development, manufacture, assembly and delivery of arms, munitions and other equipment for the JAF. They were considered to be part of the JAF and subject to 'the appropriate competent bodies of member states'.[56] Quality guarantees for enterprise output were to be fixed under the contractual obligations of both parties. A representative of a customer might be provided at the manufacturer plant for signing contracts.

[54] *Military News Bulletin* (Moscow), Mar. 1992, p. 2 (in English).
[55] See note 54.
[56] See note 54.

The agreement also left unchanged old normative–technical documents on the standardization and unification of arms and munitions. However, it was resolved to prepare new regulations on arms standardization.

As for logistic support, the JAF had to proceed from the rates of supply, accumulation and distribution of arms, munitions and other *matériel* then in use. The JAF Commander-in-Chief was responsible for proposing specific procedures to implement this requirement for approval by the Council of Heads of State.

The document urged the member states to introduce measures for the top-priority supply of material resources, including consumer goods, to the JAF via state deliveries and on the basis of contracts.

R&D work undertaken for the JAF must be carried out on the basis of a joint development programme and contracts with appropriate plants and research institutions. The co-signatories agreed to conduct a constant exchange of scientific and technical information on specimens of arms and *matériel* in development and exploitation. R&D for the JAF was funded on the basis of a separate item in the common defence budget.

On the face of it, these arrangements created the basis for military–technical cooperation between the CIS countries. However, in reality this cooperation developed slowly and in a sporadic manner. A number of factors prevented cooperation, including the decline of the defence industrial base in the CIS countries, the lack of a proper legal basis for economic and military cooperation, mistrust of the CIS and its institutions and suspicion about the objectives of military integration.

The next step in developing the legal and regulatory framework occurred when the CIS leaders met in Ashkhabad, Turkmenistan on 23 December 1993 and signed the Agreement on the General Conditions and Mechanism for Support of the Development of Production Cooperation of Enterprises and Sectors of Commonwealth of Independent States Participating States. The protocol on the mechanism for the realization of this agreement was signed in Moscow on 15 April 1994.[57] These documents paved the way for both economic and defence industry cooperation at the enterprise and sector levels.

At the Almaty CIS summit meeting of 10 February 1995 a Concept of Collective Security of Participating States was adopted. The document proposed three stages for forming the CIS system of collective security. A programme of military and military–technical cooperation among participating states was one of the elements to be included in the first stage.[58]

At the level of institutions, the CIS also took a series of decisions. In December 1993, the CIS Council of Defence Ministers created a CIS Military Cooperation Coordination Headquarters (MCCH) in Moscow, with 50 per cent of the funding provided by Russia.[59]

[57] *Rossiyskaya Gazeta*, 5 May 1995, p. 14 (in Russian).
[58] *Diplomaticheskiy Vestnik*, Mar. 1995, pp. 36–37 (in Russian).
[59] Olcott (note 20), p. 358.

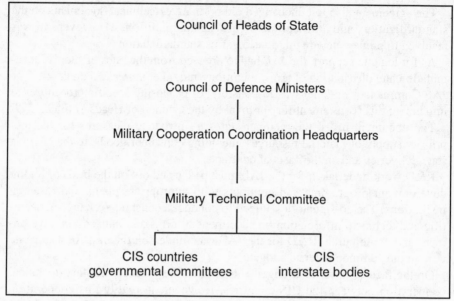

**Figure 9.1.** CIS organs involved in military–technical cooperation

As shown in figure 9.1, a Military–Technical Committee (MTC) was established within this body. Its sessions were usually attended by deputy defence ministers responsible for armaments along with representatives of other bodies coordinating work in the military–technical sphere (such as, in Russia, the then State Committee for Defence Industries, the State Committee for Military–Technical Policy, now dissolved, and the Ministry for Economic Cooperation with Members of the CIS) and officials from CIS interstate organs (such as the Interstate Economic Committee of the Economic Union of the CIS).[60]

According to some reports, the immediate priority tasks for military–technical cooperation are to support: (*a*) the rehabilitation of armaments, military equipment and ammunition that are currently unfit for use, and (*b*) the development of a concept for providing an information, command and control system for the armed forces of CIS countries.[61] The MTC drafted a document setting out the basis elements of this cooperation and referred it to the Council of Defence Ministers for endorsement in June 1996.

## VI. The CIS joint air defence system

Section V suggests that a legal and institutional basis for military–technical cooperation at the CIS level has been created. However, in reality most countries have preferred to develop bilateral channels for this cooperation. This is caused by at least two factors. First, as mentioned above, some CIS member

---

[60] *Krasnaya Zvezda*, 26 Aug. 1995, p. 3 (in Russian).

[61] *Novoe Vremya*, no. 8 (Feb. 1995), pp. 8–9 (in Russian) in FBIS-SOV-95-058-S, 27 Mar. 1995, pp. 13–15.

states are still afraid of the implications of too rapid a tempo in CIS integration. There is a suspicion that either Russia or some new supranational organ will undermine their newly gained sovereignty. Second, in practice the CIS bodies that have been created are rather ineffective in carrying out agreements and joint programmes. By 1996 about 400 of the roughly 500 agreements concluded by the CIS had not entered into force because one or more parties had not ratified them.[62] For that reason many countries regard bilateral relations as a more effective instrument for cooperation than the CIS.

The CIS member states are likely to combine collective and bilateral efforts to develop their military–technical ties in future. The creation of the CIS unified air defence system exemplifies this approach.

After the disintegration of the Soviet Union, the assets and forces of the Soviet air defence system were divided between the former Soviet republics. What had been an integrated system became less efficient.[63] Moreover, the division of *matériel* and armaments was not conducted in an orderly way everywhere. In some places, components of the system such as missile fuel, parts of anti-aircraft guided missiles and aircraft were abandoned.[64]

In the view of Colonel-General Viktor Prudnikov, Commander-in-Chief of the Russian Air Defence Forces, the non-Russian CIS states are unlikely to be able to build credible air defence forces without help from Russia. Russian air defence personnel are currently serving in Azerbaijan, Belarus, Kazakhstan, Latvia, Tajikistan, Turkmenistan and Uzbekistan, none of which has the independent capacity to train replacement cadres of personnel. Air defence installations are manufactured and for the most part repaired in Russia. At the strategic level, the monitoring of airspace depends on a network of assets scattered over Armenia, Azerbaijan, Belarus, Kazakhstan, Russia and Ukraine.[65]

The Russian Ministry of Defence began to study the question of creating a unified air defence system for the CIS member countries in 1994 and this initiative was immediately backed by Kazakhstan. Finally, the Agreement on the Creation of a Unified Air Defence System of CIS Member States was signed by the CIS heads of state on 10 February 1995 in Almaty.[66] CIS countries with the exception of Azerbaijan and Moldova have decided to pool their efforts to protect the common CIS airspace and to assign means and forces from each state to the combined air defence system. It is expected to have a coordinating committee to include air defence commanders from each member state, their deputies and other high-ranking officials. Military–technical cooperation should provide for the delivery of *matériel*, repairs to armaments and training of per-

---

[62] *Budushchee Postsovetskogo Prostranstva* [The future of the post-Soviet space], (Council on Foreign and Defence Policy: Moscow, 1996), p. 11 (in Russian).

[63] Col-Gen. Viktor Prudnikov, Commander-in-Chief of the Russian air defence troops and then Commander-in-Chief of the CIS allied air defence system. ITAR-TASS, 17 Feb. 1995 (in English) in FBIS-SOV-95-034, 21 Feb. 1995, p. 1.

[64] Karatchenya, I., CIS Executive Secretary, *Narodnaya Gazeta*, 6 Dec. 1995, p. 1 (in Russian) in FBIS-SOV-95-238, 12 Dec. 1995, p. 21.

[65] FBIS-SOV-95-238, 12 Dec. 1995 (note 64).

[66] The text is reproduced in *Rossiyskaya Gazeta*, 25 Feb. 1995, p. 5 (in Russian) in FBIS-SOV-95-040, 1 Mar. 1995, pp. 2–4.

sonnel. Article 6 of the agreement states that 'air defence armaments and military hardware shall be supplied on the basis of bilateral agreements between the member states' governments, while repairs of air defence armaments and military hardware shall be effected under the procedure laid down by the CIS Council of Heads of State'.[67] The agreement thus emphasized bilateral relations rather than unified CIS structures as far as military–technical cooperation was concerned. Training of specialist personnel for the unified air defence forces should also be provided for by bilateral agreements.

Elements of a multilateral export control system were also set up under the same agreement. Member states pledged not to sell or transfer air defence armaments and military hardware defined on a list ratified by the CIS Council of Heads of State to states which are not party to the 10 February agreement.[68]

The details of cooperation were not specified in the agreement and were to be worked out later during meetings of air defence commanders. According to General Prudnikov, the CIS joint air defence force will initially concentrate its efforts on air surveillance and the exchange of information. In the first stages neither anti-aircraft rocket launchers nor combat aircraft are expected to be under the command of the joint air defence force. Each CIS member will decide which units and which equipment will be dedicated to air defence.[69]

At the January 1996 CIS summit meeting Russia agreed to finance upgrades of the air defence forces of Armenia, Georgia, Kazakhstan, Kyrgyzstan, Tajikistan and Uzbekistan. According to Colonel-General Sergey Sapegin, First Deputy Commander of the Russian Air Defence Forces, Russia planned to provide Georgia with 10 billion roubles ($2.1 million) for its air defences.[70]

In August 1996 the CIS Interstate Economic Committee approved the establishment of the Granit international financial–industrial group to execute air defence projects. Under this arrangement 10 Russian and four Ukrainian companies as well as enterprises from Armenia, Belarus, Georgia and Kazakhstan formed a group to provide the joint air defence system with armaments, service and repairs.[71]

The integrated air defence system began operations in the spring of 1996. On 1 April Russia and Belarus started joint patrols of the western air border of the CIS and on 1 May Russian and Kazakh air defence troops began joint patrols of the southern border. On 1 June Russian and Georgian air defence forces began joint patrols of the airspace along the border with Turkey and by the end of 1996 it was anticipated that other member states would be involved in joint patrols.[72]

In the view of many Russian defence experts there is no alternative to the creation of a unified air defence system even though this is a costly project for

[67] See note 66.
[68] See note 66.
[69] See note 63.
[70] OMRI Daily Digest, no. 28, part 1 (8 Feb. 1996).
[71] Krasnaya Zvezda, 21 Aug. 1996, p. 1 (in Russian).
[72] Moscow Mayak Radio Network, 27 Mar. 1996, p. 8 (in Russian) in FBIS-SOV-96-061, 28 Mar. 1996, p. 8.

Russia. It will help settle many problems in stabilizing the conditions of national air defence forces and consolidate the sovereignty of CIS member states in the face of external threats to the group.[73]

Military analyst Nikolay Plotnikov has identified four arguments in favour of the project. First, unified and centralized control over all air defence forces is in the interests of each state and the CIS as a whole in the event of an external threat. Second, the comprehensive employment of forces in the framework of a single concept and the preservation of a unified system for reconnaissance, tactical surveillance, and command and control will provide timely information to military and political leaders. Third, a coordinated military–technical policy and standardization of weapons and equipment can bring considerable financial savings. Fourth, a unified training programme can both raise personnel standards and reduce costs.[74]

## VII. Bilateral cooperation

Bilateral relations between Russia and each of the members of the CIS remain the main channel for military–technical cooperation within the CIS. However, the scope and nature of and the motives for cooperation vary from country to country.

Several different levels of bilateral cooperation can be identified. In some countries fully fledged cooperation covers arms transfers and defence industrial cooperation (perhaps even at the level of joint design and co-production). This type of cooperation is inbuilt in relations between Russia and Belarus and between Russia and Kazakhstan. The second type is based on a preference for defence industry cooperation. Ukraine is interested in restoring former defence industrial links with Russia but opposes other forms of military–technical cooperation. Third, Russia is assisting in the creation of national armies through arms transfers and the provision of services and training. This type of cooperation does not imply broad defence industrial cooperation or technology transfer. Central Asian republics such as Kyrgyzstan, Tajikistan and Uzbekistan are involved in this kind of cooperation with Russia. A fourth type is limited to repair and maintenance needed to sustain capabilities such as air defence networks or equipment inherited from former Soviet arsenals. This is an option for countries which are not happy about full military rapprochement with Russia: Armenia and Georgia exemplify this type of relations with Russia. Finally, a number of CIS countries are indifferent or hostile to military–technical cooperation with Russia but for various reasons see no alternative. Azerbaijan, Moldova and Turkmenistan are in this category.

Table 9.3 offers an index of the interest among CIS countries in cooperation with Russia.

---

[73] Prudnikov (note 63).
[74] *Nezavisimaya Gazeta*, 21 Feb. 1995, p. 2 (in Russian) in FBIS-SOV-95-036, 23 Feb. 1995, p. 3.

**Table 9.3.** Index of the level of military–technical cooperation between Russia and the CIS member states in select areas

| | Armenia | Azerbaijan | Belarus | Georgia | Kazakh. | Kyrgyz. | Moldova | Tajik. | Turkmen. | Ukraine | Uzbek. |
|---|---|---|---|---|---|---|---|---|---|---|---|
| Defence industrial cooperation | Low | None | Very high | High | Very high | None | None | Low | None | Very high | High |
| Conversion | Low | None | Very high | Low | High | None | None | Low | None | Very high | High |
| Arms transfers | High | Low | High | High | High | Low | High | Very high | High | Low | High |
| Division of Soviet arsenals | Very high | Very high | High | Very high | High | Low | Very high | Very high | High | High | Low |
| Air defence | High | None | Very high | Very high | Very high | High | Low | High | None | High | Very high |
| Russian military bases | High | None | None | High | None | None | None | Very high | None | None | Low |
| Russian military advisers | Low | None | Low | High | High | Low | None | Very high | None | None | High |
| Military training | High | Low | Very high | High | Very high | High | Low | Very high | Low | Low | Very high |

*Notes:* Kazakh. = Kazakhstan; Kyrgyz. = Kyrgyzstan; Tajik. = Tajikistan; Turkmen. = Turkmenistan; Uzbek. = Uzbekistan.

Military relations between Russia and *Belarus* have always been relatively good. Belarus was and is in favour of military integration with Russia. On 6 January 1995 the two countries signed a number of agreements on military cooperation which covered some military–technical aspects.[75] This was the first step in what seemed to be an accelerating dialogue in this field. The Russian–Belarussian Treaty on Friendship and Cooperation of 21 February 1995 stipulated that the two countries would develop industrial cooperation, including cooperation between their defence industries.[76]

On 8 December 1995 Russian Defence Minister Grachev visited Minsk and, according to the concluding press release, discussed issues related to military–technical cooperation. At a final plenary session the Russian and Belarussian military delegations signed a number of documents raising cooperation to a qualitatively new level. Belarussian President Alexander Lukashenko noted that there were no difficulties in their military and military–technical cooperation and underlined that the two sides 'have decided to cooperate and cooperate very seriously!'[77]

The main avenues of bilateral military–technical cooperation have been defined in the first stage. Contracts between defence industry enterprises will lead to the supply of spare parts, the repair and servicing of military equipment and the refurbishing of munitions unfit for combat use. Another element of the agreement is how to resolve the problem of the existing debts owed to enterprises by the Russian and Belarussian defence ministries for past repairs of arms and military equipment. For the future, Russian defence analysts believe that cooperation in the production of attack helicopters in Belarus is the most attractive for Russian industrialists.[78]

According to some reports, the two countries also agreed to conduct a joint arms trade policy.[79] For Russia this is an issue of some importance because Belarus has been active in exporting second-hand arms to other countries: it has, for example, sold China tanks and ammunition and China has apparently shown interest in purchasing air defence and optical weapon systems as well as repairing aircraft in Belarus. China has sent officers for training in the Belarussian Military Academy.[80] This policy cannot continue indefinitely since the arsenals of Soviet-made weaponry will be exhausted. However, it can undermine the Russian policy of not exporting single items or technical documentation where these can be used by the defence industry of the importing country.

Since Belarus has no complete circle of military production, in the longer term it also has an interest in cooperating with Russia in arms export policy.

The Russian–Belarussian Treaty on Confederation of 2 April 1996 confirmed the importance of military cooperation between the two countries. It included

[75] *Novoye Vremya*, no. 9 (Mar. 1995), p. 14 (in Russian).

[76] *Diplomaticheskiy Vestnik*, Mar. 1995, p. 40 (in Russian).

[77] *Krasnaya Zvezda*, 10 Dec. 1995, p. 1 (in Russian). Emphasis in original.

[78] *Novoye Vremya*, no. 18–19 (May 1996), p. 27 (in Russian).

[79] Minsk BTK TV Network, 10 Dec. 1995 (in Belarussian) in FBIS-SOV-95-238, 12 Dec. 1995, p. 60.

[80] *Nizhegorodskiye Novosti*, 27 Apr. 1996 (in Russian).

special provision on common principles of military construction and use of military infrastructure.[81]

The development of military–technical cooperation between Russia and *Ukraine* has been hindered by a number of unresolved problems related to nuclear weapons and strategic forces and the division of the Black Sea Fleet.

As of mid-1996 the issues of the Black Sea Fleet had not been finally resolved. Nevertheless, under the presidency of Leonid Kuchma conditions for developing military–technical cooperation between Russia and Ukraine have become quite favourable. Under his administration, joint ventures with Russia have gained in favour. On 28 March 1996, the then defence ministers of Russia and Ukraine, Pavel Grachev and Valeriy Shmarov, finalized 10 documents on expanding military–technical cooperation. These included agreements on defence industry cooperation to repair equipment, but no further details were disclosed.[82]

The two countries have begun to use financial–industrial groups intensively as instruments of cooperation between defence industries. In April 1996 they established the International Aircraft Engines group which consists of 50 Russian and Ukrainian enterprises, design bureaux and banks. The total number of employees is 215 000, of which 140 000 are in Russia and 75 000 are in Ukraine.[83] The group has received special concessions to ease its operations. For example, both value-added tax and customs tariffs are waived for trans-actions within the group and the debts of the enterprises participating in this group have been liquidated. The priority of the group is the development of two new aircraft engines designated the D-27 and D-436.

Another financial–industrial group, the International Aircraft Project, was formed in August 1995 to develop and produce civilian and military transport aircraft. Production of the Tu-154M, Tu-156, Tu-334, Tu-354, Tu-230, An-72, An-74, An-70 and An-140 is the responsibility of this group. It was established by leading Russian and Ukrainian organizations including the Tupolev and Antonov design bureaux, the Aviakor industrial association, the Kiev Aircraft Plant (Aviant), the Kharkov State Aircraft Enterprise, the Inter-regional Volgo-Kama Bank of Reconstruction and Development, Prominvestbank and the Savings Bank of Ukraine.[84]

Russia and Ukraine both regard the creation of joint financial–industrial groups of this kind as a promising method of developing cooperation not only in the military–technical field but also in the economic and financial sphere.

The division of the property—including hardware, weapons and munitions—of the Russian 14th Army in *Moldova* became a central issue in relations between Russia and Moldova after the decision that the army would withdraw.[85] This issue has not been finally resolved. Along with the fact that Moldova has

---

[81] *Diplomaticheskiy Vestnik*, May 1996, p. 40 (in Russian).
[82] *OMRI Daily Digest*, no. 64, part II (29 Mar. 1996).
[83] *Izvestiya*, 21 Feb. 1995; and *Krasnaya Zvezda*, 20 July 1996, p. 4 (in Russian).
[84] *Vybor*, 7 Sep. 1995 (in Russian).
[85] According to some reports there were around 400 000 tonnes of supplies that would require *c.* 3000 freight-car trips to transport them to Russia. *Izvestiya*, 21 Feb. 1995, p. 2 (in Russian).

not shown any desire to develop significant armed forces of its own, this has effectively blocked dynamic military–technical cooperation with Russia.

Military–technical cooperation between Russia on the one hand and *Armenia* and *Azerbaijan* on the other was prevented for a number of years by the conflict between those countries and Russia's role in it. The military rapprochement between Azerbaijan and Ukraine was one more obstacle to Russian military–technical cooperation with them both.

As noted above, immediately after *Georgia* became independent the issue of military–technical cooperation was subordinated to other issues confronting the new state. Apart from the war against Abkhazian separatism, internal changes in the Georgian armed forces also prepared the ground for rapprochement between Russia and Georgia in this area. Shevardnadze dissolved the units loyal to Kitovani and Ioseliani and formed a regular army. General Varden Nadibaidze, the former Deputy Commander of the Transcaucasus MD, became the new Georgian Defence Minister. Nadibaidze—who had been responsible for logistics and armaments in the Transcaucasus MD—had participated in the creation of the Georgian armed forces and was a personal friend of Grachev.[86]

The military–technical cooperation between Russia and Georgia has three elements: defence industrial cooperation, arms transfers, and agreements on bases and support facilities.

The elements of defence industry cooperation between Russia and Georgia are to be found under Article 15 of the Russian–Georgian Agreement on Trade and Economic Cooperation of March 1996, which states that 'the parties will undertake measures to develop cooperation between defence enterprises on a mutually advantageous basis'.[87] Russian designers have assisted the Tbilisi Aviation Industrial Association to develop a two-seater trainer version of the Su-25UB aircraft.[88] However, the prospects for cooperation between enterprises are limited because the plants in Georgia have been almost completely destroyed since 1992.

In March 1996, Russia and Georgia concluded an agreement on sending Russian military advisers and specialists to Georgia to train Georgian servicemen and repair military hardware.[89]

In February 1994 Russia and Georgia signed a Treaty of Friendship and Military Cooperation. The agreement allowed Russia to maintain three military bases in Georgia and called for Russian forces to help train and equip a new Georgian Army.[90] However, the leaders of every faction in the Russian State Duma, citing domestic instability in Georgia, signed a letter to President Yeltsin opposing the treaty and warning that it might not be ratified.[91] The Russian

[86] *Novoye Vremya*, no. 27 (July 1996), p. 19 (in Russian). After Grachev's resignation in 1996 some analysts noted that Russia had lost an important channel of communications with and influence on Georgia.

[87] *Diplomaticheskiy Vestnik*, Apr. 1996, p. 55 (in Russian).

[88] *Krasnaya Zvezda*, 11 July 1996, p. 2 (in Russian).

[89] *Diplomaticheskiy Vestnik*, May 1996, pp. 47–48 (in Russian).

[90] *International Herald Tribune*, 2 Feb. 1994, p. 1.

[91] This treaty was ratified by Georgia in Feb. 1996.

President said he would not submit the treaty to the State Duma until two ethnic conflicts involving tiny breakaway republics within Georgia but on Russia's border (Abkhazia and South Ossetia) had been resolved.

In March 1995 the Russian and Georgian defence ministers signed an agreement on airfield technical support services and discussed the details of future military–technical cooperation. Details of their discussions were not disclosed.[92] However, it took another year and half to conclude a special treaty on military bases. This was signed during the visit of Russian Prime Minister Viktor Chernomyrdin to Tbilisi in September 1995. The treaty provided for Russian control over military bases at Akhalkalaki, Vaziani, Batumi and Gudauta for 25 years. In exchange Georgia would receive economic assistance, especially in the energy sector, and support in regaining control over Abkhazia.[93]

Anti-Georgian forces in the Russian Duma have also blocked the ratification of this treaty and, as of late 1996, neither Russia nor Georgia had ratified it.[94]

The issues of arms transfers and Russian access to bases and facilities have become intertwined. According to some reports, Grachev promised to transfer to Georgia about 100 T-72 tanks in exchange for assistance from Nadibaidze in concluding the bilateral agreement on military bases.[95] Two additional issues have complicated military–technical cooperation between Russia and Georgia. First, all decisions on both deployments and arms transfers must be consistent with the 1990 CFE Treaty.[96] Second, because of the fighting going on in Georgia, some equipment declared under the CFE information exchange has been destroyed or is not fit for use.[97]

Alongside Belarus, *Kazakhstan* has been the most eager for economic and military integration with Russia. The legal basis for military–technical cooperation was laid on 25 May 1992 when Presidents Yeltsin and Nursultan Nazarbayev signed the Treaty of Cooperation and Mutual Assistance. Kazakhstan strongly supported the 1992 Tashkent Treaty on Collective Security and since late 1993 cooperation between the armed forces of Kazakhstan and Russia has gradually assumed greater importance. On 28 March 1994 the two countries signed the Treaty on Military Cooperation. Russia and Kazakhstan agreed to pursue a coordinated policy in the areas of joint design, production, repair and supply of arms, military equipment, and material and technical resources. This was to include efforts to preserve and establish cooperation between enterprises designing and manufacturing arms and military equipment.

---

[92] *Krasnaya Zvezda*, 25 Mar. 1995, p. 1 (in Russian).

[93] *New Europe*, 24–30 Sep. 1995, p. 40.

[94] *Krasnaya Zvezda*, 27 July 1996, p. 3 (in Russian).

[95] *Novoye Vremya*, no. 27 (July 1996), p. 19 (in Russian).

[96] If the Russian side accumulates treaty-limited items in excess of its permitted quotas in the North Caucasus MD (which includes Chechnya), Russian forces stationed in Georgia will automatically be deprived of this quantity of equipment. The treaty allows up to 220 tanks, 220 armoured combat vehicles, 100 combat aircraft, 50 combat helicopters, and over 280 artillery systems of a calibre of over 100 mm to be stationed in Georgia. One possible solution is for Georgia to transfer part of its quota for treaty-limited equipment to Russian forces on a temporary basis. The Georgian Government has indicated that it can accept this solution. *Krasnaya Zvezda*, 24 Nov. 1992, p. 2 (in Russian).

[97] *Krasnaya Zvezda*, 25 Mar. 1995, p. 1; *Krasnaya Zvezda*, 10 Feb. 1996, p. 2; and *Pravda*, 15 Nov. 1995, p. 2 (in Russian).

In order to implement a coordinated policy, Kazakhstan and Russia established an interstate commission for military–technical cooperation.[98] However, this document was a declaration of intent rather than a concrete programme. Each provision relating to military–technical cooperation needed additional agreements and further detailed work. It was agreed that supplies and services should be provided duty-free at prices agreed by the parties and specified in each case in a separate agreement. Specific issues of coordinating policy and the supply of work and services should be determined on a project basis.

The two countries decided to cooperate in defence industry research and on experimental and design work. Each side would retain and develop existing specializations. They also agreed to cooperate in such fields as training officers and military transport.

On 24 December 1994, after nearly two years of negotiation and hard bargaining, the prime ministers and defence ministers of the two states signed a number of additional documents of a technical nature: Procedures for the Maintenance and Use of the Balkhash Missile Warning System in Kazakhstan; the Agreement on Air Defence Facilities of the Russian Federation and Kazakhstan, and their Joint Operation; and the Agreement on Issues of Joint Planning of the Armed Forces in the Interest of the Mutual Security of the Russian Federation and Kazakhstan.

At a meeting on 20 January 1995 between Presidents Yeltsin and Nazarbayev military cooperation between the two countries was placed on a long-term footing. Of 17 the documents which they signed, 8 dealt specifically with military cooperation. The two countries decided to start forming joint armed forces on the basis of common armaments. Defence industrial cooperation will develop alongside programmes of standardization.[99]

As noted above, Kazakhstan has a significant defence industrial potential inherited from the Soviet past. In mid-March 1995 senior officials of the Defence Industry Committee of Kazakhstan and the Russian State Committee on Defence Industries concluded an agreement on economic, scientific and technical cooperation in the defence sector. The Russian state agency Oboronresurs [Defence Resources] and the Kazakh state agency Kazkontrakt will be responsible for implementing this agreement.[100]

In July 1995, the Russian and Kazakh governments agreed on procedures to compensate for hardware and armaments withdrawn to Russian territory from Kazakhstan. Kazakhstan is transferring to Russia equipment (strategic systems) which could be better utilized by the Russian armed forces. In return, Russia is sending to Kazakhstan a number of types of arms, including MiG-29 fighter aircraft, which would be more effectively used under Kazakh conditions.[101]

On 26 January 1996 the Russian and Kazakh defence ministers signed a package of 16 documents on cooperation in the military sphere, including

[98] See the English translation of the treaty in FBIS-SOV-94-206, 25 Oct. 1994, pp. 56–60.

[99] *Diplomaticheskiy Vestnik*, Feb. 1995, p. 41 (in Russian).

[100] Woff, R., 'Kazakh–Russian relations: an update', *Jane's Intelligence Review*, Dec. 1995, p. 568.

[101] *Izvestiya*, 13 July 1995 (in Russian).

agreements on the organization of communications, procedures for reciprocal payments, cooperation in the air defence system and collaboration in military science.[102] In essence, Kazakhstan will receive supplies of Russian arms and equipment in exchange for allowing Russia to use test ranges in Kazakhstan. There is also a programme of assistance for the creation of a Kazakh naval base on the Caspian Sea.

Kazakh maritime forces are centred on the naval base opened on 17 August 1996 in Aktau. Russia has transferred to Kazakhstan two coastal defence vessels of the Sunkara Class, and a further Griff Class vessel is under construction. This programme involved contracts with 800 Russian plants.[103]

At the same time Russians have expressed concern that alongside its military ties with Russia Kazakhstan is also developing defence cooperation with Western countries. Nazarbayev has irritated Russia with his firm support for the NATO Partnership for Peace programme.[104] Russia was also worried about the implications of the Charter of Democratic Partnership between the Republic of Kazakhstan and the United States of America according to which the United States promised 'to support Kazakhstan's efforts to meet its legitimate defence requirements'.[105]

Another subject for concern for both countries is the development of illegal arms traffic. Some corrupt high-ranking military officials have been involved in illegal arms transfers to a number of Central Asian countries. In 1995, two senior Kazakh officials, former Deputy Defence Minister General Valeriy Sapsayev and a Ministry of Defence official, Colonel Zhailaubai Sadibekov, were jailed for illegally exporting weapons to an unidentified country. Newspaper reports also suggested that a Russian general was involved in the $2 million deal.[106]

Like Azerbaijan, *Turkmenistan* has been reluctant to participate in the process of CIS military integration. Turkmenistan has not signed the Treaty on Collective Security within the CIS and generally keeps aloof from the other former union republics. However, it has a small army which is in need of combat equipment and cannot end military cooperation with Russia entirely. In 1994 Rosvooruzhenie concluded a contract to supply weapons in exchange for 5 billion cubic metres of natural gas to be supplied to southern regions of Russia, but this agreement has not been implemented and the Russian Government has since renounced its commitment to modernize the Turkmen air force and air defences.[107]

---

[102] Almaty Kazakh TV, 26 Jan. 1996 (in Russian) in FBIS-SOV-96-019, 29 Jan. 1996, p. 56.

[103] E.g., the diesel engines for the Griff Class were built at the Zvezda plant in St Petersburg. *Krasnaya Zvezda*, 20 Aug. 1996, p. 3.

[104] In particular, Nazarbayev's statement that active Kazakh participation in the programme 'will give us great assurances as to Kazakhstan's future as a sovereign state'.

[105] Kortunov *et al.* (note 16), pp. 307–308.

[106] *Asian Defence Journal*, Dec. 1995, p. 138.

[107] This may have been a response to the statement of President Saparmurat Niyazov that Turkmenistan wishes to be a neutral state. *Rossiyskaya Gazeta*, 18 May 1995, p. 7 (in Russian).

In May 1995 the Russian and Turkmen presidents signed a package of documents on military cooperation including agreements on military–technical cooperation, air defence and military interstate transport.[108]

In contrast with Turkmenistan, *Uzbekistan* favours military integration within the CIS. It has signed all the major agreements on military cooperation between CIS member states and accepted Russian assistance in restoring and upgrading the joint air defence system. During the visit of Prime Minister Chernomyrdin to Tashkent an agreement on Russian–Uzbek military–technical cooperation was signed. The most important element of defence industrial cooperation is the decision to establish a transnational financial–industrial group, Ilyushin, to produce civil and military aircraft.[109]

The needs of the Uzbek Army are quite modest and it is not a major market for Russian arms transfers. This country does not intend to develop large-scale armed forces and the scope for Russian–Uzbek cooperation is not wide.

## VIII. Conclusions

A number of important factors have dictated the need to develop military–technical cooperation between the members of the CIS. These include the need to restore elements of defence industry cooperation and supplies of certain raw materials, the dependence of the other CIS countries on Russia for continued supplies of arms and spare parts and for repairs and maintenance of equipment in their inventories. Russia has had some success in re-establishing an integrated air defence system with its associated infrastructure on parts of the territory of the former Soviet Union and several CIS states (Georgia, Kazakhstan, Tajikistan and Uzbekistan) have created new national armies.

Since 1992 the legal and institutional basis for cooperation has been established in Russia and at the CIS level. Russian interest in and commitment to the further development of the CIS have grown steadily and Russia now places great importance on the organization. It has used both the CIS framework and bilateral relations to develop cooperation with its new neighbours. The approach used by Russia to stimulate interest in cooperation and the precise nature of the programmes undertaken vary from country to country. Some important defence industrial links have been restored. Financial–industrial groups have proved to be an effective instrument for resuming and developing defence industrial relations between CIS member states.

Military–technical cooperation within the CIS faces many problems and is far from ideal. The legislation in this field is often different in different CIS countries, which makes cooperation more difficult. Defence industrial cooperation still exists only in embryo. Joint conversion projects currently exist only on paper. Providing military assistance to the other CIS member states is a heavy burden for Russia to carry given its own economic problems. At the

---

[108] *Diplomaticheskiy Vestnik*, June 1995, p. 36 (in Russian).
[109] *Diplomaticheskiy Vestnik*, Aug. 1995, p. 20 (in Russian).

political level, Russia has often shown a tendency to follow the old imperialist principle of 'divide and rule' in managing its relations with its new neighbours.

The development of the CIS will challenge decision makers both inside and outside the member states for the foreseeable future. It is still unclear whether the objective of military–technical cooperation is to restore Russian power or to create stability in the post-Soviet strategic space through cooperation. It is also important to find a level of CIS military integration which could help to restore natural and traditional ties between the members and meet their legitimate defence needs without threatening the security balance or undermining relations with countries in adjacent regions.

# 10. Military–technical cooperation between Russia and countries of East–Central Europe

*Irina Kobrinskaya and Peter Litavrin*

## I. Introduction

The years 1994–95 appeared to be a period in which the underlying forces shaping Russian policy on arms transfers and defence production were beginning to stabilize—at least in comparison with the complete turmoil which accompanied the dissolution of the WTO and the disintegration of the Soviet Union at the beginning of the 1990s. This does not mean that there is now complete clarity in the future path of development. It is true that decision-making processes are still in a state of flux across the entire space occupied by the post-Soviet independent states and East–Central European countries.[1] However, in 1996 it is possible to identify some of the main tendencies that are likely to define further developments in military–technical cooperation.

In spite of the differences in scale of the problems facing Russia and the countries of East–Central Europe, there are certain similarities between them as regards their defence industrial structures. This chapter is confined to a discussion of arms transfers and military–technical cooperation between Russia and the non-Soviet countries which were members of the WTO[2] but, because of the structural similarities of state socialist command economies, some of the observations in this chapter probably apply across East–Central Europe.

The present state of arms procurement, arms transfers and arms production in the Russian and Central European states is closely linked to the tectonic shifts that have occurred in Europe in the last decade. One of the primary characteristics of this change has been the significant and asymmetrical cuts in defence expenditure that have occurred in Europe.[3] These reductions have not been fully compensated for by increased spending in other potential markets. At the same time, the countries of East–Central Europe retain significant arms production capacities. The current circumstances could therefore change if developments in the international arena led to increases in military expenditure and arms procurement.

[1] In this book East–Central Europe is defined as those non-Soviet countries that were members of the WTO—Bulgaria, the Czech Republic, Hungary, Poland, Romania and Slovakia—but excepting Albania.
[2] Military–technical cooperation and arms transfers between Russia and other members of the Commonwealth of Independent States are discussed in chapter 9 of this book.
[3] For example, while military expenditure among the European members of NATO between 1990 and 1995 fell by 14% in real terms, Poland's (the former WTO country for which the most reliable data are available) declined by 36%. George, P. *et al.*, 'Military expenditure', *SIPRI Yearbook 1996: Armaments, Disarmament and International Security* (Oxford University Press: Oxford, 1996), table 8A.2, pp. 365–66.

Domestic economic, political and social developments in these 'transitional societies' are also of tremendous importance for the future national policies on military–technical cooperation, arms transfers and arms production.

The following basic questions have to be answered before a full understanding of the factors which define military–technical cooperation between Russia and East–Central European countries can be achieved:

1. Is there a final understanding of the structure, priorities and needs of the defence industries in the post-socialist countries within the frameworks of national security policy?

2. What role will the defence industries play as a sector in the post-command economies? and

3. Will the pattern of future relations between the defence industries of what was an integrated WTO production system be characterized by cooperation and reintegration or by disintegration and competition?

None of the post-socialist countries has resolved the questions what structure and size of defence industry can best meet their defence needs or what role the defence industries should play in national economic policy.

## II. The legacy of the WTO

One element of the WTO's work was procurement of equipment for the armed forces. With the ending of the WTO the bureaucratic mechanisms for managing the relationships between governments, armed forces and industrial enterprises broke up. This breaking of ties did not occur according to a gradual and phased timetable but was sudden and abrupt. The peaceful end of the WTO was possible because of the strategic decisions taken by the then leaders of the Soviet Union in 1990 not to use force (either directly or in cooperation with certain elements in former allies) to prevent the disintegration of the alliance. However, the dissolution of the WTO reflected decisions taken at the initiative of the governments of Czechoslovakia, Hungary and Poland in particular.[4] These decisions were taken without consultation with, for example, their armed forces or representatives of their domestic arms industries.

The break-up of the common trade and payment system managed by the CMEA also had a major impact on the defence industries of both the former Soviet Union and East–Central Europe.[5] The ending of the trading system based

---

[4] The decision to dissolve the military structure of the WTO was finally taken after a meeting of WTO states in Budapest in Feb. 1991. However, throughout 1990 a series of decisions—such as bilateral agreements on the withdrawal of Soviet troops from various former allies and the statement of the WTO Consultative Political Committee after their meeting in Moscow on 7 June 1990—highlighted the accelerating pace of political change in Europe in that decisive year. Most of the relevant documents are reproduced in Rotfeld, A. D. and Stützle, W. (eds), SIPRI, *Germany and Europe in Transition* (Oxford University Press: Oxford, 1991).

[5] The CMEA was dissolved in two stages in Jan. and July 1991. At the end of 1990 its membership consisted of Bulgaria, Czechoslovakia, the German Democratic Republic, Hungary, Poland, Romania, the USSR, Cuba, Mongolia and Viet Nam.

on material quotas and a clearing system for payments led to an immediate reduction in intra-regional trade of over 40 per cent in 1991. The impact was particularly severe in Bulgaria (where the reduction in trade within the CMEA was the equivalent of a contraction in GDP of 10 per cent in one year) and in what was then Czechoslovakia.[6]

The sudden break in relations with the Soviet Union caused great disruption in the system of military–technical cooperation and arms transfers. This was made worse by the consequences of the disintegration of the Soviet Union only a year later. Then, in 1992, the new Russian Government began to formulate a foreign policy in which relations with the United States in particular and the Western countries in general were given central place. In this phase little or no attention was paid to relations with countries in East–Central Europe.

The state of military–technical cooperation between Russia and East–Central European states can be explained by a number of factors on both sides. The problems that the two sides faced in 1990 had some common characteristics, but the fact that they all tried, for political reasons, to define their own independent and separate solutions aggravated their economic and financial conditions.

The lack of attention and, if anything, unwillingness on the Russian side to increase military–technical cooperation with East–Central European states in the first half of the 1990s can be explained by three factors.

First, in the Russian military there was a predominant feeling of distrust towards its East–Central European counterparts after the dissolution of the WTO. This unwillingness to cooperate with former allies probably stemmed more from psychological factors than from objective arguments.

Second, the possibility of short-term profit from arms transfers was a new phenomenon for Russia and the long-term possibilities that could come from continued cooperation with traditional partners were not sufficiently taken into consideration. Obviously, the immediate attractiveness of the East–Central European market was less than that of Asia or the Middle East, and these regions received most attention.

If these reasons for indifference have proved to be erroneous, a third was perhaps more valid: Russian arms producers were wary about agreements such as issuing production licences to their former East–Central European 'brothers' if that meant creating or sustaining potential competitors in third-country markets.

In the East–Central European countries themselves, the initial reactions to the breakdown of relations with the Soviet Union and then Russia among managers and the government ministries and departments responsible for the defence industries were not always the same. However, in many cases there was anger and frustration at what was seen as sudden abandonment by the responsible authorities.[7] Many of these industrial managers and government officials

[6] 'Trading patterns and trade policies', European Bank of Reconstruction and Development, *Quarterly Review*, 30 Sep. 1992, pp. 4–10.

[7] Kiss, Y., SIPRI, *The Defence Industry in East–Central Europe: Restructuring and Conversion* (Oxford University Press: Oxford, 1997).

regarded the breaking of relations with their counterparts in the former Soviet Union as damaging and counter-productive since it deprived them in many cases of their most important suppliers, technologies and customers. In most cases the governments of East–Central Europe broke their old ties with no clear alternative security policy or defence industrial policy. The only policy was to hope for rapid integration into Western political, military and economic structures.

In the period since 1991 there has been a gradual recognition in some of the countries of East–Central Europe that managing the consequences of the decisions taken in 1990 will be easier if they do not rule out all forms of cooperation with Russia. By 1996 all had re-established some form of military–technical cooperation or arms transfer relationship with Russia. However, the nature of these relations has been different in different countries depending on their specific national conditions. One country—the Czech Republic—has been less interested in re-establishing military–technical cooperation with Russia than, for example, Bulgaria and Slovakia. Hungary, Poland and Romania fall somewhere between the positions of these three other states.

The policies of the East–Central European states can be grouped according to two basic motivations. First, although the non-Soviet WTO states had some arms production capacity, their armed forces are still dominated by equipment of Soviet origin. Second, they have realized that cooperation with Russian partners can help East–Central European producers to be successful in future projects. To this could be added disappointment with the extent of cooperation achieved with new partners in the West and elsewhere.

Similarly, from the Russian side it is possible to see the gradual development of greater interest in military–technical cooperation and arms transfer relationships with former allies.[8] Russian foreign and security policy has become more multi-dimensional and less centred on relations with the United States and the West. The view that it is unwise to abandon markets in which it has advantages has become more widely held. Undoubtedly, however, by 1995 the position of Russia in the East–Central European market had been undermined even though the principal feature of the armed forces of those countries has been and still is the dominance of Soviet arms and military technology.

## General tendencies in Russian defence industries and export policy

Apart from aspects which are specific to Russia's relations with East–Central Europe, the development of military–technical and arms transfer relations is also affected by overall Russian policy priorities.[9] This approach has been characterized by Zinoviy Pak, at one time head of the Ministry of Defence Industry, as 'the state turning its face to the "oboronka" [defence establish-

---

[8] This change can probably be dated to mid-1993. de Weydenthal, J., 'Russia mends fences with Poland, the Czech Republic and Slovakia', Radio Free Europe/Radio Liberty, *RFE/RL Research Report*, vol. 2, no. 36 (10 Sep. 1996), pp. 33–36.

[9] These developments are described in detail in chapters 5, 6 and 7 of this book.

ment]'.[10] This new approach is motivated primarily by the need to use arms transfers to help achieve both economic and military reform objectives—specifically, to provide the armed forces with modern arms and equipment and to shift the accent in Russian exports from raw materials and energy to a more broadly based mix of goods.

The reasoning behind this policy is based on premises that never lost their popularity among either the military or the techno-scientific élite, who have always regarded the defence sector as a 'locomotive' able to pull the Russian economy out of crisis.[11] Historically, in Russia the bulk of the scientific, technical and industrial potential has been concentrated in the defence industry and this sector is still believed to account for 60–65 per cent of total national R&D. The new attitude towards the defence industry is motivated to a significant extent by the belief that: (a) Russia's future scientific and technological position will be decided by the fate of this sector, and (b) the sector cannot be preserved during the process of market transformation without direct state intervention.

More subjectively, there is also a view, in Moscow at least, that strategic decisions are increasingly motivated by the struggle between various political lobbies. The relative importance of the two most powerful economic complexes in Russia—the oil and gas industry and the defence industry—as a political power base is the subject of much discussion. This focus sharpened in 1996 as the political profile of two individuals—the Prime Minister, Viktor Chernomyrdin, and the National Security Adviser to the President (later presidential candidate), Alexander Lebed—was raised.[12]

In other words, it is still true that in Russia there is a real danger that the national authorities will 'put the cart before the horse'. Policies on military–technical cooperation and arms exports are being made before any coherent programme is in place that defines the size and shape of the future Russian armed forces on the basis of a comprehensive national security doctrine. Rather, the new tendency seems to be to provide the defence industry with orders for modern weapons and military equipment and then to reform the armed forces on the basis of the outcome of these programmes. For example, the Ministry of Defence and General Staff have drafted a new long-term defence programme for developing the armed forces up to the year 2005 which is intended to ensure

---

[10] At the time the new ministry was created the president issued a decree 'On urgent measures to support the Russian Federation defence complex enterprises', 8 May 1996. *Delovye Lyudi*, June 1996, pp. 24–27 (in Russian).

[11] See, e.g., Kokoshin, A., 'Defence industry conversion in the Russian Federation', eds T. P. Johnson and S. E. Miller, *Russian Security after the Cold War*, CSIA Studies in International Security no. 3 (Brassey's: Washington, DC, 1993) as well as many other articles and interviews published in the period 1993–95 by Deputy Defence Minister Kokoshin.

[12] Prime Minister Chernomyrdin is considered to represent the interests of the oil and gas complex and in particular the largest corporation, Gazprom. During his period as National Security Adviser to the President, Gen. Lebed emphasized the need to increase taxes on the oil and gas branches of industry as well as exerting tighter state control over exports of all raw materials. At the same time, Lebed emphasized the need to stimulate new development and production of high-technology products. Voloshin, V., 'Driving the economy', *Business in Russia*, no. 65 (1996), p. 63; and *Kommersant Daily*, 26 June 1996 (in Russian).

'adequate military might based on developed defence industry and science'.[13] At the same time, a comprehensive military reform programme has still not been elaborated, neither is a new military doctrine in place.[14]

There has also been a move in Russia towards widespread acceptance that the state has an important and legitimate role to play in managing the economy. A former Minister of Defence, Igor Rodionov, wrote in 1995 that 'our market economy should be forced—I am not afraid of the word—forced to work for defence'. Defining the priorities of military reform, Rodionov stated that one of them should be 'keeping and enlarging the military–economic potential of the country'.[15]

Thus as an intermediate conclusion it may be said that at present in Russia several domestic political and economic tendencies as well as foreign policy priorities give strong ideological grounds for developing the defence industrial complex.

Apart from these general tendencies, what specific elements in Russia's relations with East–Central Europe determine the future of military–technical cooperation and arms transfers?

## III. Military–technical cooperation between Russia and East– Central European states

As noted above, for several years relations between Russia and the East–Central European countries in the sphere of military–technical cooperation were mostly neglected by Russia. Starting in 1993 Russia began to take cautious steps to see whether it was possible to make up for lost opportunities. From 1995 the pace of these contacts increased. However, old fears of and prejudices about Russia among the countries of East–Central Europe and the fact that at least three countries—the Czech Republic, Hungary and Poland—are to become members of NATO make it difficult to achieve new understandings and agreements.

The visit of the newly elected Polish President, Alexander Kwasnewski, in the spring of 1996, before the presidential elections in Moscow, marked something of a culmination to this renewed dialogue. Paradoxically, it could be that the knowledge that these countries are to achieve their long-expected NATO membership might make Poland and perhaps also other East–Central European countries more open-minded in considering cooperation with Russia. However, this is by no means assured.

Since 1995 the Russian approach to East–Central Europe has undergone certain changes. The commercial aspects of military–technical cooperation have become very significant elements of Russian policy. Boris Kuzyk, adviser to the president on military–technical cooperation, has underlined that in formu-

[13] Interview with First Deputy Defence Minister Andrey Kokoshin, ITAR-TASS (in English) in Foreign Broadcast Information Service, *Daily Report–Central Eurasia* (hereafter FBIS-SOV), FBIS-SOV-96-144, 25 July 1996, p. 24.

[14] *Komsomolskaya Pravda*, 4 June 1996 (in Russian).

[15] *Nezavisimaya Gazeta*, 22 Apr. 1995 (in Russian).

lating its new arms transfer policy Russia has carried out a market survey which examines the cycles of rearmament and modernization of the armed forces as well as projections for future defence budgets in the countries that are at present of interest to Russian arms exports. On the basis of this study, as well as an evaluation of the political and economic situation in different regions of the world, seven individual programmes were elaborated. These were for Latin America, the Middle East, South-East Asia, India, China, Western Europe and East–Central Europe.[16]

If this rather pragmatic approach were to be the sole basis of Russian policy, the import prospects of East–Central European states would have low priority as prospective markets. There is general agreement that in 1995 and 1996 Russia began to increase the volume of its arms transfers in spite of the continued overall contraction of the global arms market.[17] Not only are the prospects for modernization of the armed forces of East–Central European states limited; they are also unlikely to be able to make contracts on a direct payment basis. In 1995, according to statements by Russian officials, 75 per cent of Russian arms transfers were concluded on a direct payment basis. As the General Director of Rosvooruzhenie, Alexander Kotelkin, has said, Russia has stopped all philanthropy in the arms market. Weapons and military equipment are sold exclusively on commercial terms, although taking into account Russian strategic interests.[18]

At the same time it is acknowledged that penetrating markets may require significant investment, perhaps in the form of credits. Russia should not retreat from its traditional markets in East–Central Europe. Although these countries want to become part of the West, they do not have the money for full-scale rearmament in the near future. For this reason it may be worthwhile for Russia to consider more flexible forms of trade, including barter deals, which might give it access to consumer goods produced in East–Central Europe in exchange for military equipment.

This statement underlines that recent negotiations between Russia and East–Central European countries have demonstrated that, although both sides realize the existence of certain limits, they have overcome prejudices for the sake of finding mutually profitable solutions to problems—albeit on a more primitive and less ambitious level than was once hoped.

In the view of another official, Sergey Svechnikov, former Chairman of the State Committee on Military–Technical Policy (Gosudarstvenny komitet po voyenno-tekhnicheskoy politike, GKVTP), Russia would even be ready in principle to cooperate with East–Central European states in transforming their armies to NATO technical standards. According to Svechnikov, however, 'that doesn't mean Russia is going to transfer to Western standards itself. But if

---

[16] *Nezavisimaya Gazeta*, 25 Apr. 1996 (in Russian).
[17] The current trends in the global arms market are described in chapter 2 of this volume.
[18] *Izvestiya*, 27 Dec. 1995 (in Russian); and *Military Review*, no. 5 (14 Apr. 1996).

anybody needs cheaper weapons and equipment that they are more in the habit of using but which are no worse than those of the West, we are ready to do it'.[19]

In 1996 particular attention began to focus on the possibility that East–Central European countries will introduce new models of fighter aircraft into their air forces.[20] According to some estimates, the market for new fighter aircraft in the Czech Republic, Hungary, Poland and Slovakia might reach as many as 600 aircraft in future. Russian officials have stressed that there is no reason why possession of Russian fighter aircraft should prevent a country from cooperating in NATO operations. In particular, attention is drawn to the positive experience of the German Air Force in using the MiG-29 aircraft taken over from the former German Democratic Republic.[21]

Increased attention is being paid by officials and experts to the new relationships emerging between former partners in East–Central Europe and in particular to the identification of a new niche in the world market—the modernization of old Soviet equipment. Some have estimated that this market may be worth huge amounts and that Russian enterprises should win a substantial share of contracts.[22] For example, about half of all tactical combat aircraft currently in service in the world are either Soviet-made or based on Soviet designs.[23] This should give Russia opportunities to dominate the supply of spare parts, maintenance, infrastructure support, staff training and other services—provided that it can improve its performance in post-shipment services.[24]

East–Central European states, along with partners in Western countries and in Israel, have been very active in developing technical approaches to this market niche. East–Central European states are using components and sub-systems of Western origin—such as communication or electronics systems—on platforms of Soviet origin such as tanks. As the point of origin of the weapons is Russian, the question is whether this market development should be seen by Russian industry as a way of widening cooperation with foreign partners or whether it is stimulating competition.[25] One of the reasons for a reformulation of Russian approaches to military–technical cooperation may in fact be a desire to neutralize the danger of growing competition from East–Central European states acting together with Western companies.

As was mentioned above, the particular state of Russia's military–technical and arms transfer relations with the East–Central European countries depends in part on the domestic situation in the countries themselves and their attitude to cooperation with Russia. As far as new contracts for major equipment are concerned, in some cases Russia seems to be the loser even before any tender is

---

[19] *Financial Izvestiya*, 13 Feb. 1996.

[20] Kolyadin, S., 'Russian fighter planes on international arms markets: new realities', *Military Parade*, July–Aug. 1996, pp. 23–25.

[21] *Segodnya*, 11 Oct. 1995 (in Russian) in FBIS-SOV-95-199, 16 Oct. 1995, pp. 45–46.

[22] Puchov, R., [Russian at the world arms market], *Export Obychnych Vooruzenii* [Conventional arms transfer], no. 1 (May 1996), pp. 9–13 (in Russian).

[23] Dreger, P., 'Selling Russian combat aircraft: an unbiased assessment', *Military Technology*, Aug. 1996, pp. 10–18.

[24] *Bulletin of the CIS States Staff for Military Cooperation*, no. 43 (23–29 Dec. 1995).

[25] Puchov (note 22).

announced. The preference of many East–Central European countries seems to be to modernize their armed forces with equipment of Western origin in spite of the economic difficulties of such an approach.

## IV. The defence industry and arms transfer policies of East–Central European states towards Russia

The situation in the defence industries of Central Europe shares many characteristics with that in Russia. These countries also experienced the negative consequences of the collapse of the socialist structure. Difficulties of the transition period have included large budget deficits, the lack of a legal basis and institutions to implement many actions, problems of privatization, domestic political turmoil, clashes between particular interest groups, the formulation of national security and defence doctrines, and other problems of military reform. At the same time, not only has each of the East–Central European countries followed a different path in post-socialist development; they also have different levels of engagement in the process of joining Western structures, first and foremost NATO.

*Bulgaria* presents a special case. It is geographically close to some of the most vulnerable and conflict-prone regions in Europe—such as Macedonia—where a clash of Russian and US interests is not excluded. It is also a country where Russian strategic interests have not been lost as they have in most of East–Central Europe. Bulgaria may become the East–Central European country where there are the greatest prospects for military–technical cooperation. At present, the forms of this cooperation are only at the stage of being elaborated. An important factor in this process may be the domestic political struggles going on in Bulgaria between proponents of closer alignment with the West and those who place greater emphasis on cooperation with Russia. The forms can also be fairly diverse, ranging from a Soviet-style 'philanthropy' to barter deals or joint ventures producing equipment for third countries.

Russia and Bulgaria established a bilateral committee on military–technical cooperation in May 1994 to work out 'legal, economic and financial conditions for mutually beneficial cooperation in defence industry'.[26] The representation on the committee is at the level of deputy prime minister (on the Bulgarian side) and the deputy chairman of the State Committee on Defence Industries (as it then was) on the Russian side.[27]

Under a June 1995 agreement, in mid-1996 Russia began to deliver 100 T-72 tanks and 100 BMP-1 armoured personnel carriers (APCs) to Bulgaria as military assistance.[28] These transfers were reported to the responsible authorities in

---

[26] *Balkan News International and East European Report*, 5–11 June 1994, p. 9.
[27] ITAR-TASS, 22 May 1995 (in English) in FBIS-SOV-95-099, 23 May 1995, p. 8.
[28] Open Media Research Institute, *OMRI Daily Digest*, no. 141, part II (23 July 1996), URL <http://www.friends-partners.org/friends/news/omri/1996/07/960723II.htmlopt-tables-mac-english->; and 'Bulgaria will receive free Russian armour', *Jane's Defence Weekly*, 31 July 1996, p. 9. During discussions between Bulgaria and Russia the idea of transferring 12 Mi-24 attack helicopters was also raised.

accordance with the 1990 CFE Treaty.[29] According to Western reports, Bulgaria is acquiring 10–12 new MiG-29 fighter aircraft as well as additional T-72 tanks from Russia in a barter deal that apparently includes the transfer of ownership of property in Bulgaria to Russia.[30]

Bulgaria is also involved in some of the alternatives for developing the system of oil distribution from Russia (the Novorossiysk–Burgas–Alexandropolis pipeline) and the current 'strategic pipe game' suggests additional reasons why Russia can be expected to be fairly active in developing its cooperation with Bulgaria.

The most successful country where military reform is concerned has been the *Czech Republic*. It is reported that the restructuring required to meet NATO standards is almost complete. With bilateral technical assistance from NATO, the Czech armed forces have received the information necessary for modifying their logistical and supply systems and the Czech military hopes that by about the year 2000 it will be able to begin exchanging military equipment and weapons and using NATO standards. However, lack of financing is a serious obstacle. The main priority is still to get to know the alliance systems for logistics and supply so that the Czech armed forces can function alongside NATO units.[31]

The Czech Republic seems to have the least interest of all the East–Central European states in cooperation with Russia.

In *Poland*, the largest of the East–Central European countries, military reform is still far from complete either in the armed forces or in the defence industry. Poland has the biggest defence industry of East–Central Europe and intends to retain significant capacities in the future. In 1995, 31 enterprises made up the core of Polish arms production capacity, accounting for 90 per cent of its Ministry of Defence orders. A programme for defence industrial restructuring adopted by the government in April 1996 will keep the core enterprises in full government ownership.[32]

The remaining 10 per cent of orders placed by the Polish Ministry of Defence are shared between roughly 120 enterprises which are civilian in character but whose products can have military applications. Many of these enterprises have been privatized using the model of transferring ownership to a state-owned bank or financial institution (usually through a debt-for-equity swap). As a result, the government and leading political parties still have very significant

However, this was not accepted by the Bulgarian side because the costs of preparing the aircraft for use would have been too high. Presse und Informationsamt der Bundesregierung, *Stechworte zur Sicherheitspolitik*, Bonn, July 1995, p. 56 (in German); and *Air Force Monthly*, Apr. 1996, p. 3.

[29] Interfax, 22 July 1996 (in English) in FBIS-SOV-96-142, 23 July 1996, p. 8; and *Izvestiya*, 24 July 1996 (in Russian) in FBIS-SOV-96-144, 25 July 1996, p. 6. This equipment would allow Bulgaria to take out of service older equipment including T-34 tanks built in the 1950s. Interfax, 30 Jan. 1996 (in English) in FBIS-SOV-96-021, 31 Jan. 1996, p. 20.

[30] *Military Procurement International*, 15 Apr. 1996, p. 6; and *OMRI Daily Digest*, no. 138, part II (18 July 1995), URL <http://www.friends-partners.org/friends/news/omri/1995/07/950718II.htmlopt-tables-mac-english->.

[31] *Bulletin of the CIS States Staff for Military Cooperation*, no. 23 (15–21 June 1996), pp. 3–4.

[32] Kiss (note 7), pp. 115–16.

influence in the strategic direction of the sector. The issue of privatization remains unfinished business and is likely to continue to be politically controversial. The management at some of the enterprises which remain subject to specific defence-related regulations would prefer to be fully privatized.

In addition to these enterprises, there are 13 agencies under the direct supervision of the Ministry of Defence which provide repair and maintenance services to the armed forces. Like Russia, Poland has certain towns where defence enterprises are totally dominant in the local economy, mainly in the former central production region.

Poland lacks coherent plans for equipment procurement. The military has given priority to two strategic programmes—the Loara programme to develop a modern air defence system and the Huzar programme to develop the W-3 Sokol helicopter into an anti-tank attack helicopter. These programmes are regarded as potential 'locomotives' that will pull the most technically capable defence enterprises out of crisis.[33] However, the Polish Parliament, the Sejm, did not receive a list of priorities in 1996 that would have made possible a budget to support these programmes in 1997. Recent levels of military expenditure in Poland do not allow for significant new procurement. In 1994 and 1995 the state did not provide any new orders to six of the 31 defence enterprises that are considered to be the most important. The head of the financial department of the Polish Ministry of Defence, Tadeusz Grabowski, has stated that the arms procurement priorities of Poland include armoured vehicles, modern communication systems and combat aircraft. However, the PT-91 Twardy tank which is entering series production is being ordered by the army in very low numbers, so that the production facility where it is constructed is operating at around 5 per cent of capacity. If Poland is to buy a new fighter aircraft—which would be by far the largest spending commitment of the present plans—a separate decision will have to be taken outside the regular defence budget. No decision had been taken by the end of 1997.

In addition, Poland has not yet adopted a law on military–technical cooperation or what is termed trade in 'special production'. In May 1996 a special commission of the Sejm was established to improve the current draft of the law.

From a Russian perspective this law could have some implications. Up to the present Polish companies have re-exported Russian arms without regard to earlier agreements on re-transfer and are acting practically without any control. This issue was raised by the Russian side in talks between Russian former Deputy Prime Minister Oleg Davydov and Polish Minister for Foreign Trade Jacek Buchacz during talks in Warsaw in April 1996.[34] A Polish law could benefit Russia if it ended this practice. On the other hand, there are strong concerns that any centralization of military–technical cooperation could further reduce the prospects for privatization of the defence industry.

---

[33] *Rzeczpospolita*, 5 Jan. 1996 (in Polish).
[34] Buchacz agreed that in future contracts between Poland and Russia there would be end-user commitments. *New Europe*, 14–20 Apr. 1996, p. 19.

Military–technical cooperation with Russia remains a most controversial political issue in Poland. Almost every word uttered by Minister of Defence Stanislaw Dobrzanski in his talks with the then Russian Defence Minister, Pavel Grachev, in Moscow in April 1996 was discussed and criticized in Poland. According to the Polish mass media, Dobrzanski spoke about strengthening military–technical cooperation. Grachev reportedly mentioned in this connection that the Polish minister was interested in spare parts for MiG-29 and MiG-21 fighter aircraft. *Gazeta Wyborcza* added that Dobrzanski proposed establishing Polish–Russian joint ventures to manage repairs and production.[35]

Commenting on Dobrzanski's visit, a member of Poland's Defence Commission, former Deputy Defence Minister Bronislaw Komorowski, said that Poland should not take on risky obligations.[36] Some cooperation will probably remain necessary from a Polish perspective: around 80 per cent of the equipment in the Polish armed forces is of Russian origin and the Polish defence industry is mostly producing weapons developed in Russia. However, there is a view that Poland should try to replace industrial cooperation with Russia by imports of spare parts from other post-communist countries, notably Slovakia or Ukraine. As early as March 1994 former Polish Defence Minister Piotr Kolodzejczik signed an agreement in Kiev worth $150–200 million for the repair of T-72 tanks as well as MiG and Sukhoi combat aircraft in Ukraine.[37]

In general, many representatives of the Polish political establishment do not regard stronger cooperation with Russia as a secure investment. It is also true that, in the ongoing political struggle in the country, anti-Russian statements have become rather popular. Accusations of cooperation with Russia or the former Soviet Union have become common—most notably the accusations of espionage against the former Prime Minister Josef Oleksy.

In comments on military–technical cooperation with Russia it is often mentioned that both in Poland and in Russia the interests and intentions of the lobbies that support the defence industry and those that are interested in the development of economic relations between the countries are the same. In 1995 Russia offered Poland a package deal to modernize the Polish Air Force with MiG-29 fighters including building a manufacturing facility in Poland.[38] Komorowski noted that the plan would engage Poland in technological and even political cooperation with Russia at a time when, from the point of view of its role in East–Europe and NATO, Poland's main partner should be the United States. No decision on the offer has yet been made.

---

[35] *Gazeta Wyborcza*, 18 June 1996 (in Polish).

[36] *Rzeczpospolita*, 23 May 1996 (in Polish).

[37] *Kommersant Daily*, 12 Apr. 1994 (in Russian).

[38] Interview with Polish Minister of Defence Zbigniew Okonski, *Defense News*, 13–19 Nov. 1995, p. 70. The MiG-29 fighter was offered in competition with the US F-16 and F-18 fighters, the French Mirage-2000-5 and the Swedish JAS-39 Gripen. According to the Polish press another proposal was to develop a new aircraft, designated the M-2000, which would be based on the latest version of the MiG-29 but with upgraded engines and Western avionics. *OMRI Daily Digest*, no. 169, part II (30 Aug. 1995), URL <http://www.friends-partners.org/friends/news/omri/1995/08/950830I.htmlopt-tables-mac-english->.

One possible form of cost-effective cooperation could be increased exchanges between Poland and Russia's Kaliningrad region. Poland could provide the region with agricultural and other products while getting spare parts in return. Reportedly, this was proposed by Dobrzanski in Moscow. Nevertheless, the comments on this proposal in the Polish media were negative here also.

Military–technical cooperation between Russia and Poland is based on five agreements achieved and signed by the ministries of defence of the two countries in July 1993. These agreements are valid for five years and, so long as there are no objections, are automatically prolonged for the next five years. Nevertheless, Polish military officials characterize this cooperation as being on an extremely low level and stress that much more is needed. In the view of these officials, this cooperation does not conflict with Polish plans for NATO membership, nor with Russian security interests. The Polish Ministry of Defence reportedly makes 20–40 proposals to its Russian counterparts annually and laments that the Russian Ministry of Defence does not encourage greater bilateral military cooperation—sometimes explaining that this is because of its own financial problems.

As this suggests, for Poland the issue of military–technical cooperation is a 'two-way street'. Poland has been interested in repairing aircraft engines and Russian ships built at Polish shipyards and establishing joint ventures for these purposes.[39] From the point of view of some Polish experts, creating joint industrial entities would make the task of cooperation easier. At the same time, taking into account the political sensitivity of the issue and the dominant public mood, Polish state officials prefer to omit the question of defence industrial joint ventures in their discussions with the domestic media. They prefer to reiterate that technical cooperation is necessary as long as Poland uses Russian equipment. Meanwhile, they reject the idea that cooperation will prolong this period of dependence. This will gradually be reduced, depending on the characteristics of the equipment. For instance, it is estimated that aircraft of Russian origin will be phased out between the years 2000 and 2005. In general the Polish Chief of Staff and military experts consider that the framework trade agreements signed by ministers Buchacz and Davydov in Warsaw correspond to Polish interests.

The dominant aim of Polish foreign and security policy has been to become a member of NATO. Leading Polish experts consider that this should be the most important factor in dealings with Russia, especially in the military–technical sphere. There should therefore be no obstacles to purchasing certain types of arms in Russia if they correspond to those of NATO members or permit cooperation with NATO. At the same time Russian arms should be competitive in performance and Poland should get consent for third-party sales of any arms or spare parts produced with Russian partners. In certain cases Poland should even consider the confidentiality of Russian military secrets. However, according to

[39] Vesti newscast, Moscow Russian Television Network, 3 Apr. 1996 (in Russian) in FBIS-SOV-96-065, 3 Apr. 1996, p. 15.

this view, any licences that place restrictions on Poland are not in its interests. It should be stressed that Russia is not eager to give these licences to Poland.

The general improvement in Russian–Polish relations, the first signs of which appeared at the end of 1995, may promote mutually profitable military–technical cooperation between the two countries. However, in the short term it is likely that controversies over the enlargement of NATO will dominate relations between them and block their further development.

Up to 1996 Russia carried on military–technical cooperation and arms transfers with *Hungary* on a comparatively stable basis. The cooperation was based predominantly on clearing Russian debts as assessed at the time of the dissolution of the payments system within the CMEA.

In April 1994 Russia and Hungary reached agreement on a package of equipment including the BTR-80 APC and 28 new MiG-29 fighter aircraft to offset debts. In 1995 a follow-on agreement included transfers of additional BTR-80 vehicles, 20 Smerch rocket artillery systems and spare engines for MiG-29 aircraft.[40] According to some sources, the total value of this arms-for-debt swap was calculated to be around $1.7 billion of which military equipment could account for $900 million.[41]

From the beginning of 1996, however, Hungary seemed to be beginning to take serious steps towards diversifying the equipment of its armed forces. The Hungarian Government stated its intention to launch an international tender worth an estimated $1–1.2 billion to replace ageing MiG-21 fighter aircraft.[42] Initially this would include 30 aircraft but eventually around 40 more would be needed. This potential order has attracted interest from many companies including the Swedish Saab group, Dassault Aviation in France, and Lockheed Martin and McDonnell Douglas in the United States which produce, respectively, the JAS-39 Gripen, Mirage 2000-5, and F-16 and F-18 fighter aircraft.

As in Poland, the idea of such a significant programme of modernization for the air force has led to disagreements between the Hungarian Ministries of Finance and Defence over national priorities. In May 1996 Finance Ministry officials called for a shelving of the international tender because of financial difficulties.[43] No call for tenders has been issued by the Ministry of Defence but Hungarian Air Force experts have held preliminary negotiations with several foreign companies.[44]

[40] Interfax, 6 Mar. 1995 in FBIS-SOV-95-045, 8 Mar. 1995, p. 8; *Balkan News and East European Report*, 12–18 Mar. 1995, p. 33; *Baltic Independent*, 4–10 Aug. 1995, p. 6; *Jane's Defence Weekly*, 5 Aug. 1995, p. 13; and *New Europe*, 13–19 Oct. 1996, p. 18. Apart from military equipment, Russia also supplied Hungary with oil and gas storage tanks and agricultural machinery under this agreement.

[41] *Military Technology*, Sep. 1995, p. 11.

[42] *Kommersant Daily*, 14 May 1996 (in Russian).

[43] *OMRI Daily Digest*, no. 92, part II (13 May 1996), URL <http://www.friends-partners.org/friends/news/omri/1996/05/960513II.htmlopt-tables-mac-english- >. At the same time, Israel Aircraft Industries has tried to interest Hungary in the idea of rebuilding and modernizing MiG-21 fighter aircraft which could be accomplished for around $130–150 million, one-tenth of the cost of new aircraft. *OMRI Daily Digest*, no. 96, part II (17 May 1996), URL <http://www.friends-partners.org/friends/news/omri/1996/05/960517II.htmlopt-tables-mac-english- >.

[44] Until a call for tenders is announced, the Hungarian Defence Ministry can only conduct negotiations with possible suppliers over prices with the special authorization of the cabinet or parliament. *Magyar*

Like their Western counterparts, Russian aircraft manufacturers are reported to be ready to submit bids. Reportedly, both MiG-MAPO (with the MiG-29) and Sukhoi (with the Su-27) may do so. In negotiations, the financial aspects of acquiring the MiG-29, which Hungary already uses in its air force, would include some element of debt clearing. If Hungary refuses the MiG-29 and chooses a Western aircraft in spite of the difference in cost, this will represent a serious setback for Russian hopes in the East–Central European market.

The prospects for military–technical cooperation with *Slovakia* seem more complicated. On the one hand, it will not be among the first new members of NATO. This (along with the pattern of economic reform in Slovakia known as 'people's capitalism') makes it more interested in closer cooperation with Russia, including military–technical cooperation.[45] On the other hand, in terms of numbers of tanks, armoured vehicles, artillery systems, fighters and bombers permitted under the CFE Treaty, the Slovak arms market is much smaller than those of the Czech Republic or Poland. From this point of view, Slovakia is less interesting to Russia. It has reportedly been active in attempts to establish close cooperation ties with Russia in defence production. According to Slovak experts' calculations the cost of modernizing weapons in cooperation with Russia will be 7–10 per cent of the cost of a transition to Western models.[46]

Russia and Slovakia also negotiated agreements in 1993 and 1994 to use military equipment transfers to settle bilateral debts. According to Oleg Lobov, Secretary of Russia's Security Council at the time of these negotiations, the items transferred by early 1996 included one Il-76 transport aircraft and 13 MiG-29 fighter aircraft.[47]

In March 1995 Russian and Slovak officials agreed in principle to set up joint ventures in the aviation sector involving three Slovak enterprises and Yakovlev and Klimov on the Russian side.[48] The Yak-130D trainer aircraft will use an engine of Slovak design—the DV-2 developed by Povazhska Stroyanye—which will be produced in Russia by Klimov under the designation RD-35. It is not known whether this will lead to any industrial joint venture beyond the sale of the production licence.

Slovakia has also sought to develop its military–technical cooperation with Ukraine—a factor which might increase the competition between Russia and Ukraine. However, in mid-1996 it was reported that the Slovak side was not satisfied with the course of the negotiations and from 1997 relations in this sphere would be downgraded.[49]

*Hirlap*, 21 May 1996 (in Hungarian); and *OMRI Daily Digest*, no. 98, part II (21 May 1996), URL <http://www.friends-partners.org/friends/news/omri/1996/05/960521II.htmlopt-tables-mac-english->.

[45] In Oct. 1996 the Chief of the Russian General Staff, Mikael Kolesnikov, said in Bratislava that Russia and Slovakia would also develop closer military-to-military ties. *OMRI Daily Digest*, no. 191, part II (2 Oct. 1996), URL <http://www.friends-partners.org/friends/news/omri/1996/10/961002II.htmlopt-tables-mac-english->.

[46] *Nezavisimaya Gazeta*, 4 June 1996 (in Russian).

[47] *Balkan News and East European Report*, 19–25 Mar. 1995, p. 42; *Military Technology*, Nov. 1995, p. 72; *Air Force Monthly*, Dec. 1995, p. 13; and *New Europe*, 10–16 Dec. 1995, p. 22.

[48] *Jane's Intelligence Review*, June 1996, p. 247.

[49] UNIAN (Kiev), 31 Aug. 1996 (in Ukrainian) in FBIS-SOV-96-171, 3 Sep. 1996, p. 39.

As noted above, Poland has also developed some ties with Ukraine. In the early stage of independence, some nationalist-minded Ukrainian politicians (such as a former Ambassador to Canada, Levko Lukyanenko) appealed to Poland to form an anti-Russian alliance.[50] Poland has kept firmly to the good-neighbour principle in relations with the two countries to its east. However, in Polish political history the concept of exploiting problems between Russia and Ukraine is well known. A policy of this kind on the part of Poland and Slovakia is not likely but cannot be completely excluded. Military–technical cooperation could serve as one effective instrument.

Although the mandatory UN arms embargo against all the countries that were created after the collapse of the *former Yugoslavia* was lifted in 1996, the prospects for Russian arms transfers there remain problematic. President Yeltsin signed a decree that laid out a three-phase lifting of Russian national export restrictions against the republics of the former Yugoslavia in March 1996.[51] However, financial difficulties hinder these new states from buying large quantities of arms. It is more likely that Bosnia and Herzegovina, Croatia and the Federal Republic of Yugoslavia (Serbia and Montenegro) will try to develop their defence industries and become exporters of second-hand arms. Under the Agreement on Sub-Regional Arms Control of 14 June 1996, the three countries accepted ceilings on their inventories of five categories of armament—battle tanks, armoured combat vehicles, artillery, combat aircraft and attack helicopters. According to article VI of this agreement, up to 25 per cent of the reductions required can be achieved through exports.[52]

The chances for Russian arms exports mostly seem to be connected to clearing Soviet debts in accordance with the agreement on military–technical cooperation that was signed before the sanctions were lifted.[53] One transfer that may occur under this programme is reported to involve 20 MiG-29 fighter aircraft.[54]

## V. Conclusions

Some important decisions were taken in mid-1997 regarding the enlargement of NATO. However, while the Czech Republic, Hungary and Poland have begun formal discussions with NATO about accession, for the other countries the question of eventual membership is not resolved. This is a serious limitation on

[50] *Zycie Warszawy*, 24 Sep. 1992 (in Polish).

[51] Interfax, 12 Mar. 1996 (in English) in FBIS-SOV-96-050, 13 Mar. 1996, p. 9.

[52] Details of these exports must be notified to the other parties and to the Personal Representative of the Chairman-in-Office of the Organization for Security and Co-operation in Europe. OSCE, Agreement on Sub-regional Arms Control, 14 June 1996, OSCE document INF/98/96, 18 June 1996, article VI, reproduced in *SIPRI Yearbook 1997: Armaments, Disarmament and International Security* (Oxford University Press: Oxford, 1997), pp. 517–24.

[53] Former Russian Defence Minister Pavel Grachev and Federal Republic of Yugoslavia Defence Minister Pavle Bulatovic met in Moscow in Feb. 1996 to discuss renewed military–technical cooperation. ITAR-TASS, 27 Feb. 1996 (in English) in FBIS-SOV-95-039, 28 Feb. 1995, p. 11; and *Kommersant Daily*, 20 June 1996 (in Russian).

[54] *New Europe*, 6–12 Oct. 1996, p. 18.

Russian military–technical cooperation with and arms transfers to East–Central Europe. If the expansion of NATO takes on an anti-Russian character, for whatever reason, or (as is fairly likely) if it continues to be perceived as such by Russia, the prospects for military–technical cooperation and arms transfers between Russia and East–Central Europe will deteriorate. The competition between the East–Central European countries along with the West on the one side and Russia on the other will increase, particularly in third markets.

Under this scenario it also cannot be excluded that Ukraine would prefer cooperation with the East–Central European and Western states to closer ties with Russia. This would further aggravate the political situation in the European part of the post-Soviet space. Such a development would also pose certain technical problems for parts of the Russian defence industry.

If the transformation of NATO takes place alongside the development of more formalized relations between Russia and NATO in the interests of both sides, the present tendencies in military–technical cooperation and arms transfers between Russia and East–Central Europe will have chance to develop further. In either case, however, the rivalry between Russia and companies from the West will not diminish. It is more likely to intensify—a development that corresponds to the general pattern in global arms transfers.

In the East–Central European region in general, Russia is likely to pay most attention to strengthening its positions in South-Eastern Europe, developing in particular its military–technical relations with Bulgaria. However, even under the best-case scenario, East–Central Europe is not likely to play a leading role in Russia's military–technical cooperation and arms transfers as measured by commercial value. Other parts of the world will be more important for a considerable time to come.

# 11. Sino-Russian military–technical cooperation: a Russian view

*Alexander A. Sergounin and Sergey V. Subbotin**

## I. Introduction

The post-cold war era is replete with uncertainties and paradoxes. Yesterday's foes become friends and rivalry is growing between former allies. The Sino-Russian economic, diplomatic and military rapprochement exemplifies this paradox in a period of transition and rapid change.

During the early and mid-1950s, the Soviet Union provided the People's Republic of China with a wide array of military hardware. This period of Sino-Soviet strategic cooperation, however, gave way to an era of enmity by the early 1960s. All military cooperation between the two communist countries ceased.[1] During the late 1980s, Soviet President Mikhail Gorbachev started the normalization of Sino-Soviet relations. The two governments opened negotiations on a series of agreements including reciprocal force reductions, demarcation of disputed borders, the resumption of military-to-military exchanges and greatly expanded economic relations. It was Russian President Boris Yeltsin, however, who concluded the most extensive military agreements with China since the 1950s, promising, after a visit to Beijing in December 1992, to sell to China 'the most sophisticated armaments and weapons'.[2] In May 1995, during his visit to China, the then Russian Defence Minister, Pavel Grachev, confirmed that arms transfers would remain an essential element in bilateral relations. China is now emerging as one of Russia's most important arms purchasers.

This chapter examines the motives, purposes and major programmes involved in the new Sino-Russian military cooperation since its resumption at the beginning of the 1990s. Why have China and Russia agreed to resume military ties? What do the Russian Government and defence industry hope to gain from this relationship? The discussion below addresses these questions.

The Russian policy- and decision-making system for arms exports is still in a period of transition and a stable balance between political, economic and military priorities has not yet been found. All three of these issue areas can be seen as playing a role in the renewed Sino-Russian relationship. However, in con-

[1] Hickey, D. V. and Harmel, C. C., 'United States and China's military ties with the Russian republics', *Asian Affairs*, vol. 20, no. 4 (winter 1994), p. 241.

[2] Boulton, R., 'Yeltsin hails new era in Russian relations with China', *The Independent,* 18 Dec. 1992, p. 11.

* This chapter was prepared with the help of a fellowship research grant from the United States Institute of Peace, Washington, DC.

trast with Sino-Soviet military cooperation in the 1950s, when the USSR generously shared weapons and military technology with China, current Russian policy is more heavily influenced by economic than by strategic or ideological considerations.

## II. Economic incentives for cooperation

The first and most obvious rationale for Russia in seeking arms exports is to provide financial support to the defence industry. Since the break-up of the Soviet Union the formidable Soviet defence industry has found itself in a deep economic decline. This dramatic development raised some concerns in the West. In December 1991 Robert Gates, then Director of the US Central Intelligence Agency (CIA), warned that

the former Soviet defence industries, enterprises involved in special weapons and missile programmes that face cuts in military funding may well try to stay in business by selling equipment, materials and services in the international market place. The hunger for hard currency could take precedence over proliferation concerns, particularly among republic and local governments with high concentrations of defence industry and little else that is marketable.[3]

Those fears proved well-founded. In October 1993 the CIA reported that 'Russia has been actively promoting military sales to China this year to secure needed hard currency and to help defence industries cope with declines in domestic procurement'.[4]

The Russian leadership has many times underlined the need to keep production facilities, technicians and scientists employed lest massive unemployment and falling investment ruin the sector and undermine readiness and technological competitiveness. In 1992 alone, military procurement was cut by 70 per cent.[5] According to the estimates of the Moscow-based economics agency Novecon, defence production fell by 33.4 per cent in 1993.[6]

The Russian Government was often unable to pay defence enterprises for weapons ordered for its own use. In 1993 more than 100 new MiG-29s worth an estimated $2 billion were parked, unclaimed and unpaid for, at an assembly plant near Moscow.[7] In 1994–95 the Ministry of Defence paid for only 23 per cent of an order to the Fakel (Torch) enterprise, builder of air-defence missile systems. The government owed the defence enterprises in and around the city of

---

[3] Testimony of Robert Gates in US Congress, House of Representatives, *Potential Threats to American Security in the Post-Cold War Era*, Hearings before the Defense Policy Panel of the Committee on Armed Services, 102nd Congress, 1st session, 10, 11, 13 Dec. 1991 (US Government Printing Office: Washington, DC, 1992), p. 9.

[4] Nai-kuo, H., 'Russia promoting military sales to mainland China', Central News Agency (Taipei), 12 Oct. 1993; and Hickey and Harmel (note 1), p. 244.

[5] *Chancen und probleme der rustungs-konversion in der GUS* [Prospects and problems of defence conversion in the CIS], (Bonn International Center for Conversion: Bonn, 1995), p. 4.

[6] Quoted in Beaver, P., 'Russian industry feels the cold', *Jane's Defence Weekly*, 7 May 1994, p. 30.

[7] Kogan, E., 'The Russian defence industry: trends, difficulties and obstacles', *Asian Defence Journal*, Oct. 1994, pp. 43–44.

Nizhniy Novgorod 150 billion roubles by the end of 1994.[8] According to Viktor Glukhikh, at that time the Chairman of the State Committee on Defence Industries (Goskomoboronprom), by the end of 1993 the government owed the defence industry 8 trillion roubles.[9] By the end of 1994 about 400 defence enterprises had stopped all production, while another 1500 defence plants were working part-time.[10] In 1995 the situation became progressively worse.

Among other negative consequences, government non-payment of debts prevented the defence industry from pursuing an effective arms export policy since plants (which now often had to pay their suppliers in cash) had no money to start production of equipment ordered by foreign clients.

Under these conditions the Russian leadership turned to arms exports in the hope of saving the slowly dying defence industry. As President Yeltsin noted, 'the weapons trade is essential for us to obtain the foreign currency which we urgently need and to keep the defence industry afloat'.[11]

Russian officials contend that the restoration of Sino-Russian military ties is the natural outgrowth of a broad and maturing relationship with China. Economic concerns, however, are the driving force behind it. Igor Rogachev, Russian Ambassador to China, explains: 'I think it's quite natural that we consider this [military] cooperation as an integral part of our general relationship. China has been and I hope it will be our partner. Our defence industry needs some impulse. We need hard currency. We now have a lot of economic troubles'.[12] As US analyst Norman Friedman has observed, 'they [the Russians] have one product worth buying and they are selling it'.[13]

Economic considerations have meant that there is considerable pressure to make deals with any country ready to pay in hard currency. A former Minister for Foreign Economic Relations, Pyotr Aven, has noted that Russian defence plants put formidable pressure on the government to permit arms deal with Taiwan.[14] The main block on sales of Russian arms to Taiwan was the damage that this could do to relations with China. According to Sergey Glaziev, Deputy Minister for Foreign Economic Relations, Russia was ready to issue a licence for arms merchants to sell warships, missiles and light arms to Taiwan until the leadership decided that this would do too much damage to relations with mainland China.[15] As is discussed in section III in this chapter, the decision not to proceed with military–technical cooperation illustrated that Russia has not made economic considerations the only element in its arms transfer policy.

[8] *Izvestiya*, 10 Oct. 1995, p. 5 (in Russian) in Foreign Broadcast Information Service, *Daily Report–Central Eurasia* (hereafter FBIS-SOV), FBIS-SOV-95-205-S, 24 Oct. 1995, pp. 33–37.

[9] *Komsomolskaya Pravda*, 27 Apr. 1994 (in Russian).

[10] *Segodnya*, 18 Oct. 1994 (in Russian).

[11] Interview with President Yeltsin, *Izvestiya*, 24 Feb. 1992, pp. 1, 3 (in Russian). To illustrate how military exports can be helpful, in 1993 the Russian defence industry repaid 400 billion roubles ($220 million) in loan credits from profits from export orders. Beaver (note 6).

[12] 'Russia hopes to sell more arms to Peking', Central News Agency (Taipei), 15 Dec. 1992.

[13] *Baltimore Sun*, 17 Oct. 1992.

[14] Moscow Teleradiokompaniya Ostankino, 14 Mar. 1992 (in Russian) in FBIS-SOV-92-051, 16 Mar. 1992, p. 49.

[15] Vesti (Russian Television Network), 3 Mar. 1992 (in Russian) in FBIS-SOV-92-044, 5 Mar. 1992, p. 48.

## Financing economic reform

Arms exports have been at the centre of a fundamental debate in Russia about where the resources needed to finance economic reform should come from. One school of thought favours seeking external financial support, another would rely mainly on national resources.

Russian politicians and industrialists point out that the value of Russian arms exports per year (in the region of $1.5–3.5 billion) is comparable to the level of Western financial assistance, while arms transfers do not increase the national debt.[16] According to Boris N. Kuzyk, since the autumn of 1994 assistant to the Russian President responsible for advice on arms transfers and military–technical cooperation, Russia concluded contracts worth $2.5 billion in 1995.[17]

Industrialists initially claimed that arms sales could finance conversion. The former adviser to President Yeltsin on arms transfers, Mikhail Maley, suggested that Russia must sell $5–10 billion worth of arms each year for 15 to 30 years to cover the estimated $150 billion cost of conversion.[18] President Yeltsin proposed that part of the income from the defence industry be used to finance social programmes for armed forces personnel.

Arms export policy as an instrument of economic reform has had its opponents in Russia. Before he became Russian Foreign Minister, Andrey Kozyrev expressed reservations about the compatibility of an active arms export policy with the new principles of Russian international policy.[19] He also later argued that revenues from arms sales could not substitute for Western aid because they would be channelled to a narrow sector of the Russian economy. Vsevolod Avduevskiy, Chairman of the Russian Commission for Conversion, has voiced primarily political objections, referring to the financing of conversion through exports as a dirty business which he likened to adding kerosine to local conflicts. In his view, such a policy would simply prolong the agony of breaking with a militarized economy by postponing difficult decisions that must be faced. Moreover, there is no guarantee that the proceeds from arms exports will not simply boost the defence industry.[20]

## Providing the Russian people with consumer goods

China has tended to conclude deals with Russia only on condition that the financial arrangements include a significant element of barter. However, the Russian defence industry is willing to sell China sophisticated armaments in the hope of getting at least some hard currency in its present economic situation.

[16] *Jane's Defence Weekly*, 9 July 1994, p. 28.

[17] *Rossiyskaya Gazeta*, 17 Oct. 1995 (in Russian), p. 3.

[18] *Asian Defence Journal*, no. 3 (1994), p. 74.

[19] *Izvestiya*, 20 Feb. 1990 (in Russian).

[20] Cited in Cooper, J., *The Soviet Defence Industry: Conversion and Economic Reforms* (Royal Institute of International Affairs/Council on Foreign Relations Press: New York, 1991), pp. 65–66. See also *Izvestiya*, 7 Feb. 1990, p. 2 (in Russian) in FBIS-SOV-90-031, 14 Feb. 1990, pp. 117–21; and Avduevskiy, V., 'Conversion and economic reforms: experience of Russia', *Peace and the Sciences*, Mar. 1992, pp. 7–10.

If arms transfer deals with China cannot be paid for on a purely cash basis many Russian politicians and industrialists believe that military cooperation could still contribute to the resolution of the problem of the consumer goods deficit in Russia. China can offer Russia a range of goods—such as toys, some electronics, textiles, shoes, leather and tea—and, although some are of relatively poor quality, it has adapted itself to the Russian market much better than other developing (and even developed) countries. Moreover, China is able to offer a much wider variety of goods in barter deals than other arms trade partners such as India and Malaysia.

### Improving Sino-Russian economic relations

The Russian leadership has also pointed out that arms and technology transfers promote other forms of economic cooperation between the two countries. The value of Sino-Russian bilateral trade was $7.68 billion in 1993 and $5.1 billion in 1994.[21] The value of their trade in these two years added together was greater than that of the two decades of 1950–69. In the framework of discussions about military–technical cooperation a number of Russian regions and particular enterprises have established direct contact with Chinese counterparts. In turn, this commercial and industrial infrastructure serves as an additional spur to the development of Sino-Russian cooperation in various areas.

## III. Political and strategic considerations

Economic incentives are not the only reasons for Sino-Russian military–technical cooperation. Political and military motivations have been identified in the resumption of the relationship.

### Framing a new security complex on the Eurasian continent

According to the Russian draft foreign policy doctrine of 1993, China is not a very high priority for Russia. In early 1993 the Asia–Pacific region was ranked sixth on a list of 10 priorities in Russian foreign policy, behind relations with the CIS, arms control and international security, economic reform, relations with the United States and relations with Europe. At the same time, it ranked higher in priority than South and West Asia, the Near East and Latin America.[22] However, this order of priorities probably no longer reflects China's significance for Russia.

The place of China in the new Russian world-view must be seen in the light of the search for a new Russian identity and international role. According to the Russian leadership, Russia should become a focal point of a new Eurasian security complex. During his visit to India in December 1992 President Yeltsin

[21] *Rossiyskaya Gazeta*, 17 Aug. 1995 (in Russian).
[22] 'Kontseptsiya vneshney politiki Rossiyskoy Federatsii' [The foreign policy concept of the Russian Federation], *Diplomaticheskiy Vestnik*, Jan. 1993 (special issue), pp. 15–16 (in Russian).

emphasized Russia's Eurasian identity by pointing out that the greater part of Russia's territory (10 million out of 17 million km$^2$) lay in Asia and that most Russian citizens live in the Asian part of Russia.[23] During a visit to South Korea in November 1992 he stressed that Russia's foreign policy was turning from the West to the Asia–Pacific region.[24] Along with India and Kazakhstan, China is perceived by Russia as an important pillar of this new security system.

In recent years Russian leaders have faced a continuous temptation to play the 'Chinese card' against Japan and the USA in order to end Russian isolation from the principal economic and security institutions gradually being developed in the Asia–Pacific region. Russia could influence the regional power balance through its arms transfer policies towards China. Given Russia's weakening economic, political and military position in East Asia, it views a strong China as a counterweight to Japan and the United States.

At the present time commercial rationales sometimes challenge strategic considerations when a potential recipient is ready to pay for arms in cash. In East Asia, Taiwan has both a stated need for modern weapons and foreign exchange reserves of around $97 billion. However, the temptation to sell weapons to Taiwan was resisted, in part for strategic reasons.

## Promoting Sino-Russian relations

Russian leaders believe that Chinese interest in military cooperation with its northern neighbour will help in the further development of a stable bilateral relationship and lead to greater flexibility in the resolution of common problems. Despite some security concerns among elements of the Russian military and some politicians, most prominently from the Yabloko Party, the Yeltsin Government and a majority of defence experts are confident that China will not use its growing military potential against Russia.[25] In spite of the fact of military confrontation between China and Russia in the recent past, Pavel Grachev, then Russian Defence Minister, used his visit to China in May 1995 to state his view that China would never pose a military threat to Russia again.

## Fear of Islamic fundamentalism

The Russian leadership believes that the Chinese military threat has disappeared for the foreseeable future and Russia no longer plans for a general war to preserve its territorial integrity against a potential Chinese invasion. In the post-cold war international environment, however, there remain elements of unpredictability. Some contend that Russia is selling military equipment to China as

[23] Singh, A. I., 'India's relations with Russia and Central Asia', *International Affairs* (Moscow), vol. 71, no. 1 (1995), p. 71.

[24] Gill, B., 'North-East Asia and multilateral security institutions', *SIPRI Yearbook 1994* (Oxford University Press: Oxford, 1994), pp. 156–57.

[25] Afanasiev, E., 'Russia–China relations: from normalization to partnership', *Far Eastern Affairs*, no. 1 (1994), pp. 3–8.

part of an effort to remain vigilant against two potential threats to the integrity of Russia: the possible growth of Islamic fundamentalism within Russia and the resurgence of regional powers with an Islamic background.

According to some observers, the strategic and economic alignments emerging in Central Asia (mainly among Islamic peoples) will shape the strategic balance of Asia in the coming years.[26] China and Russia have a mutual interest in monitoring the activity of Islamic peoples in Central Asia and in adjacent countries such as Iran. Joint action to prevent any potential threat will be a solid basis for future Sino-Russian ties and a lasting feature of their policies.

## IV. The management of arms transfers to China

The dislocation of the Soviet decision-making system and the development of a new export control system in Russia are described in earlier chapters of this book. Discussions on military–technical cooperation with China were already under way in the late Soviet period, and deals in progress or under negotiation by individual enterprises were disrupted by the sudden changes in the rules.[27]

After 1991 Oboronexport (the predecessor of Rosvooruzhenie) and the Central Engineering Directorate (Glavnoye inzhenernoye upravleniye, GIU) within the Ministry of Foreign Economic Relations (MFER) were the chief negotiators and points of contact for discussions with China. Suddenly after November 1993 the new state company Rosvooruzhenie became the chief negotiator for major arms deals.

Chinese requests for Russian military equipment and related technology were usually relayed to specialist agencies through the Ministry of Foreign Affairs and the MFER. Applications were first considered by the licensing authority— at that time the Interdepartmental Commission for Military–Technical Cooperation between the Russian Federation and Foreign States (Komitet voyenno-tekhnickeskogo sotrudnichestva, KVTS). Afterwards the state authority (now Rosvooruzhenie) together with the State Committee on Defence Industries would identify specific defence plants which might be interested to produce the arms which had been licensed.

If an enterprise was the only producer of a given system a choice was very easy. For example, for a time the Nizhniy Novgorod Krasnoye Sormovo plant was the sole producer of the Kilo (Varshavyanka) Class submarine.[28] The Irkutsk Aviation Production Association (IAPO) is the main builder of the Su-27 aircraft. However, if a deal could involve several exporters Rosvooruzhenie was subjected to heavy lobbying: the MiG-29 fighter aircraft offered to China, for instance, could have been transferred by either the Moscow

[26] Rumer, B. Z., 'The gathering storm in Central Asia' and Malik, M. J., 'India copes with the Kremlin's fall', *Orbis*, vol. 38, no. 1 (winter 1993).

[27] Author's interview with Sergey L. Zimin, formerly Director, Volga Innovation Company, Nizhniy Novgorod, 11 Sep. 1995.

[28] Early in 1996 the Admiralty shipyard in St Petersburg decided to begin construction of the Kilo Class submarine. *Jane's Intelligence Review Pointer*, Oct. 1996, p. 1.

Aviation Production Association (MAPO) or the Sokol plant in Nizhniy Novgorod. Orders were sometimes allocated to the plant which was in the worst economic situation and sometimes to the one able to produce the best equipment. In the case of China the equipment was usually taken 'off the shelf' from stocks produced for the Russian Ministry of Defence or a foreign customer but never paid for. According to Nikolay Zharkov, Director of the Krasnoye Sormovo plant, two Kilo Class submarines which were sold to China were initially designated for Poland and Romania but they had refused to pay for them at the last moment.[29] Fifty T-80 tanks were sold to China by the Kirov Plant in St Petersburg after the Russian Ministry of Defence refused to pay for them.[30]

After the decision on exporters was taken, the state authority formed a mixed team which included both state officials and representatives of the producer to negotiate price and payment schedules with the Chinese. Representatives of enterprises had no major voice in the negotiating process, usually playing more the role of consultants. On a number of occasions negotiations were carried out only by the state agency. For example, the Kilo submarine deal (including prices and other financial conditions) was concluded by Rosvooruzhenie without any participation of the Krasnoye Sormovo plant.[31] At the same time an enterprise might have some opportunities to renegotiate some technical aspects of the contract (such as the delivery and payment schedule or the shares of hard currency and barter payments): representatives of the Krasnoye Sormovo plant succeeded in increasing the hard currency element of the submarine deal through direct negotiations with China after the general contract had been signed.[32] The producer was also responsible for adapting the weapon system to the specific requirements of the Chinese forces and for after-sales service.

Since the dissolution of the Soviet Union, some regional governments in Russia have also had a voice in arms export decisions. The regional governments of, for example, Yekaterinburg, Irkutsk, Nizhniy Novgorod, St Petersburg and Tula all promoted the establishment of local arms trading firms and issued export licences to defence plants. The regional governments gave plants which allocated some arms export proceeds to conversion programmes exemption from certain taxes or privileged tax status.[33] Local governments received Chinese delegations to assure them of their interest in and support for the Russian defence industry.[34]

[29] *Izvestiya*, 10 Sep. 1994 (in Russian).

[30] *Izvestiya*, 30 Mar. 1994 (in Russian).

[31] *Delo*, 24–30 Mar. 1995, p. 7 (in Russian).

[32] See note 31.

[33] Author's interview with Andrey A. Khudin, Director, Nizhniy Novgorod Division, Institute of Economy and Conversion of Military Production, 16 Oct. 1995; and interview with Vladimir A. Andreyev, formerly Head, Planning Division, Department of Conversion, Nizhniy Novgorod Regional Government, 25 Oct. 1995.

[34] Author's interview with Igor V. Moskayev, former Head, Department of International Relations, Nizhniy Novgorod Regional Government, 23 Oct. 1995.

## Management of the production process

The management of the production process for arms to be exported depends on the specific conditions laid down in the contract. The producers manage the programmes themselves, including coordinating the many levels of production and suppliers of equipment and sub-systems. This represents a fundamental change compared with previous practice by which this coordination was managed by central ministries. Since 1993 many defence enterprises have become joint-stock companies largely independent from central government.[35] For example, by April 1995, 22.5 per cent of shares of the Krasnoye Sormovo plant had been sold through auction and 35 per cent more were to be auctioned.[36] In July 1995 the Sokol plant also sold 22 per cent of its shares through auction.[37] After this the plants applied for local and central government support only where they experienced troubles with their subcontractors.

Some problems were revealed in the functioning of the relationship between design bureaux and production enterprises. A number of design bureaux complained about neglect of their copyright in particular weapon systems. One of the chief designers at the Sukhoi Design Bureau (designer of the Su-27 fighter sold to China) complained in March 1992: 'Our situation violates the laws of the market . . . As soon as the design bureau transfers [design] documentation to the aircraft production plant, the production plant becomes sole proprietor of the aircraft it produces. They forget the enormous intellectual effort invested in the aircraft, which ought to bring the designers definite dividends'.[38]

In most cases designers and manufacturers have moved to form production associations that would assure them both a share of the proceeds from the sales of a given weapon system. For example, the Krasnoye Sormovo plant works in close contact with its design bureau, which is in St Petersburg. In 1994–95 the plant used the facilities of the design bureau to test submarines in the open sea before they were transferred to China.[39] The MiG Design Bureau agreed to sign final contracts together with the assembly plants of MAPO and Rosvooruzhenie. This 'triple signature' is also intended to reassure the potential customer that the Russian Government is in full control of the deal. The general designer (MiG) will be responsible for aircraft modifications, while the director of the assembly facility (MAPO) will be responsible for maintenance, technical service and supply of spare parts.[40] In 1995, the MiG Design Bureau, MAPO, the Sokol plant and a number of banks took a further step and formed the MiG financial–industrial group to qualify for privileges regarding taxation, tariffs, credits and orders.[41] Outside the aircraft building sector, the Admiralty shipyard

[35] Author's interview with Andrey A. Khudin, Director, Nizhniy Novgorod Division, Institute of Economy and Conversion of Military Production, 16 Oct. 1995.

[36] *Delo*, 7–13 Apr. 1995 (in Russian).

[37] *Birzha*, 14 July 1995, p. 4 (in Russian).

[38] *Krasnaya Zvezda*, 21 Mar. 1992 (in Russian).

[39] See note 36.

[40] Kogan (note 7), p. 45.

[41] *Nizhegorodskaya Yarmarka*, no. 32 (1995), p. 3 (in Russian).

along with a leading Russian commercial bank, Inkombank, established a financial–industrial group to produce Kilo Class submarines for export.

## Payment

The financial aspects of arms export deals with China remain an unresolved problem from the Russian point of view. China usually pays in hard currency for only 20–30 per cent of contract value; the rest is paid for by shipments of consumer goods. For example, Russia received hard currency payments to cover one-third of the value of the first Su-27 Flanker fighter aircraft contract and 35 per cent of the contract to provide ship-borne guns of 77-mm calibre produced by the Mashzavod machine-building plant in Nizhniy Novgorod.[42]

In some cases Russia has succeeded in signing more advantageous contracts with China. For example, in the deal to transfer the Kilo Class submarine, the producer was able to shift the ratio of hard currency and barter to 50 : 50. However, the Krasnoye Sormovo plant remained unsatisfied since, after paying taxes and commissions, its hard currency payment would be only 8–10 per cent of contract value.[43] The programme to supply Su-27 aircraft is likely to extend over several phases. Russia would like the ratio shifted to 50 : 50 in future deals and would like the barter goods to be of sufficient quality for re-export.[44] Some sources have reported that this hard currency ratio has been achieved in the second phase of the programme.[45]

The Chinese consumer goods received in part payment are not popular in Russia because of their low quality and are already available from many other businesses. The lack of clarity in the contracts has been used by China to propose an assortment of goods that left their Russian counterparts no choice and could not be effectively resold in Russia or re-exported.

In time this problem may be solved if both China and Russia move towards fully convertible currencies. At present, however, the Russian defence industry has no alternative but to accept such business practices. For many plants exports remain the only way to survive and readjust production. According to some reports, the Kilo Class submarine deal in September 1994 prevented the financial collapse of the factory and a strike in the Krasnoye Sormovo plant.[46]

Industrial leaders are nevertheless very critical of what they call the 'banana approach' to arms exports and put pressure on the Russian Government and Rosvooruzhenie to try to modify it.[47] In their view, this kind of financing can serve only as a temporary tactic to survive a transitional period.

---

[42] See note 31; *Birzha*, 14 Apr. 1995, p. 3 (in Russian); and *Far Eastern Economic Review*, 3 Sep. 1992, p. 21.

[43] See note 31.

[44] *Jane's Defence Weekly*, 22 Jan. 1994, p. 3.

[45] Other sources even report that a more likely split will be 70 : 30 in favour of hard currency. *Jane's Defence Weekly*, 6 May 1995, p. 3.

[46] See note 31; and note 36.

[47] This term became popular after the Philippines offered bananas in exchange for Russian weapons. *Moscow News*, 19 Mar. 1993.

A number of complaints have been lodged against Rosvooruzhenie by industrialists and arms-trading companies, many addressed to the way in which Rosvooruzhenie handles its business. In particular, they have blamed Rosvooruzhenie for neglecting the economic interests of the weapon producers.

As a state company, Rosvooruzhenie has often established the prices and financial conditions of agreements on the basis of political and strategic considerations rather than commercial reasons. This is similar to the Soviet practice. For instance, Kilo Class submarines were sold at substantially reduced prices for that class of vessels—$90 million each—while Germany is believed to have sold submarines recently for in excess of $200 million each.[48]

Second, industrialists are not content with the size of the commission charged by Rosvooruzhenie for its services. Under current legislation, Rosvooruzhenie is allowed to take 5–10 per cent of the value of sales in commission. The rest of the after-tax proceeds should, theoretically, go to the producers. In reality the commission was often 15–20 per cent of the value of the deal.[49] In one case— the deal to supply China with T-80 tanks—the commission was 25 per cent.[50] Other government and private trading companies have limited themselves to commissions of up to 5 per cent.

While deals with China have usually been concluded on the basis of barter, Rosvooruzhenie took its commission in hard currency regardless of the conditions of the contract. Naturally, this evoked resentment in enterprises which were left with the bulk of their proceeds in the form of goods. Moreover, plants which succeeded in getting some part of their payments in hard currency were required, under the prevailing rules, to sell 50 per cent of their currency to the Russian Central Bank at an artificial exchange rate.

Industrialists also insist that in future a general arms export contract should be signed after or at least simultaneously with an agreement on the method of payment. This could increase the bargaining power of Russia both in deciding the hard currency portion of a deal and regarding the specific selection of Chinese goods accepted as barter.[51]

Russian industrialists and trading firms were not content with the idea that Rosvooruzhenie should enjoy a monopoly on foreign trade contacts. Following a government decree of 6 May 1994, a number of defence enterprises and local trading companies (for example, Aviaexport and Promexport in Moscow, Russkoye Oruzhiye in Tula and the Volga Innovation Company in Nizhniy Novgorod) received licences to engage in arms export operations directly.[52] Trading companies offered enterprises much better financial conditions and

[48] *Izvestiya*, 10 Sep. 1994 (in Russian).

[49] Author's interview with Sergey L. Zimin, former Director, Volga Innovation Company, Nizhniy Novgorod, 11 Sep. 1995.

[50] *Izvestiya*, 30 Mar. 1994 (in Russian).

[51] Author's interview with Sergey L. Zimin, former Director, Volga Innovation Company, Nizhniy Novgorod, 11 Sep. 1995.

[52] Decision of the Government of the Russian Federation on granting the enterprises of the Russian Federation the right to participate in military–technical cooperation with foreign countries, no. 479, 6 May 1994, reproduced in appendix 3 in this volume as document 10.

services than Rosvooruzhenie, especially with regard to the commission that they charged. However, the export efforts of trading firms were met with a very hostile reaction from Rosvooruzhenie, which accused them of incompetence or even damaging national security.[53]

As described in chapter 6 of this volume, the nature of Russian Government control over arms exports has been continuously shifting since 1994 and remains very uncertain. From the defence enterprises' point of view, by effectively retaining its monopoly over arms exports, the government has returned to the Soviet practice of subordinating commercial to political and strategic objectives. The centralization of bureaucratic procedures is also seen as neglecting the interests of producers. However, whereas the Soviet period was characterized by stable administrative procedures, in the new environment producers are forced to deal with a constantly changing group of government agencies.[54]

## V. Major bilateral programmes

Since the resumption of Sino-Russian military cooperation, Russia has become China's biggest arms supplier. According to former Prime Minister Yegor Gaidar, Russian arms sales to China totalled $1.8 billion in 1992.[55] Few agreements were reached in 1993, but Sino–Russian arms trade was reactivated in 1994 and 1995. New contracts were agreed for a total of $1 billion by the end of 1994.[56]

Russia is assisting with the modernization of the Chinese ground forces, air force and navy as well as transferring some military technologies.

### Ground forces

China is estimated to have 10 000 tanks in its inventory, mostly Chinese versions of Soviet-designed main battle tanks.[57] Most of these are of outdated designs, having been developed from the Soviet T-54/55 series of the early 1950s. Moreover, many Chinese tanks are believed to be non-operational. The need to modernize the tank fleet became obvious by the end of the 1980s.

According to Russian military sources, in 1992 China agreed to purchase about 50 T-72 tanks and 70 BMP-1 armoured infantry fighting vehicles at a cost of c. $250 million.[58] According to some reports, Russia delivered these

---

[53] *Izvestiya*, 10 Oct. 1995 (in Russian). A government audit and investigation of Rosvooruzhenie itself, led by presidential representative Marshal Ye. Shaposhnikov, after numerous accusations against this company revealed many violations of the law (including corruption). Subsequently many employees were made redundant and legal proceedings were begun against Gen. Viktor Samoylov, the former head of Rosvooruzhenie, and a number of his colleagues. *Ponedelnik*, no. 42 (1994), pp. 2–3 (in Russian).

[54] Author's interview with Sergey L. Zimin, former Director, Volga Innovation Company, Nizhniy Novgorod, 11 Sep. 1995.

[55] Sismanidis, R., 'China and the post-Soviet security structure', *Asian Affairs*, vol. 21, no. 1 (spring 1994), p. 51.

[56] *Izvestiya*, 22 Sep. 1994 (in Russian).

[57] *World Defence Almanac 1993–94*, vol. 18, issue 1 (1994), p. 222.

[58] *Far Eastern Economic Review*, 8 July 1993, p. 26; and *Washington Post*, 31 Mar. 1993.

**Table 11.1.** Deliveries of major conventional weapons to China, 1990–94

| Seller | 1990 | 1991 | 1992 | 1993 | 1994 | Total |
|---|---|---|---|---|---|---|
| France | 30 | 10 | 5 | 2 | 2 | **49** |
| Italy | 10 | 8 | 0 | 0 | 0 | **18** |
| USSR/Russia | 86 | 133 | 951 | 677 | 1 000 | **2 847** |
| **Total** | **126** | **151** | **956** | **679** | **1 002** | **2 914** |

*Source:* SIPRI arms trade database.

tanks at the end of 1993.[59] The version of the T-72 involved in the deal was an improved version of the T-72M1, among the most modern of this series. If the T-72 were to replace the immense inventory of older tanks used by China this would represent a major increase in capability.[60] China and Russia have also apparently discussed the transfer of more modern BMP-3 armoured infantry fighting vehicles including a licence to manufacture these vehicles in China.[61] Moreover, China and Russia continue to explore other areas of possible cooperation. During President Yeltsin's visit in April 1996 Russian and Chinese specialists apparently discussed the modernization of older Chinese tanks with new fire-control systems and the possible transfer of the BTR-80 armoured personnel carrier.[62]

In October 1992, the People's Liberation Army (PLA) became the first export customer to receive the Russian S-300 (the NATO designation is the SA-10B Grumble) surface-to-air missile.[63] China has bought three complexes, one apparently for deployment near Peking, one for Wuhu air base in Anhui Province, and one for training purposes.[64] Since the PLA had no system equivalent to the S-300, this represents a significant upgrade in air defence capability.[65]

## The air force

Before China resumed military cooperation with Russia, it had a fleet of 5000 obsolete combat aircraft, most of them based on old Soviet designs such as the

---

[59] *Izvestiya*, 30 Mar. 1994 (in Russian).

[60] Bain, W., 'Sino-Indian military modernization: the potential for destabilization', *Asian Affairs*, vol. 21, no. 3 (fall 1994), pp. 133–34. It is also reported that China has received T-80U tanks from Russia, although this is denied by Russian experts. *Jane's Intelligence Review*, Sep. 1996, p. 9.

[61] *Rossiyskaya Gazeta*, 5 Oct. 1996, p. 12 (in Russian), in FBIS-SOV-96-196, 5 Oct. 1996.

[62] *Jane's Defence Weekly*, 24 Apr. 1996, p. 10.

[63] Dantes, E., 'Changing air power doctrines of regional military powers', *Asian Defence Journal*, Mar. 1993, p. 43.

[64] *World Defence Almanac 1993–94* (note 57); *Izvestiya*, 5 Mar. 1993 (in Russian); and Gill, B., 'Trade, production and control of conventional weapons in East Asia', 1995, p. 21. Unpublished manuscript.

[65] The S-300 is a local-area air-defence system that was developed to defend against attacks by low-flying aircraft such as the US F-111 or the British Tornado. Later versions of the S-300, designated the SA-12 Gladiator by NATO, have limited capabilities to defend against ballistic and cruise missile attacks. However, it is not thought that this is the version bought by China. Tai Ming Cheung, 'Sukois, sams, subs', *Far Eastern Economic Review*, 8 Apr. 1993, p. 23.

MiG-21 and MiG-19 fighter aircraft and the Tu-4 bomber. Chinese helicopters are also mostly based on Soviet designs, the Mi-4 and Mi-8/17 series.[66]

The PLA Air Force (PLAAF) has made a major investment in trying to modernize its equipment by domestic means, but with limited success. In 1990, China introduced the F-8II Finback. However, this aircraft is derived from the Soviet MiG-21 Fishbed and is not comparable to contemporary Western or Russian aircraft.[67] The failure of the Finback programme forced the PLAAF to seek alternative aircraft and the dramatic reduction in tension between China and Russia made Russia an obvious choice as supplier. In 1992 China received 26 Su-27 Flanker fighter aircraft—Russia's most advanced air superiority fighter—including two trainer versions.[68]

The Su-27 is designed for air-to-air combat, equipped with Russia's most advanced avionics and capable of carrying the most advanced weapons.[69] It has, among other features, multiple-target engagement and look-down/shoot-down capabilities and a combat radius of approximately 1600 km, which could be extended if China can acquire in-flight refuelling capability—an acquisition priority. The Su-27s are currently based at the Wuhu air base and will primarily be used as interceptors. If deployed in southern China (probably on Hainan Island), the aircraft could operate over the South China Sea.[70]

A further batch of 24 Su-27 aircraft (including two twin-seater trainers) was acquired in 1995–96. In April 1996 Yeltsin apparently agreed to the transfer of a third batch of 18 Su-27s and in principle to begin producing the aircraft under licence in China.[71]

The Su-27 deal was followed in 1992 by a contract for 100 Klimov RD-33 aircraft engines, which Russia uses to power its MiG-29 fighter. China will employ these to upgrade its export-oriented Super F-7 fighter.[72]

It has been reported that China is prepared to buy between 24 and 36 Russian-produced MiG-31 fighter aircraft. There are also reports that it is prepared to buy 40 MiG-29 fighter aircraft and 12 Su-24 fighter bombers.[73] According to

[66] *World Defence Almanac 1993–94* (note 57).

[67] The F-8II was at the centre of the Sino-US 'Peace Pearl' programme which involved fitting 50 Finback aircraft with a Westinghouse radar and fire-control computer and a Litton inertial navigation system. The programme was cancelled, among other reasons because the PLAAF determined that the F-8II would not meet performance requirements. Bin Yu, 'Sino-Russian military relations', *Asian Survey*, vol. 33, no. 3 (1993), p. 305; and Jencks, H., *Some Political and Military Implications of Soviet Warplane Sales to the PRC* (Sun Yat-Sen Center for Policy Studies: Kaohsiung, 1991), pp. 5–6. On recent Chinese modernization efforts, see Gill, B. and Taeho Kim, *China's Arms Acquisitions from Abroad: A Quest for 'Superb and Secret Weapons'*, SIPRI Research Report no. 11 (Oxford University Press: Oxford, 1995).

[68] Tai Ming Cheung (note 65).

[69] Taylor, J., 'Gallery of Soviet aerospace weapons', *Air Force Magazine*, Mar. 1990, p. 75.

[70] Fulghum, D. and Proctor, P., 'Chinese coveting offensive triad', *Aviation Week & Space Technology*, 21 Sep. 1992, p. 21.

[71] *Jane's Defence Weekly*, 24 Apr. 1996, p. 10; and *Defense News*, 9 Dec. 1996, p. 26.

[72] Gill (note 64); and *Jane's Defence Weekly*, 22 Jan. 1994, p. 3; and 19 Feb. 1994, p. 26.

[73] Dantes (note 63), p. 43; Anthony, I. *et al.*, 'Register of the trade in and licensed production of major conventional weapons in industrialized and developing countries, 1992', *SIPRI Yearbook 1993: World Armaments and Disarmament* (Oxford University Press: Oxford, 1993), p. 501; International Institute for Strategic Studies, *The Military Balance 1993–1994* (Brassey's: London, 1993), p. 148; Bin Yu (note 67), pp. 308–10; *Asian Security 1994–95* (Brassey's: London, 1994), p. 15; and *Military and Arms Transfers News*, 17 June 1994, p. 5.

**Table 11.2.** Deliveries of major conventional weapons to China from the former Soviet Union, 1990–97

| Supplier (S) or licenser (L) | No. ordered | Weapon designation | Weapon description | Year of order/ licence | Year(s) of deliveries | No. delivered/ produced | Comments |
|---|---|---|---|---|---|---|---|
| **S:** | | | | | | | |
| USSR | (2) | Ka-27PL Helix-A | ASW helicopter | (1991) | | .. | For Navy |
| | 24 | Mi-17 Hip-H | Helicopter | 1990 | 1990–91 | (24) | Deal worth $700 m. (offsets 40%); incl. 4 Su-27UB trainer version |
| | 24 | Su-27 Flanker | Fighter aircraft | 1991 | 1992 | (24) | |
| Russia | (288) | AA-11 Archer/R-73M1 | Air-to-air missile | (1991) | 1991–92 | (288) | For 24 Su-27 fighter aircraft |
| | .. | AA-8 Aphid/R-60 | Air-to-air missile | 1991 | 1991–92 | 96 | For 24 Su-27 fighter aircraft |
| | 1 | Il-28 Beagle | Bomber aircraft | 1992 | 1993 | 1 | Ex-Russian Air Force; exchanged for canned fruit |
| | 7 | Il-76M Candid-B | Transport aircraft | 1992 | 1993 | 7 | Barter deal worth $200 m. (offsets 60%) |
| | 2 | Su-27 Flanker | Fighter aircraft | 1992 | 1992 | 2 | Original order for 12 Su-27 fighter aircraft reduced to 2 Su-27UB trainer version |
| | 24 | Su-27 Flanker | Fighter aircraft | 1995 | 1996 | 24 | Deal worth $2.2 b.; incl. 6 Su-27UB trainer version |
| | 4 | AK-130 130-mm | Naval gun | 1996 | | .. | On 2 Sovremenny Class destroyers |
| | 15 | SA-15 SAMS | AAV(M) | (1996) | 1997 | 15 | |
| | (200) | T-80U | Main battle tank | 1993 | 1996 | (200) | |
| | (1) | 36D6 Tin Shield | Surveillance radar | 1992 | 1993 | (1) | For use with SA-10c/S-300PMU SAM systems |
| | (1) | 76N6 Clam Shell | Surveillance radar | 1992 | 1993 | (1) | For use with 4 SA-10c/S-200PMU SAM system |
| | 4 | Bass Tilt | Fire control radar | 1996 | | .. | On 2 Sovremenny Class destroyers; for use with AK-630 30-mm guns |
| | 12 | Front Dome | Fire control radar | 1996 | | .. | On 2 Sovremenny Class destroyers; for use with SA-N-7 ShAMs |
| | 2 | Kite Screech | Fire control radar | 1996 | | .. | On 2 Sovremenny Class destroyers; for use with AK-130 130-mm guns |
| | 6 | Palm Fond | Surveillance radar | 1996 | | .. | On 2 Sovremenny Class destroyers |
| | (4) | SA-10c/S-300PMU | SAMS | 1992 | 1993–97 | (4) | |
| | 4 | SA-N-7 ShAMS/Shtil | ShAM system | 1996 | | .. | On 2 Sovremenny Class destroyers |
| | 2 | SS-N-22 ShShMS | ShShM system | 1996 | | .. | On 2 Sovremenny Class destroyers |

| | | | | | | Comments |
|---|---|---|---|---|---|---|
| 2 | Top Plate | Surveillance radar | 1996 | 1996 | .. | On 2 Sovremenny Class destroyers |
| (288) | AA-11 Archer/R-73M1 | Air-to-air missile | 1996 | 1995 | (288) | For 24 Su-27 fighter aircraft |
| (1 200) | AT-11 Sniper/9M119 | Anti-tank missile | 1993 | 1995 | (1 200) | For 200 T-80U tanks |
| (144) | SA-10 Grumble/5V55R | SAM | 1992 | 1993–97 | (144) | For 4 SA-10c/S-300PMU SAM systems |
| (255) | SA-15 Gauntlet/9M330 | SAM | (1996) | 1997 | (255) | For SA-15 SAM system |
| (88) | SA-N-7 Gadfly/Smerch | ShAM | 1996 | | : | For 2 Sovremenny Class destroyers |
| (32) | SS-N-22 Sunburn/P-80 | ShShM | 1996 | | : | For 2 Sovremenny Class destroyers |
| 2 | Kilo Class/Type 636 | Submarine | 1993 | 1997 | 1 | |
| 2 | Kilo Class/Type 877E | Submarine | 1993 | 1995 | 2 | Originally built for Poland and Romania but cancelled |
| Ukraine | | | | | | |
| 2 | Sovremenny Class | Destroyer | 1996 | | .. | Originally ordered by USSR, but cancelled |
| .. | AA-10a Alamo/R-27R | Air-to-air missile | 1991 | 1991–94 | (144) | For 24 Su-27 fighter aircraft |
| (144) | AA-10a Alamo/R-27R | Air-to-air missile | 1995 | 1996 | (144) | For 24 Su-27 fighter aircraft |
| 1 | Fedko Class | Tanker | 1992 | 1993 | 1 | Chinese designation Fusu or Nancang Class; delivered incomplete and fitted out in China |
| Uzbekistan | | | | | | |
| 15 | Il-76M Candid-B | Transport aircraft | 1993 | 1994–95 | (15) | Ex-Uzbek Air Force |
| L: | | | | | | |
| Russia | | | | | | |
| 120 | Su-27SK Flanker | Fighter aircraft | 1996 | | .. | Incl. assembly from kits; Chinese designation F-11 (J-11) |

*Notes:* This register lists major weapons on order or under delivery, or for which the licence was bought and production was under way or completed in or after 1990. 'Year(s) of deliveries' includes aggregates of all deliveries and licensed production since the beginning of the contract. ASW = Anti-submarine warfare; SAM = Surface-to-air missile; AAV = Anti-aircraft vehicle; ShAM = Ship-to-air missile; ShShM = Ship-to-ship missile.

*Source:* SIPRI arms trade database.

some accounts, in July 1994 China's State Council approved an additional $5 billion-worth of armament imports from Russia including an unspecified number of Su-30MK and Su-35 fighters.[74] Apparently Russia refused to sell the advanced Su-35 but offered the Su-27 and Su-30 aircraft as an alternative.[75]

Table 11.2 summarizes recent Chinese imports of Russian aircraft and missile systems. The Su-27 provides the PLAAF with an instant qualitative boost. The acquisition of MiG-29s and Su-24s would, if confirmed, also give China a further qualitative leap: the MiG-29 has dual-role air superiority/attack capabilities, while the Su-24 is a highly capable attack aircraft.[76]

In addition to these fighter aircraft, Russia has apparently offered to modernize China's bomber fleet, replacing obsolete H-6 bombers with newer models. The supersonic Tu-22M Backfire (with a 4000-km range without refuelling) has been mentioned in press reports, although it should be stressed that these are not confirmed.[77] In 1996 it was reported that China may order four Tu-26 bombers from Russia along with 118 air-to-surface missiles. However, this report is also unconfirmed.[78] China has also expressed interest in developing an airborne warning and control (AWAC) aircraft, perhaps modelled on the Russian A-50 Mainstay aircraft and long-range early-warning radar systems.[79]

The PLAAF has taken delivery of four Ilyushin Il-76 heavy transport aircraft which should prove to be a particularly important addition, since until now its transport fleet has only had light cargo aircraft. A further seven Il-76s are said to be on order.[80]

In October 1990 the first significant Chinese post-*détente* military purchase was made from the then Soviet Union—24 Mi-17 HIP-H transport helicopters.[81]

If these programmes are all completed, the addition of sophisticated Russian equipment will represent a spectacular improvement over current PLAAF hardware. Aircraft such as the Flanker and the Backfire would give the Chinese a credible tool for military intervention beyond its borders. In lieu of actual combat, such aircraft would stand as a symbol of Chinese power and prestige and offer an effective deterrent. Modern military aircraft will also help PLAAF efforts to develop an effective combined arms capability.[82]

[74] *Asian Recorder*, 27 Aug.–2 Sep. 1994, p. 24192.

[75] *Military and Arms Transfers News*, 26 Aug. 1994, p. 5.

[76] Taylor (note 69), pp. 74, 76.

[77] Bain (note 60), p. 135; Blank, S., *Challenging the New World Order: The Arms Transfer Policies of the Russian Republic* (Strategic Studies Institute: Carlisle Barracks, Harrisburg, Pa., 1993), pp. 53–60; Blank, S., 'Russia arms exports and Asia', *Asian Defence Journal*, Mar. 1994, p. 78; and Davis, M., 'Russia's big arms sales drive', *Asia–Pacific Defence Reporter*, Aug.–Sep. 1994, p. 12.

[78] *Jane's Intelligence Review*, July 1996, p. 330. Another report has suggested that China may acquire from Russia the AS-15 air-launched cruise missile, although the status of this report is also uncertain. Allen, K., Krumel, G. and Pollack, J. D., *China's Air Force Enters the 21st Century* (RAND: Santa Monica, Calif., 1995), p. 159.

[79] *Far Eastern Economic Review*, 8 July 1993, p. 26; and *Air Force Magazine*, July 1993, p. 59.

[80] *World Defence Almanac 1993–94* (note 57).

[81] Jencks (note 67), p. 15.

[82] Bellows, M. (ed.), *Asia in the 21st Century: Evolving Strategic Priorities* (National Defense University Press: Washington, DC, 1994), p. 95; Sismanidis (note 55); Hickey and Harmel (note 1), pp. 241–53; Bain (note 60), pp. 131–47; Afanasiev (note 25), pp. 3–8; Taylor, R. I. D., 'Chinese policy towards the Asia–Pacific region: contemporary perspectives', *Journal of the Royal Society for Asian Affairs*, vol. 25,

## The navy

Russia has also contributed greatly to the development of the PLA Navy (PLAN). The PLAN includes seven ex-Soviet and Chinese Romeo Class submarines (although these are probably no longer operational), 20 former Soviet Kronstadt Class patrol craft and 23 Soviet T-43 Class ocean minesweepers.[83]

The dispute over the Spratly Islands and the growth of Japanese sea power have provided China with an immediate incentive to modernize its naval forces. Chinese leaders are said to have 'attached a high priority to modernizing China's navy'.[84] Naval modernization includes the introduction of a new class of destroyer, new conventional and nuclear-powered submarines and substantial talk of acquiring an aircraft-carrier. According to some assessments, the eventual objective of this programme is to move from a brown-water coastal navy to one that is capable of projecting power into the Pacific and Indian oceans.[85]

Until recent years, the priority in Chinese shipbuilding was laying down large numbers of hulls which were not equipped with sophisticated sensors or weapons. This trend has only begun to change in the 1990s with the acceptance, in 1993, of the first Luhu Class (Type 052) guided-missile destroyer.[86]

The PLAN is also acquiring an upgraded version of the Luda Class destroyer, a new class of frigates (the Jiangwei Class) and new classes of re-supply and amphibious assault ships. Once completed, this programme will allow for sustaining operations further from shore. China has also enhanced the air base and anchorages on Woody Island (Lin-tao) in the Paracel Islands.[87]

In 1991 Russia sold two Ka-27 Helix-A anti-submarine warfare (ASW) helicopters to the PLAN. There are also reports that China may purchase from Russia several Sovremenny Class destroyers. These ships have formidable air defence and anti-ship capabilities and can accommodate ASW helicopters.[88]

In February 1994 the Mashzavod plant in Nizhniy Novgorod signed a contract with the PLAN to supply three ship-borne 77-mm calibre automatic artillery systems. In March 1995 Chinese specialists were trained at the Mashzavod plant to use these guns which were to be delivered by the end of the year.[89]

The PLAN is also working to modernize its submarine fleet. China has purchased four Kilo Class submarines from Russia and apparently intends to obtain

part 3 (Oct. 1994), pp. 259–69; Shambaugh, D., 'The insecurity of security: the PLA's evolving doctrine and threat perceptions towards 2000', *Journal of Northeast Asian Studies*, vol. 13, no. 1 (spring 1994), pp. 3–25; and Munro, R., 'China's waxing spheres of influence', *Orbis*, vol. 38, no. 4 (fall 1994), pp. 585–605.

[83] *World Defence Almanac 1993–94* (note 57), p. 221.

[84] Glaser, B., 'China's security perceptions: interests and ambitions', *Asian Survey*, vol. 33, no. 3 (1993), p. 265.

[85] Bain (note 60), p. 136.

[86] Tai Ming Cheung, *Growth of Chinese Naval Power* (Institute of Southeast Asian Studies: Singapore, 1990), p. 22; and *The Military Balance 1993–1994* (note 73), p. 148.

[87] Ball, D., 'A new era in confidence building: the second-track process in the Asia/Pacific region', *Security Dialogue*, vol. 25, no. 2 (June 1994), pp. 159–60.

[88] Preston, A., 'Russian weapons and ships in the Asia–Pacific region', *Asian Defence Journal*, Dec. 1992, p. 60.

[89] *Birzha*, 14 Apr. 1994, p. 3 (in Russian).

the rights to build additional vessels in China.[90] The two submarines, produced in Nizhniy Novgorod, were delivered to China in 1995.[91] Some reports contend that China may ultimately obtain up to 22 Kilos, but sources in Beijing with a closer knowledge of the programme dismiss this.[92] It was reported in March 1995 that China had struck a new deal with Russia for the purchase of six more submarines.[93] The Kilo is considered to be an advanced conventionally-powered vessel that is extremely effective in the coastal defence role. With a range of 9650 km and the ability to remain at sea for up to 45 days, these vessels represent a significant addition not only to the PLA's coastal defence but also to its offensive potential.

According to Tai Ming Cheung, the clearest sign of China's blue-water aspirations is its plan to acquire an aircraft-carrier.[94] There have been frequent although unconfirmed reports that the leadership has decided to acquire one[95] and that China was interested in the Ukrainian ship *Varyag*—a large unfinished carrier that is part of the disputed Soviet Black Sea Fleet. It now appears that China will not purchase the *Varyag* but will either acquire a smaller Russian carrier or build a 30 000- to 48 000-tonne vessel domestically.[96] Although there is no confirmation of China's intentions here either, another indicator of genuine interest in an aircraft-carrier has been the attention paid to the Yak-41 vertical/short take-off and landing (V/STOL) naval fighter aircraft. Bin Yu also argues that it is no coincidence that China purchased the Su-27 for the PLAAF as it can be modified for use on board an aircraft-carrier.[97] Numerous sources have suggested that after the completion of the existing Su-27 Flanker deal a follow-on purchase may include the Su-27K, the naval variant specially designated for carrier-based operations.[98] Moreover, if the PLAN were also to purchase the naval variant of the S-300 (NATO designation SA-N-6), it would possess the foundation for building an adequate defensive and escort force for an aircraft-carrier.

Although Chinese officials can cogently argue that the Spratly Islands dispute demands that China modernize its naval capabilities, this is probably not the underlying or fundamental reason for Chinese aspirations for a blue-water navy. The cost of systems such as the Han Class nuclear-powered attack submarine (SSN) or an aircraft-carrier with its associated escort vessels and air wing make

---

[90] Tai Ming Cheung (note 65), p. 23; and Tai Ming Cheung, 'China's buying spree', *Far Eastern Economic Review*, 8 July 1993, p. 26.

[91] *Delo*, 7–13 Apr. 1995 (note 36).

[92] *Jane's Defence Weekly*, 13 May 1995, p. 18.

[93] *Asian Recorder*, 26 Mar.–1 Apr. 1995, p. 24672; and *Jane's Defence Weekly*, 18 Mar. 1995, p. 3.

[94] Tai Ming Cheung (note 86), p. 27.

[95] *East Defence & Aerospace Update*, May 1993, p. 3; Ball (note 87), pp. 159–60; and Lin, C., 'Chinese military modernization: perceptions, progress and prospects', *Security Studies*, vol. 3, no. 4 (summer 1994), p. 731.

[96] *New York Times*, 11 Jan. 1993; and Ryan, S., 'The PLA Navy's search for a blue water capability', *Asian Defence Journal*, May 1994, p. 30. If this is true, this domestic programme would take many years to complete.

[97] Bin Yu (note 67), pp. 302, 308.

[98] Dantes (note 63), p. 43; Ackerman, J. and Dunn, M., 'Chinese airpower revs up', *Air Force Magazine*, July 1993, p. 59; and *Military and Arms Transfers News*, 7 Oct. 1994, p. 4.

them expensive tools for use against countries like the Philippines or Malaysia with extremely weak navies. Chinese military interests in the Spratlys would probably be better served by warships covered by long-range and air-refuelled aircraft from the Paracel Islands. An aircraft-carrier, as well as China's nuclear submarines, would be much better suited for use on the open seas rather than in the relatively shallow and constricted waters of the South China Sea.[99]

## Military technology transfer

Chinese military technology is as much as 20 years behind that of the West. Past efforts to resolve this problem through reverse engineering (often of Soviet-made equipment) have not overcome this gap. China's defence industry has a history of problems with reverse-engineered systems and some Chinese copies of foreign-designed weapons never reached production—for example, the Chinese copies of the Soviet T-62 tank and MiG-23 fighter-bomber.

Since resuming military cooperation with Russia, China has been extremely cautious in signing deals to purchase Russian military hardware. Chinese officials would prefer to purchase technology and production licences rather than buying equipment 'off the shelf'. This reluctance to place large orders is probably partly because of budget restrictions and partly because of the fear of the potential political consequences of over-dependence on any one supplier. China would prefer to modernize its defence industry. In its pursuit of defence industrial cooperation, China has found Russia a more willing partner than Western countries. Russia has been prepared to consider transfers of advanced technology even at the risk of a long-term adverse impact on the regional balance of power. This willingness stemmed from the desperate economic straits of the Russian defence industry and pressure from the defence industry to overrule the objections of opponents in the Russian Government.

In late 1995, Russia agreed to produce the Su-27 aircraft in China. The licensed production deal is covered by a letter of intent that should be finalized once the second batch of the Su-27s is delivered and paid for. A two-stage programme is proposed, the first being assembly in China from kits produced in Russia, and the second full production in China (probably at the Shenyang Aircraft Factory). The licence is thought to cover annual production of 90–100 aircraft, but most observers say that production will probably be half that number, beginning at a rate of 10–20 aircraft per year.[100] Sukhoi Design Bureau officials also reportedly proposed co-production of the Su-35 in China on condition that China purchase close to 120 of the aircraft produced.[101]

---

[99] Bain (note 60), p. 137.

[100] *Jane's Defence Weekly*, 6 May 1995, p. 3.

[101] Taeho Kim, *The Dynamics of Sino-Russian Military Relations: An Asian Perspective* (Chinese Council of Advanced Policy Studies: Taipei, 1994), p. 19; and Taeho Kim, 'The Russian factor in China's arms acquisition: implications for China's evolving security relations in the Asia–Pacific region', Paper prepared for presentation at the 5th Annual Staunton Hill Conference on China's People's Liberation Army (PLA), sponsored by the American Enterprise Institute, Staunton Hill, Va., 17–19 June 1994, p. 11.

It has also been suggested that China has had negotiations with Russian officials for a technology exchange programme involving the MiG Design Bureau that could include production of an advanced fighter aircraft, probably the MiG-31.[102]

Another recent report suggests that Russia has offered to develop a brand-new fighter for the PLAAF for as little as $500 million. Senior Russian Ministry of Defence officials have said that there have been negotiations over a deal which could see Russian aerospace firms providing up to two-thirds of the required technical and design work, as well as avionics and an engine, for a new fighter based on the Xinjian J-10 airframe. China is supposedly planning to produce the new aircraft at the rate of 100 per year according to Russian statements.[103]

At present, however, it is unclear what type of aircraft (if any) China plans to co-produce with Russia. One analyst has suggested that, if pursued, such a programme would provide China with its 'first step towards a new manufacturing capability' that could both replenish and modernize the air force's obsolete fleet of aircraft, as well as compete with Western manufactures in the lucrative Asian arms market.[104] However, in order to pursue such a programme China would need technical assistance in the areas of aircraft engines and stealth technology. China is also attempting to purchase Kilo Class submarines, ASW technology and technical data on the design and construction of airframes.

Military technology transfers have been combined with exchanges of personnel and expertise. According to Russian Ministry of Defence sources, 'more than 1000 Russian defence scientists and technicians have travelled to China since 1991 on defence-industrial exchanges [and] . . . there are around 300–400 Chinese defence specialists in Russia'.[105] Some of the Russian scientists now believed to be based permanently in China are apparently experts 'in the fields of cruise missiles, ASW, missile launching experiments and nuclear explosions'.[106] Chinese defence scientists and technicians are working at Russian aerospace institutes, including some in Moscow, Ryazan, Samara and Saratov. Some are studying at organizations such as the Central Institute of Aircraft Dynamics in Moscow.[107]

Some sources have suggested that Russia's chaotic economic, social and political conditions have also permitted China to recruit scientists and acquire technology without official approval. However, as Shulong Chu puts it, 'the

---

[102] Tai Ming Cheung, 'China's buying spree' (note 90), p. 24; Dantes (note 63), p. 43; and Bin Yu (note 67), pp. 308–10. It has been suggested that 150–300 MiG-31 Foxhounds could be made in China over an 8-year period. This aircraft is a high-altitude interceptor with superior extended-range radar and multiple target-engagement capabilities.

[103] Gallaher, M., 'China's illusory threat to the South China Sea', *International Security*, vol. 19, no. 1 (summer 1994), p. 175; and *Jane's Defence Weekly*, 19 Feb. 1994, p. 28.

[104] Mecham, M., 'China updates its military, but business comes first', *Aviation Week & Space Technology*, 15 Mar. 1993, p. 58.

[105] Tai Ming Cheung, 'China's buying spree' (note 90), p. 24.

[106] 'Peking recruits Russian weapons experts: report', Central News Agency (Taipei), 29 Dec. 1992; *Wall Street Journal*, 14 Oct. 1993; and Shulong Chu, 'The Russian–US military balance in the post-cold war Asia–Pacific region and the "China threat"', *Journal of Northeast Asian Studies*, spring 1994, pp. 89–90.

[107] *Jane's Defence Weekly*, 19 Feb. 1994, p. 28.

reports that China recruits thousands of Russian weapons experts may come from speculation, because there is no governmental source from China or Russia of such exchange programme. The Russian government has lost a lot in controlling its society, but it has not lost everything'.[108]

## Defence industry conversion

China and Russia have also agreed to strengthen their cooperation in conversion programmes. China has been engaged in a programme intended to convert its defence industry to civilian production.[109] A document has been signed similar to that signed by China and the United States during the visit of US Secretary of Defense William Perry to Beijing in October 1994.[110]

A number of Sino-Russian joint ventures were set up to develop conversion programmes. The companies involved in the first Sino-Russian venture are Xing-Yui-Ju (Beijing), Yuilang Trading (Hong Kong), the Nizhniy Novgorod-based Impex and the Institute of Applied Physics at the Russian Academy of Sciences.[111] The joint venture will take electro-optical defence items and re-configure the designs to create commercial laser, electro-optic and optical devices for sale in the Middle East. Russia will provide research personnel and expertise, leaving the manufacturing and marketing to the Chinese.

The Sungari Sino-Russian joint venture, set up by the Ural Device-Building Plant (Yekaterinburg) and the Kharbin Commercial Trade Company, has started production of cassette tape recorders for motor cars at the Lazur former defence plant in Nizhniy Novgorod.[112]

# VI. Conclusions

This chapter has argued that Russia has vital interests in the resumption and development of military cooperation with China. The immediate background for the reopening of the military relationship has been the need for Russia to support its defence industry. At the same time, China and Russia hope that their bilateral military ties will provide them with a strategic counterweight to a number of threats and challenges of the post-cold war era. These might include US hegemonism, the rise of Japanese power, or a militant Islam. These considerations have helped China and Russia to overcome the hostility which characterized their relations until recently and develop close military ties despite the cautious reaction (or overt opposition) of other major players in the region.

To date it appears that actual deliveries of Russian military equipment to China have been much more modest than has sometimes been reported. The

---

[108] Shulong Chu (note 106), p. 91.
[109] *Asian Economic News*, 4 July 1994; and *NOD & Conversion*, no. 30 (Sep. 1994), p. 42.
[110] *Izvestiya*, 19 Oct. 1994 (in Russian).
[111] Beaver (note 6), p. 30.
[112] *Nizhegorodskiye Novosti*, 14 Apr. 1995 (in Russian).

military potential acquired by China through recent purchases is still not sufficient to make China a leading military power in the region.

In future, Sino-Russian joint and collaborative efforts could be restored to the levels of the 1950s, when Soviet technology effectively armed the PLA, if the financial arrangements can be worked out to the satisfaction of both sides. If this happens, Chinese force modernization could be achieved at lower cost than through imports from Western suppliers. However, it is also obvious that Russia's arms transfers to China have had a destabilizing effect on the regional security debate and have been used as an excuse by countries throughout East Asia to justify their own acquisition programmes. This is not surprising given the low levels of trust between the governments of the region.

Sino-Russian arms deals and defence industrial cooperation have become the focal point of China's efforts to engage Russia in a substantive long-term military relationship, obtain advanced military equipment and technology to modernize the PLA inventory, enhance its air force and naval capabilities, and advance Chinese power projection in East and South-East Asia.

How dangerous Sino-Russian military cooperation will be in the regional context will depend on the extent to which the major regional players include China and Russia in the evolving Asia–Pacific community, reducing their temptation to form a separate strategic coalition.

# 12. Illicit arms transfers

*Ian Anthony*

## I. Introduction

Most of the preceding chapters of this book deal with the issue of arms transfers as an act of state policy. In recent years growing attention has been focused both in Russia and in the international community on those arms transfers which are not acts of state policy—illicit weapon sales, as they are called. Several authors suggested before the end of the cold war that transfers of this type were of increasing importance.[1] However, it is the new conditions after the end of the cold war which have boosted the attention paid to this issue.[2]

This increased attention is connected in many ways with developments in East–Central Europe in the period after 1989. There was a great fear in Europe in the early 1990s that rapid political changes would produce conflicts of interest of various kinds that could not be managed by peaceful means. The rapid disintegration of the military structure of the WTO and then the disintegration of the armed forces of the Soviet Union itself added a new dimension to the question how to control arms. Suddenly the future disposition of enormous quantities of arms and military equipment became very difficult to predict.

This combination of escalating conflict and widespread availability of arms seemed to contain the ingredients for a serious breakdown of peace and order. In the event, the pattern of development after 1990 was not uniform across Europe. In places such as Azerbaijan, Georgia, Moldova, Russia (in the Republic of Chechnya) and in the former Yugoslavia the most pessimistic predictions were fulfilled. In many others—for example, the Baltic states, the former Czechoslovakia, Romania and Ukraine—the pattern of post-cold war development has been more or less peaceful.

The danger that illicit arms transfers could fuel armed conflicts was one concern in the early 1990s. In addition, it was feared that a combination of domestic developments—collapsing economies and the erosion of central authority with the elimination of the decisive role of communist parties—would create conditions in which criminality would thrive. Some external observers believe that this is in fact what has happened in Russia. In the United States, Director of

---

[1] Laurance, E. J., 'The new gunrunning', *Orbis*, spring 1989, pp. 25–37; and Karp, A., 'The trade in major conventional weapons', *SIPRI Yearbook 1988: World Armaments and Disarmament* (Oxford University Press: Oxford, 1988), pp. 188–94.

[2] In the growing literature on light weapons trade relatively little has been written about Russia. One exception is Gonchar, Ks. and Lock, P., 'Small arms and light weapons: Russia and the former Soviet Union', eds J. Boutwell, M. Klare and L. Reed, *Lethal Commerce: The Global Trade in Small Arms and Light Weapons* (Committee on International Security Studies, American Academy of Arts and Sciences: Cambridge, Mass., 1995).

Central Intelligence John Deutch told the Congressional Committee on International Relations that Russian organized crime groups 'exploit corruption, poor living conditions and chronic late wages in the Russian military to gain access to weapons and other stocks. Theft and illegal sales of these items have become routine. Military officers purchase weapons and smuggle contraband, including weapons and narcotics, via military transport, which cannot be searched by Russian law enforcement officials'.[3]

This chapter attempts to describe the extent and forms of illicit arms transfers both into and out of Russia in the context of these political and economic developments. As noted above, illicit arms transfers are defined as those that are unauthorized by the state. This chapter does not discuss transfers which seem to have taken place with the knowledge and consent of the state—for example, to sub-state groups in certain members of the CIS—which are discussed in chapter 9 of this book. The discussion is made more difficult by the fact that comprehensive data on illicit arms transfers are by definition unavailable. Some data on the volume of weapons intercepted by law enforcement agencies are available. However, what percentage of the total trade this represents can only be the subject of speculation.

The descriptive and anecdotal information which is available in public sources is also difficult to evaluate. The public sources may give an impression of events but it is likely that some of what is reported is wrong (even in some cases deliberate disinformation) and that most of what is reported represents only a part of the truth. While this is a general problem with public information on arms transfers, it is undoubtedly worse in the case of the illicit arms trade.

Interpreting the data available is made more complicated by the fact that the issue of the illicit arms trade has become an element of the fierce domestic political debate as the new state of Russia develops new government structures. There have been conflicts between the high command of the Russian armed forces, the Federal Security Service (the successor to the Committee of State Security, Komitet gosudarstvennoy bezopasnosti, KGB), the paramilitary forces of the Ministry of the Interior and the border security forces over how responsibility for certain security-related tasks should be divided. In this 'turf war' allegations of criminality or incompetence have been made by one service against another and the illicit arms trade has often featured in these allegations. Similarly, there has been competition between different agencies about who should have the responsibility of implementing Russia's arms transfer policy.[4] Here, too, public allegations that one or other state agency is either corrupt or incompetent in its management of the arms trade have played a role in an inter-agency competition for power.

---

[3] Prepared statement of John Deutch, Director of Central Intelligence, The Threat from Russian Organized Crime, Hearing before the US House of Representatives Committee on International Relations, 30 Apr. 1996 (US Government Printing Office: Washington, DC, 1996), p. 49.
[4] These are described in more detail in chapters 6 and 7 of this volume.

## Some observations on illicit arms transfers

At the outset it is useful to comment on the relationship between illicit arms transfers and the law. Arms transfers can be seen as a policy issue or as a legal issue. From a policy perspective, there is scope for disagreement about the wisdom of any individual arms transfer and some observers are likely to regard any given transfer as a bad policy choice. However, these arms sales are not illegal. It is also true that a sale may take place which is regarded by the government of the exporter or importer as undesirable but there may be no law that gives the authority to prevent this arms transfer. For this reason, in recent years the UN has encouraged all member states to enact laws defining all the conditions for legal arms import and export.[5] Finally, there may be transactions which take place without the knowledge or consent of governments which have national laws establishing the conditions for legitimate arms transfers. These transfers are illegal.

The notion of an illicit arms sale is therefore wider than the notion of an illegal arms sale because it covers those cases where there are no laws. There is widespread agreement that the state should exercise control over arms and military equipment because of the special dangers that these goods pose to people and property. For the purposes of this chapter, therefore, the definition of an illicit arms transfer is one which is not undertaken as a conscious act of policy by either the government of the country from which supplies originate or the government of the eventual end-user.

Legitimate arms transfers—that is, those conducted with the knowledge and approval of governments—are usually considered to be part of the military dimension of international relations. This being so, the actions and intentions of national governments are the central focus of the analysis.[6] This focus on governments remains valid even in countries where defence manufacturing is conducted by private companies rather than by the state. In countries with significant private arms industries there also tend to be strict regulations governing sales to foreign customers.

Where illicit transfers are concerned, however, the study of government decisions is not the only avenue of inquiry. By definition, the relationships on which the illicit arms trade depends include at least one non-government actor.[7] This is a key distinction. It is widely acknowledged that there are some circumstances in which the use of violence by the state can be legitimate and necessary—in the domestic context of maintaining order and enforcing the law or in

---

[5] For instance, this was recently restated in the Guidelines for International Arms Transfers in the context of General Assembly Resolution 46/36H of 6 Dec. 1991, adopted unanimously by the UN Disarmament Commission at its meeting of 22 Apr.–7 May 1996. UN document A/CN.10/1996/CRP.3, 3 May 1996.

[6] Krause, K., 'Military statecraft: power and influence in Soviet and American arms transfer relationships', *International Studies Quarterly*, vol. 35 (1991).

[7] Phil Williams and Stephen Black tried to develop a conceptual approach to what they call the 'grey area phenomenon' of efforts by states to regulate transnational interactions. Williams, P. and Black, S., 'Transnational threats: drug trafficking and weapons proliferation', *Contemporary Security Policy*, vol. 15, no. 1 (Apr. 1994), pp. 127–51.

the international context of defence against external aggression or participation in UN operations.

There is not the same degree of consensus behind the idea of legitimate use of force by non-state actors. Assaults on other persons or property for personal gain are clearly not acceptable forms of behaviour. However, there are various UN resolutions which acknowledge the rights of non-state groups to resist colonial occupation. Equally, it is acknowledged that within a political commonwealth the rights of the sovereign power are not unlimited. If power is exercised by the sovereign in a way that violates the conditions on which the commonwealth is based then the authority of the sovereign is likely to be challenged. In other words, there are ambiguities and matters of judgement which inevitably surround the notion of an illicit arms transfer.

A special category of international arms transfers are those that occur with the knowledge and consent of one government (which could be at either the supplier or the recipient end of the transaction) but without the knowledge and consent of the other. As described in chapter 3, there were many such arms transfers during the cold war when both the Soviet Union and the United States regularly used military assistance to sub-state groups as an instrument of policy. These cases may lead to criminal acts being committed in either the supplier country (violations of export laws) or the recipient country (violations of import laws) but not both.[8]

As far as the specific case of Russia is concerned, the production, possession, import and export of arms are all regulated. Article 218 of the Criminal Code of the Russian Federation establishes procedures for the legal acquisition, storage and sale of firearms in Russia. The procedures for export regulation are described in chapter 6 of this volume.

## II. The extent and forms of illicit arms transfers

During the Soviet period unauthorized possession of arms was not unknown in Russia. In 1988 the deputy minister of civil aviation said that each year routine security checks at airports produced 'hundreds' of cases of smuggling of firearms and other weapons (usually hand-grenades).[9] However, it seems clear that the number of cases and the volume of unauthorized weapons in circulation are higher than before.

The possession of firearms is regulated but not prohibited in Russia. According to Russia's national submission to the international study on firearms regulation organized by the UN Commission on Crime Prevention and Criminal Justice, there are estimated to be 3.2 million gun-owners in Russia. The great

---

[8] For example, in the 1980s much attention was paid to the arms procurement efforts of the Government of Iran.

[9] *The Independent*, 4 July 1988, p. 9. One source has estimated that in 1986 there were around 80 000 unlicensed firearms in the former Soviet Union, although the basis for the estimate is not provided. *Argumenty i Fakty*, no. 5 (Jan. 1996), p. 11 (in Russian) in Foreign Broadcast Information Service, *Daily Report–Central Eurasia* (hereafter FBIS-SOV), FBIS-SOV-96-025, 6 Feb. 1996, pp. 32–34.

majority of these guns are legally registered and are not classified as military-style firearms.[10] However, the number of seizures of such weapons by police has grown in recent years. In 1997 the police in Moscow were confiscating around 1000 guns per month which were held without the required documentation.[11] In the period 1994–96 the police of the Ministry of the Interior confiscated 1250 illegal firearms, 1 million rounds of ammunition, 200 grenades and 900 kg of explosives at Russian airports.[12]

The number of crimes involving firearms in Russia grew from 3600 in 1988 to 7200 in 1990 and 22 500 in 1993. In 1993, according to the Federal Security Service, there were over 3000 non-state paramilitary and armed criminal formations in Russia which held around 200 000 automatic weapons between them.[13] At the end of 1996 the sub-unit of the Federal Security Service dealing with illegal trafficking in drugs and weapons had more than 1000 independent inquiries under way, although not all were related to arms trafficking.[14] It also seems likely that the types of weapon available without government authorization are of greater capability than those in circulation during the Soviet period. In 1993, according to the Public Order Directorate of the Russian Ministry of the Interior, almost 2000 automatic rifles, 140 machine-guns, six anti-tank missile launchers and 33 grenade-launchers were confiscated from criminals.[15] In 1993, also according to the Ministry of the Interior, 300 000 hand-grenades were stolen from Russian arms depots.[16]

In terms of the international dimension of the illicit arms trade, the indicators available consist of customs or border security service interceptions of shipments either entering or leaving Russia. Press reports of seizures of weapon shipments entering and leaving Russia have been frequent in recent years. These reports underline the fact that this kind of trade occurs. While they cannot give any definitive measure of the scale or pattern of the trade, in December 1994 the Deputy Chief of the Russian Federal Border Troops stated that seizures at the border in 1994 were sufficient to arm two anti-tank regiments.[17] There are reports of arms transfers into and out of Russia across virtually all the new state borders. However, the border between Azerbaijan and the Russian republic of Dagestan and the borders between Russia and Estonia, Latvia and Lithuania are particularly prominent in many reports. There are also press reports that arms are smuggled by sea, for example, to the Kurdish Workers' Party (PKK) in Turkey.[18]

[10] UN, Commission on Crime Prevention and Criminal Justice Division, Draft United Nations International Study on Firearm Regulation, UN document E/CN.15/1997/CRP,6, 25 Apr. 1997, p. 37.

[11] Both figures as reported in the Internet source *Johnson's Russia List*, 13 Mar. 1997, distributed from fweir.ncade@rex.iasnet.ru.

[12] ITAR-TASS, 3 June 1997 (in Russian) in FBIS-SOV-97-154, 3 June 1997.

[13] *Nezavisimaya Gazeta*, 27 Oct. 1995 (in Russian) in FBIS-SOV-95-224, 21 Nov. 1995, pp. 2–3.

[14] *Rossiyskaya Gazeta*, Weekend Edition, 20 Dec. 1996 (in Russian) in FBIS-SOV-96-246, 20 Dec. 1996.

[15] *Segodnya*, 8 June 1994, p. 1 (in Russian) in FBIS-SOV-94-111, 9 June 1994, p. 26.

[16] *Süddeutsche Zeitung*, 9 June 1994, p. 8 (in German).

[17] *Jane's Intelligence Review Pointer*, Jan. 1995, p. 7.

[18] *Milliyet*, 7 Jan. 1997 (in Turkish) in Foreign Broadcast Information Service, *Daily Report–West Europe* (hereafter FBIS-WEU), FBIS-WEU-97-007, 7 Jan. 1997.

The problem of monitoring imports and exports in the post-Soviet geographical space in any systematic way is immense. Neither the legal basis for cooperation in managing trade nor the physical disposition of customs posts has been brought into line with Russia's new borders. Gary Bertsch and Igor Khripunov point out that in recent years the Russian Customs Service has grown from 7000 employees to 54 000 (around three times the size of the US Customs Service). However, Russia has 25 000 km of frontiers, many of which have only recently become international borders.[19] Russian customs officials also continue to operate in what are now independent states to assist in monitoring trade and enforcing different export laws. However, what these officials can and cannot do is regulated by the specific laws in the host country.

Given these realities, export control authorities in Russia are particularly dependent on the voluntary compliance of industry. However, there is evidence that customs officials are the targets of criminals who offer bribes in order to escape from the existing legal framework for exports.[20]

Sometimes arms shipments are stopped because of legal violations or irregularities related to the documentation required for trans-shipment rather than because they are illicit. For example, a shipment of Russian arms to Angola was detained in the UK because it lacked the documentation required by British port authorities. The transfer was authorized by both the Russian Government as a supplier and the Angolan Government as a recipient.[21] Similarly, a shipment of artillery ammunition from Russia to Lebanon was detained at the Russian–Ukrainian border because it lacked a transit document.[22]

Other information is probably too unreliable to be used as any kind of indicator. For example, there have been many stories published in Russia and elsewhere about illicit arms transfers for which there is no evidence at all or which cannot be confirmed. Many of these unsubstantiated stories have appeared in the German media and concerned arms transfers from Russian forces in Germany to the armed conflict in the former Yugoslavia.[23]

## III. Sources of supply and demand

The sources of supply of weapons in Russia can be divided into three different categories. First, weapons have been available from the inventory of the armed

---

[19] Bertsch, G. and Khripunov, I., *Restraining the Spread of the Soviet Arsenal: Export Controls as a Long-Term Nonproliferation Tool* (Center for International Trade and Security, University of Georgia: Athens, Ga., Mar. 1996), pp. 10–11.

[20] According to Bertsch and Khripunov, in 1994 the General Prosecutor's Office identified over 1700 cases of violations by customs officials, although it is not specified how many of these related to transfers of conventional arms. See note 19, p. 10.

[21] *Campaign Against the Arms Trade News*, Feb. 1994, p. 4.

[22] *Neue Zürcher Zeitung*, 29 Sep. 1995, p. 3 (in German). The Ukrainian authorities have occasionally pointed to the issue of customs control over trains passing through their territory. *Molod Ukrayiny*, 7 Sep. 1995 (in Ukrainian) in FBIS-SOV-95-175, 11 Sep. 1995, p. 58.

[23] In Sep. 1994 a senior representative of the German security services confirmed that there was no information to support these allegations. ITAR-TASS, 30 Sep. 1994 (in Russian) in FBIS-SOV-94-192, 4 Oct. 1994, p. 3. Sergey Stepashin, Director of the Russian Federal Counter-Intelligence Service, described the stories as 'a complete fabrication'.

forces of the former Soviet Union. Second, weapons have been available from arms manufacturers in Russia and other newly independent states. Third, weapons have been imported from other states. Similarly, the demand for illicit weapons comes from three different types of user. First, there are Russian users with political motivations—for example, the irregular forces fighting in Chechnya. Second, there are foreign customers engaged in armed conflicts. Third, there are criminal elements (either in Russia or abroad) who wish either to use the weapons themselves or to act as intermediaries, supplying either of the other types of user for profit.

## Sources of supply for illicit arms

### Inventory control in the armed forces of the former Soviet Union

The speed of the disintegration of the Soviet Union created an immense problem of inventory control for the armed forces. Suddenly it was necessary to accomplish several things for which little or no planning had been undertaken. The Soviet armed forces stationed throughout East–Central Europe in the framework of the military structures of the WTO were to be withdrawn. Before this could be accomplished, the integrated Soviet armed forces were to be divided between the newly independent states as part of their attempt to create independent, national armed forces. At the same time, the armed forces also faced an extremely uncertain economic and social outlook. From 1992 sharp reductions in military expenditure began to have a direct impact on the income of servicemen. Many different figures were published by official spokesmen indicating a significant reduction in the numbers of people to be employed in the future Russian armed forces. The armed forces therefore no longer offered either adequate income or security of employment.

In the period immediately after the end of the Soviet Union the regulations that governed the new Russian armed forces were also unclear, creating some ambiguity about what was permitted and what was not permitted. In December 1991 the heads of 11 departments of the CIS Joint Armed Forces Command established a business corporation called the Military Exchange Section which was intended to 'coordinate the cooperation of the armed services and the main and central departments of the army and navy with trading and commercial structures'. Shortly afterwards the Commander of the CIS Joint Armed Forces, Marshal Yevgeniy Shaposhnikov, commissioned a report on this activity which suggested that the authority for this commercial venture was invalid since it was based on a 1991 decision of the USSR Council of Ministers.[24] Four months later all commercial transactions were prohibited by a presidential decree.

Against this background, it is not surprising that there was a degree of equipment leakage out of the inventories of the various Soviet armed formations.[25]

---

[24] *Moscow News*, no. 15 (1992), p. 8; and no. 17 (1992), p. 10.
[25] *The Guardian*, 15 Oct. 1994, p. 3.

These activities were the subject of an inquiry by the Russian Military Prosecutor's Office, which compiled a list of violations of two presidential decrees. Decree no. 361 of 4 April 1992 'On the struggle against corruption in the system of government service' made it illegal for people serving in the armed forces to engage in commercial activities. Decree no. 1513 of 30 November 1992 'On the order of the sale and use of military property being released' established regulations for the disposal of surplus equipment. The finding of the report was that these decrees were being 'systematically violated'.[26]

These activities not only involved sales of arms and other military equipment but included commercial transactions involving items of all kinds as well as unauthorized use of buildings and land.

These reports notwithstanding, the extent to which deliveries of arms from inventories of the Russian military can be called illicit is not clear. In October 1992 the Ministry of Finance approved the creation of a commercial entity, Voyentech, within the Ministry of Defence whose purpose was to generate earnings from the disposal of equipment and property to meet the social needs of servicemen. Voyentech was managed by an official with the rank of Colonel-General and with the title Deputy Minister of Defence and had both rouble and foreign currency accounts registered with appropriate authorities.[27] It is therefore questionable whether this agency could accurately be described as illicit. However, questions have been raised about whether the armed forces kept within the regulations and policy guidelines laid down for their operation. A lack of transparency and of instruments for overseeing the actions of the military has compounded the problem of controlling the disposal of assets.

The full extent to which it has actually transferred weapons and military equipment is also unclear. One of the transactions most discussed was the sale of one T-80U tank and one 2S6 Tunguska air defence system to the United Kingdom in 1992. At the time when the deal was made neither the T-80U nor the 2S6 were cleared for export. However, the negotiation of these transfers was authorized by the Deputy Prime Minister, Georgiy Khizha, and conducted within the framework of a presidential decree. In another case Voyentech applied for permission to export 2000 assault rifles and 2 000 000 rounds of ammunition to Yugoslavia (Serbia and Montenegro). However, when permission was denied the deal was not fulfilled.[28]

In April 1997 a member of parliament, Lev Rokhlin, Chairman of the Duma Defence Committee, drew attention to what he claimed was a massive diversion of equipment from the inventory of the Russian armed forces. According to Rokhlin, between 1993 and 1996 the Group of Russian Forces in the Caucasus transferred to Armenia a large amount of major equipment as well as small arms, ammunition, stores and non-lethal equipment. This was alleged to include

[26] Lt-Gen. Grigoriy N. Nosov (Acting Chief Military Prosecutor), 'Representation on the elimination of violations of the legislation forbidding military servicement to engage in commercial activities', *Moscow News*, no. 34 (26 Aug.–1 Sep. 1994), pp. 1, 3.

[27] *Moscow News*, no. 12 (28 Mar.–3 Apr. 1996), p. 4.

[28] See note 27.

**Table 12.1.** Select equipment allegedly transferred to Armenia from the Group of Russian Forces in the Caucasus, 1993–96

| Designation | Description | Quantity |
| --- | --- | --- |
| SS-1 Scud Launcher | Surface-to-surface missile launcher | 8 |
| SS-1 Scud-B | Surface-to-surface missile | 32 |
| SA-4 SAM system | Anti-aircraft vehicle (missile) | 27 |
| SA-4 Ganef | Surface-to-air missile | 349 |
| SA-8 Gecko | Surface-to-air missile | 40 |
| SA-18 Gripstock | Man-portable surface-to-air missile launcher | 26 |
| SA-18 | Man-portable surface-to-air missile | 200 |
| T-72 | Main battle tank | 84 |
| BMP-1 | Armoured personnel carrier | 4 |
| AT-4 Spigot | Anti-tank missile | 945 |
| BMP-2 | Armoured infantry fighting vehicle | 50 |
| D-30 122-mm | Towed gun | 36 |
| D-20 152-mm | Towed gun | 18 |
| D-1 152-mm | Towed howitzer | 18 |
| BM-21 | Multiple-launch rocket system | 18 |

*Source: Sovetskaya Rossiya*, 3 Apr. 1997 (in Russian) in FBIS-SOV-97-967, 3 Apr. 1997.

26 mortars, 306 sub-machine-guns, 7910 assault rifles and 1847 pistols.[29] A list of the major equipment is shown in table 12.1. A trilateral commission was subsequently established including Armenian, Azerbaijani and Russian participants in order to investigate the truth and implications of this allegation.[30]

As noted in the introduction to this chapter, the allegations of illegal or illicit arms transfers have been an element in the internal political struggle within the Russian Government and power structure, although this does not necessarily mean that the allegations are without foundation.[31]

Most public attention outside Russia has been paid to the activities of the Western Group of Forces during the withdrawal from Germany. These activities were the subject of parliamentary hearings during which the former Commander of the Western Group of Forces, Matvey Burlakov, acknowledged that some equipment had been disposed of both by legitimate transfers and through illicit sales. Former Defence Minister Pavel Grachev also acknowledged 'corruption, theft, smuggling and illegal deals' involving the Western Group of Forces.[32]

[29] *Sovetskaya Rossiya*, 3 Apr. 1997 (in Russian) in FBIS-SOV-97-967, 3 Apr. 1997.
[30] As of Aug. 1997 this commission had not yet begun its work. Turan, Baku, 16 Aug. 1997 (in Russian) in FBIS-SOV-97-228, 16 Aug. 1997; and Yerevan Snark, 19 Aug. 1997 (in Russian) in FBIS-SOV-97-232, 20 Aug. 1997.
[31] Many of the allegations have been directed at the Ministry of Defence and originated in the Presidential Security Service at the time when it was led by Gen. Alexander Korzhakov. In the most recent example, Gen. Korzhakov prepared a report alleging Ministry of Defence complicity in illegal arms sales to Croatia. *Le Figaro*, 1 Apr. 1995, p. 3 (in French).
[32] *Komsomolskaya Pravda*, 15 Sep. 1994 (in Russian) in FBIS-SOV-94-179, 15 Sep. 1994, p. 22; and Interfax, 28 Nov. 1994 (in English) in FBIS-SOV-94-229, 29 Nov. 1994, p. 29. Burlakov was dismissed

Although these have been the most widely reported activities, other armed formations have been accused of making illicit arms transfers. Russian forces in Kaliningrad and the Baltic states have been mentioned as sources of illicit arms transfers.[33] Russian troops in Georgia have often been accused of supplying arms to various armed formations in Abkhazia and Ossetia.[34] In 1996 it was also alleged that the Russian border security forces were engaged in illegal arms sales to Abkhazian forces.[35] Russian troops in the Kurgan Tyube Oblast have been accused of supplying arms to groups in Tajikistan.[36] The 14th Army, based in Moldova, has regularly been accused of transactions both with local political groups and with criminals based in Moscow and Kiev.[37]

### Sales by arms manufacturers in Russia

Allegations of illicit arms transfers from industry can be divided into two types: (a) that employees in a factory establish illegal commercial operations without the knowledge of the senior management—small groups of employees may either divert production or undertake unauthorized production to meet an order using machinery and materials available at the factory; and (b) that the enterprise managers deliberately evade the regulations on arms transfers either alone or with the cooperation of individuals in the government authorities themselves.

From the late 1980s arms manufacturers began to establish commercial trading offices under the authorization of the then Ministry of Defence Industry (a sectoral ministry subsequently incorporated into the Russian State Committee on Defence Industries, Goskomoboronprom). These entities, which initially operated legitimately, subsequently lost their rights to export arms under the revised export regulations introduced in Russia.[38] During the period of legitimate trading, commercial ties were developed with trading companies operating overseas—for example, in Cyprus. Many of the allegations of illicit arms trading suggest that, although export licences give Cyprus as the end-user of the weapons concerned, the size of the shipments makes it more likely that these will actually be re-transferred to a different destination.[39] For example, during the war in Chechnya the link between the Izhevsk Mechanical Plant in Russia

by President Yeltsin after allegations that he was involved in or, at a minimum, did too little to prevent the illicit sales. *International Herald Tribune*, 15 Nov. 1994, p. 1.

[33] *The Guardian*, 21 May 1992, p. 4.

[34] Interfax, 21 Sep. 1992 (in English) in FBIS-SOV-92-184, 22 Sep. 1992, p. 62; Interfax, 2 Oct. 1992 (in English) in FBIS-SOV-193, 5 Oct. 1992, pp. 64–65; and *New Europe*, 8–14 Oct. 1995, p. 37. In Georgia in 1992 there were also cases of local groups stealing equipment from Russian forces or coercing Russian forces to turn over arms and equipment.

[35] *Tbilisi Rezonansi*, 12–13 Mar. 1996 (in Georgian) in FBIS-SOV-96-061, 28 Mar. 1996, p. 59; and Interfax, 26 Mar. 1996 (in English) in FBIS-SOV-96-060, 27 Mar. 1996, pp. 68–69.

[36] Moscow Mayak Radio, 14 Sep. 1992 (in Russian) in FBIS-SOV-92-179, 15 Sep. 1992, p. 38.

[37] Basapress, Chisinau, 25 July 1994 in FBIS-SOV-94-144, 27 July 1994, p. 49; Basapress, Chisinau, 8 Aug. 1994 in FBIS-SOV-94-154, 10 Aug. 1994, p. 43; *Krymskaya Pravda*, 22 Feb. 1994, p. 2 (in Russian) in FBIS-SOV-95-052, 17 Mar. 1994, pp. 35–36; *Moscow News*, no. 27 (14–20 July 1995), p. 1; and *Baltic Independent*, 1–7 Sep. 1995, p. 6.

[38] This process of changing regulation is described in chapter 6 in this volume.

[39] *Literaturnaya Gazeta*, no. 16 (19 Apr. 1995), p. 13 (in Russian) in FBIS-SOV-95-084, 2 May 1995, pp. 4–6; and *Pravda*, 12–19 Jan. 1996, p. 5 (in Russian) in FBIS-SOV-96-022, 1 Feb. 1996, pp. 54–57.

and the Lora Trading Company in Nicosia was investigated by the Russian Ministry of the Interior and the Federal Security Service.[40]

Most of the reports of illicit arms transfers from arms factories seem to be connected to arms factories in Tula and Izhevsk. These are the locations of the largest factories manufacturing small arms and light weapons, which make up the largest part of the illicit arms trade, so that this is not surprising.[41]

### Illicit arms imports

The combination of the growing demand for weapons among non-state groups of various kinds in Russia and the porous borders of the new state has led to a significant volume of illicit arms imports.

As described above, there have been major challenges to both the security and the integrity of weapon stockpiles owned by the Russian armed forces and also an economic crisis in the manufacturing sector in Russia. However, in some of the countries which are Russia's new neighbours the problems are even greater. For example, in a recent poll of officers in the Ukrainian armed forces 70 per cent of respondents identified uncontrolled sales of military equipment as a serious problem.[42]

The Russian armed forces stationed in the Baltic states of Estonia, Latvia and Lithuania were often reported to be an important source of arms for criminal groups. Weapons were reported in some cases to have been stolen from bases and in some cases to have been sold by the troops.[43] There are also occasional reports of small shipments of arms from other neighbouring states, such as Azerbaijan and Georgia.[44]

---

[40] *Rossiyskaya Gazeta*, 19 Aug. 1997 (in Russian) in FBIS-SOV-97-233. 21 Aug. 1997.

[41] Tula is the main location of production facilities for portable anti-tank missiles and rockets; Izhevsk is the main location of production facilities for several lightweight automatic weapons including the Kalashnikov family of weapons. *Izvestiya*, 4 May 1994 (in Russian) in FBIS-SOV-94-087, 5 May 1994, p. 32; *Segodnya*, 8 June 1994 (in Russian) in FBIS-SOV-94-111, 9 June 1994, p. 26; *Rossiyskaya Gazeta*, 24 Nov. 1994 (in Russian) in FBIS-SOV-94-229, 29 Nov. 1994, pp. 29–30; *Rossiyskaya Gazeta*, 28 Dec. 1994 (in Russian) in FBIS-SOV-94-250, 29 Dec. 1994, p. 13; *Nezavisimaya Gazeta*, 27 Oct. 1995, pp. 1–2 (in Russian) in FBIS-SOV-95-224, 21 Nov. 1995, pp. 2–3; Tbilisi Radio, 7 Aug. 1995 (in Georgian) in FBIS-SOV-95-152, 8 Aug. 1995, p. 79; Tallin Baltic News Service, 22 May 1996 (in English) in FBIS-SOV-96-101, 23 May 1996, p. 57; and *Baltic Times*, 30 May–5 June 1996, p. 3.

[42] Open Media Research Institute, *OMRI Daily Digest*, no. 123, part II (25 June 1996). In 1992 there were already reports of Chechen arms dealers visiting Ukraine. *Molod Ukrayiny*, 3 Sep. 1992 (in Ukrainian) in FBIS-SOV-92-180, p. 39.

[43] On Estonia, see Interfax, 6 May 1994 (in English) in FBIS-SOV-090, 10 May 1994, p. 8; *Rossiyskaya Gazeta*, 26 Aug. 1994 (in Russian) in FBIS-SOV-94-169, 31 Aug. 1994, pp. 17–18; *Baltic Times*, 30 May–5 June 1996, p. 2; Interfax, 23 July 1996 (in English) in FBIS-SOV-96-143, 24 July 1996, p. 39; and *Baltic Times*, 1–7 Aug. 1996, p. 4. On Latvia, see *Defense News*, 15–21 June 1992, p. 38; *Diena*, 2 Mar. 1995 (in Latvian) in FBIS-SOV-95-048, 13 Mar. 1995, p. 86; and *Baltic Independent*, 7–13 July 1995, p. 4. On Lithuania, see Vilnius Radio, 29 Sep. 1994 (in Lithuanian) in FBIS-SOV-94-190, 30 Sep. 1994, p. 87; *Baltic Independent*, 7–13 Oct. 1994, p. 5; and *Baltic Independent*, 15–21 Oct. 1994, p. 3.

[44] Interfax, 10 May 1994 (in English) in FBIS-SOV-94-090, 10 May 1994, p. 9; and *New Europe*, 8–14 Sep. 1996, p. 47.

## IV. The case of arms supplies to Chechnya

In 1995 and 1996 a great deal of attention in Russia was paid to the questions how the irregular forces fighting in Chechnya were able to arm themselves and how they managed to re-supply themselves during the war. The information available in published sources suggests that the Chechen forces used all the sources of supply outlined above.

There have been several commissions of inquiry into the origins of the war in Chechnya, including enquiries by the Ministry of Defence and by independent investigators. While there is still some conflicting information about events, the reports of these commissions give some indication of how the armed formations operating in Chechnya were created and supplied. The main focus has been on the period of chaos in late 1991 and early 1992 as the Soviet Union was suddenly dissolved.

The Govurukhin Commission reported in February 1996 that armed formations were already being established in Chechnya in 1991. In August 1991 the National Guard of the Executive Committee of the All-National Congress of the Chechen People was formed. The Congress declared on 26 November 1991 that all military equipment stationed in Chechnya belonged to the Chechen Republic and could not be removed.[45]

In December 1991 the Soviet Union decided to close several bases in Chechnya and withdraw its forces, which had effectively become hostages in an increasingly hostile local environment.[46] In early 1992, as this was being undertaken, Chechen forces seem to have acquired large amounts of equipment of all kinds from departing Soviet forces. In some cases this acquisition was accomplished through theft but a large amount of equipment appears to have been turned over by Soviet forces on the instruction of the government.[47] In

---

[45] The commission was established by the parliament to investigate the origins of the war in Chechnya and led by Stanislav Govurukhin. Its report was published in 6 sections in consecutive issues of *Pravda*; the sections on the arming of Chechen formations are reproduced in FBIS-SOV-96-062, 29 Mar. 1996, pp. 31–38.

[46] Soviet formations in Chechnya included a training division for tank forces, an anti-aircraft defence communications and processing unit, the Ministry of the Interior 566th escort regiment, a military hospital and several smaller units. *Argumenty i Fakty*, Feb. 1996 (in Russian) in FBIS-SOV-96-029, 12 Feb. 1996, p. 7.

[47] Pavel Grachev was First Deputy Minister of Defence in the Soviet Union in Dec. 1991 and Marshal Yevgeniy Shaposhnikov was Minister of Defence. Grachev became Minister of Defence in May 1992. In Dec. 1994 and Jan. 1995 Grachev and Shaposhnikov made a series of allegations and counter-allegations accusing each other of responsibility for the loss of equipment to Chechen forces.

The formal orders to transfer equipment to Chechen forces were issued over Grachev's signature on 20 May 1992. However, a report prepared by the Ministry of Defence underlined that this reflected the circumstances inherited by Grachev from his predecessor. *Krasnaya Zvezda*, 18 Jan. 1995 (in Russian) in FBIS-SOV-95-011, 18 Jan. 1995, pp. 31–32; *Moscow News*, 20–26 Jan. 1995, p. 2; *Literaturnaya Gazeta*, nos 1–2 (11 Jan. 1995) (in Russian) in FBIS-SOV-95-015, 24 Jan. 1995, pp. 38–40; and *Argumenty i Fakty*, no. 32 (Aug. 1995) (in Russian) in FBIS-SOV-95-155, 11 Aug. 1995, pp. 36–37. This was subsequently used against Grachev by political opponents. *Komsomolskaya Pravda*, 4 Mar. 1995 (in Russian) in FBIS-SOV-95-043, 6 Mar. 1995, p. 33.

It should be noted that there are alternative versions of events. P. Shirshov, Chairman of the Committee on Security and Defence of the Federation Council, stated in an interview: 'When our troops were leaving

**Table 12.2.** Basic weapons and military equipment seized on the territory of the Chechen Republic

| Designation | No. of items stationed in Chechnya | No. of items acquired by Chechen forces |
|---|---|---|
| Tactical rocket systems | 4 | 2 |
| L-39 and L-29 jet trainer aircraft | 260 | 260 |
| Tanks | 42 | 42 |
| Infantry fighting vehicles | 34 | 34 |
| Armoured personnel carriers | 14 | 14 |
| Combat tractors | 44 | 44 |
| Other vehicles | 1 063 | 942 |
| Artillery systems | 199 | 139 |
| Anti-tank systems | 101 | 89 |
| Air-defence missile systems | 9 | 5 |
| Air-defence radars | 23 | 23 |
| Anti-aircraft guns | 9 | 9 |
| Anti-aircraft systems(gun/missile) | 18 | 16 |
| Man-portable air defence rocket-launchers | 88 | 88 |
| Firearms | | |
| automatic weapons | 35 748 | 24 737 |
| machine-guns | 1 682 | 1 682 |
| pistols | 18 715 | 10 119 |
| carbines | 946 | 895 |
| rifles | 506 | 362 |

*Sources:* Interfax, 25 Feb. 1995 (in English) in FBIS-SOV-95-038, 27 Feb. 1995, pp. 14–15; and *Pravda*, 2 Mar. 1996 (in Russian) in FBIS-SOV-96-062, 29 Mar. 1996, p. 35.

1995 documents were published in *Moscow News* indicating that the Ministry of Defence had authorized the transfers of equipment.[48]

Table 12.2 gives the estimates made by the Govurukhin Commission of the amount of equipment acquired by the Chechen forces from Soviet and Russian forces during the period December 1991–August 1992.

It has since also been alleged that Russian forces fighting in Chechnya either sold or surrendered their arms and equipment to irregular forces, although the scale of this activity seems to have been limited.[49]

During the peace negotiations in 1995 one element discussed intensively was the scale of equipment holdings among various Chechen armed formations. According to the Russian side, Chechen forces held 45 000 guns of all kinds.[50]

If this figure and those given by the Govurukhin Commission are correct, this suggests that about 38 000 (85 per cent) of the guns in the possession of

Chechnya, President Gorbachev issued an order to leave all equipment there'. Interview in *Sovetskaya Rossiya*, 24 Dec. 1994, p. 1 (in Russian) in FBIS-SOV-95-004, 6 Jan. 1995, p. 38.

[48] *Moscow News*, 20–26 Jan. 1995, p. 2.

[49] Ostankino Television, 1 Mar 1995 (in Russian) in FBIS-SOV-95-040, 1 Mar. 1995, pp. 31–32; and *Moscow News*, no. 31 (11–17 Aug. 1995), p. 3.

[50] Interfax, 19 Sep. 1995 (in English) in FBIS-SOV-95-182, 20 Sep. 1995, pp. 45–46.

Chechen forces came from the inventory of the former Soviet Union. However, as the commission makes clear, allegations of arms sales to Chechen forces by suppliers in many other countries have also been made since at least November 1991. In late 1996 analysts at the Russian Federal Security Service released a report entitled 'An analysis of supplies received by illegal Chechen formations' which apparently identified at least five countries as offering military assistance to Chechen fighters—Afghanistan, Jordan, Lebanon, Saudi Arabia and Turkey.[51] According to the report the main sources of funding for this assistance were donations from the Chechen diaspora in the Middle East (notably Iraq, Jordan and Saudi Arabia).

According to the Russian State Duma, other Russian republics have also become a source of arms supplies to Chechnya. One serious side-effect of the war in Chechnya has allegedly been the growth of arms traffic into and out of Dagestan, which is adjacent to the Chechen Republic. According to the representatives, the mass buying and selling of arms and ammunition in Dagestan has fed into a process of state-level organizations beginning to create their own paramilitary formations in order to try to cope with the problem of deteriorating law and order, a development which could in the longer term lead to a repeat of the 'Chechen scenario' in Dagestan.[52] To try to avoid this, Russia has deployed a total of 16 000 Interior Ministry troops around the external borders of Chechnya. In addition, a mixed security force including Interior Ministry troops and paramilitary policemen is manning checkpoints along the administrative border that separates Chechnya from the rest of Russia.[53]

Arms supplies to Chechen forces are also said to have originated in several CIS member states, particularly those closest to Chechnya. Azerbaijan has been named most often, although its government has denied any knowledge of or complicity in the traffic.[54] It is also alleged by some Russians that the heavier weapons of the Chechen forces (notably aircraft) which would be vulnerable to loss or capture in regular military operations have been stored at Azerbaijani bases.[55] Similarly, it has been alleged that arms have been supplied from the Abkhazia region of Georgia.[56] Among non-CIS member states Iran and Turkey

[51] Interfax, 8 Dec. 1996 (in English) in FBIS-SOV-96-237, 8 Dec. 1996.
[52] 'Appeal to the Russian Federation President and the Russian Federation Government on the adoption of measures to stabilize the situation in the North Caucasus', State Duma Decree no. 705, 18 Oct. 1996 (in Russian) in FBIS-SOV-96-212, 30 Oct. 1996.
[53] Interfax, 31 July 1997 (in English) in FBIS-SOV-97-212, 31 July 1997.
[54] *Respublika Armeniya*, 23 Feb. 1995 (in Russian) in FBIS-SOV-95-039, 28 Feb. 1995, p. 64; *Balkan News and East European Report*, 11–17 June 1995, p. 9; Turan, Baku, 5 Aug. 1995 (in Azeri) in FBIS-SOV-95-151, 7 Aug. 1995, p. 67; Interfax, 15 Aug. 1995 (in English) in FBIS-SOV-95-158, 16 Aug. 1995, p. 73; and Baku Radio Network, 25 Jan. 1996 (in Azeri) in FBIS-SOV-96-018, 26 Jan. 1996, p. 70.
[55] *Segodnya*, 5 Apr. 1996 (in Russian) in FBIS-SOV-96-068, 8 Apr. 1996, p. 6.
[56] *New Europe*, 5–11 Nov. 1995, p. 38; Tbilisi Iberia, 27 Feb. 1996 (in Georgian) in FBIS-SOV-96-039, 27 Feb. 1996, p. 60; and Tbilisi BGI, 28 Mar. 1996 (in Russian) in FBIS-SOV-96-062, 29 Mar. 1996, p. 70.

have most often been named, even by President Yeltsin, as sources of arms. Both their governments have denied all knowledge of or complicity in this.[57]

Arms are said to have come from all the Baltic states but most allegations are directed at suppliers in Estonia. Again, these allegations are denied by both the government and the dealers said to have organized the shipments.[58]

## V. Conclusions

The information available about illicit arms transfers into and out of Russia is sufficient to allow the conclusions both that such transfers take place and that this is a genuine security problem. Although the information is not adequate to support any systematic measurement of the volume or the direction of these flows, the two main centres of demand are ongoing armed conflicts and criminal gangs.

This survey of the open literature suggests that the widespread availability of arms among sub-state groups (both those with criminal and those with political objectives) has contributed to the heightened insecurity of Russian citizens. The personal security of Russian citizens has been reduced by the rise in armed criminal activity, while conflicts on Russian territory—most notably in Chechnya—have claimed many Russian lives.

Russia therefore has a strong self-interest in the success of measures (some of which are described in chapter 5 of this book) to control the illicit distribution of arms.

In terms of illicit arms supplies to other countries from Russia, it is necessary to be cautious in drawing definite conclusions about the role of the Russian authorities. For example, the extent to which the Russian Government has used arms transfers to achieve particular outcomes in the Caucasus region is difficult to quantify from the available sources.

There are also reasons to believe that the primary stocks of arms that have been traded illicitly were accumulated in 1991 and 1992. The conditions surrounding the rapid redeployments and withdrawals of Soviet and then Russian forces from Europe and elsewhere combined with the weakness of regulation and administrative arrangements after the withdrawal of the CPSU from politics to create the conditions that made illicit transfers possible. There is no strong evidence of a massive loss of control over the inventory of conventional arms owned by the Soviet armed forces. However, given that that inventory was enormous, a relatively small leakage in percentage terms probably involved significant quantities of equipment.

Although it is outside the terms of reference of this chapter, there is rather stronger evidence that control over the inventories of conventional weapons

---

[57] *Yerkir*, 22 Dec. 1994 (in Armenian) in FBIS-SOV-95-023, 3 Feb. 1995, p. 71; ITAR-TASS, 3 Oct. 1995 (in English); and Islamic Republic News Agency (Tehran), 24 Oct. 1995 (in English) in FBIS-SOV-95-206, 25 Oct. 1995, p. 24.

[58] Tallinn Baltic News Service, 17 Jan. 1995 (in English) in FBIS-SOV-95-010, 17 Jan. 1995, p. 83; and *Izvestiya*, 1 Feb. 1995 (in Russian) in FBIS-SOV-95-028, 10 Feb. 1995, p. 75.

inherited by post-Soviet states other than Russia was weak. However, it is impossible to trace the movements of weapons of Soviet origin with any precision. In conditions where Russia's external borders are still 'porous' (that is, there are no physical checks at all points of entry and exit) and cooperation within the framework of the CIS remains underdeveloped, it is likely that weapons can still move relatively freely within the post-Soviet space. It is therefore also likely that the pattern of illicit transfers will closely follow the demand generated by the various ongoing or latent post-Soviet conflicts. By extension, the primary mechanism for limiting the flow of arms will be successful conflict resolution.

# Appendix 1. The Guidelines for Conventional Arms Transfers, 1991

CLOSING COMMUNIQUÉ OF
THE MEETING OF THE FIVE
[PERMANENT MEMBERS OF
THE UNITED NATIONS SECURITY
COUNCIL] ON ARMS TRANSFERS
AND NON-PROLIFERATION,
LONDON, 17–18 OCTOBER 1991

## Guidelines for Conventional Arms Transfers, 18 October 1991

The People's Republic of China, the French Republic, the Union of Soviet Socialist Republics, the United Kingdom of Great Britain and Northern Ireland, and the United States of America,

– recalling and reaffirming the principles which they stated as a result of their meeting in Paris of 8 and 9 July 1991,
– mindful of the dangers to peace and stability posed by the transfer of conventional weapons beyond levels needed for defensive purposes,
– reaffirming the inherent right to individual or collective self-defense recognized in Article 51 of the Charter of the United Nations, which implies that states have the right to acquire means of legitimate self-defense,
– recalling that in accordance with the Charter of the United Nations, UN Member States have undertaken to promote the establishment and maintenance of international peace and security with the least diversion for armaments of the world's human and economic resources,
– seeking to ensure that arms transferred are not used in violation of the purposes and principles of the Charter,
– mindful of their special responsibility for the maintenance of international peace and security,
– reaffirming their commitment to seek effective measures to promote peace, security, stability and arms control on a global and regional basis in a fair, reasonable, comprehensive and balanced manner,
– noting the importance of international commerce for peaceful purposes,
– determined to adopt a serious, responsible and prudent attitude of restraint regarding arms transfers,
   declare that,

when considering under their national control procedures conventional arms transfers, they intend to observe rules of restraint, and to act in accordance with the following guidelines:

1. They will consider carefully whether proposed transfers will:
(*a*) promote the capabilities of the recipient to meet needs for legitimate self-defense;
(*b*) serve as an appropriate and proportionate response to the security and military threats confronting the recipient country;
(*c*) enhance the capability of the recipient to participate in regional or other collective arrangements or other measures consistent with the Charter of the United Nations or requested by the United Nations.
2. They will avoid transfers which would be likely to:
(*a*) prolong or aggravate an existing armed conflict;
(*b*) increase tension in a region or contribute to regional instability;
(*c*) introduce destabilizing military capabilities in a region;
(*d*) contravene embargoes or other relevant internationally agreed restraints to which they are parties;
(*e*) be used other than for the legitimate defense and security needs of the recipient state;
(*f*) support or encourage international terrorism;
(*g*) be used to interfere with the internal affairs of sovereign states;
(*h*) seriously undermine the recipient state's economy.

---

*Source: Disarmament*, vol. xv, no. 1 (1992), pp. 162–63.

# Appendix 2. The Organization for Security and Co-operation in Europe Criteria on Conventional Arms Transfers

DECISION BY THE OSCE FORUM
FOR SECURITY CO-OPERATION,
NOVEMBER 1993

## I.

(1) The participating States reaffirm their commitment to act, in the security field, in accordance with the Charter of the United Nations and the Helsinki Final Act, the Charter of Paris and other relevant CSCE documents.

(2) They recall that in Prague on 30 January 1992 they agreed that effective national control of weapons and equipment transfer is acquiring the greatest importance and decided to include the question of the establishment of a responsible approach to arms transfers as a matter of priority in the work programme of the post- Helsinki arms control process. They also recall their declaration in the Helsinki Document of 10 July 1992 that they would intensify their co-operation in the field of effective export controls applicable, inter alia, to conventional weapons.

(3) The participating States reaffirm:

(a) their undertaking, in accordance with the Charter of the United Nations, to promote the establishment of international peace and security with the least diversion for armaments of human and economic resources and their view that the reduction of world military expenditures could have a significant positive impact for the social and economic development of all peoples;

(b) the need to ensure that arms transferred are not used in violation of the purposes and principles of the Charter of the United Nations;

(c) their adherence to the principles of transparency and restraint in the transfer of conventional weapons and related technology, and their willingness to promote them in the security dialogue of the Forum for Security Co-operation;

(d) their strong belief that excessive and destabilizing arms build-ups pose a threat to national, regional and international peace and security;

(e) the need for effective national mechanisms for controlling the transfer of conventional arms and related technology and for transfers to take place within those mechanisms;

(f) their support for and commitment to provide data and information as required by the United Nations resolution establishing the Register of Conventional Arms in order to ensure its effective implementation.

## II.

(4) In order to further their aim of a new co-operative and common approach to security, each participating State will promote and, by means of an effective national control mechanism, exercise due restraint in the transfer of conventional arms and related technology. To give this effect:

(a) each participating State will, in considering proposed transfers, take into account:

(i) the respect for human rights and fundamental freedoms in the recipient country;

(ii) the internal and regional situation in and around the recipient country, in the light of existing tensions or armed conflicts;

(iii) the record of compliance of the recipient country with regard to international commitments, in particular on the non-use of force, and in the field of non-proliferation, or in other areas of arms control and disarmament;

(iv) the nature and cost of the arms to be transferred in relation to the circumstances of the recipient country, including its legitimate security and defence needs and the objective of the least diversion for armaments of human and economic resources;

(v) the requirements of the recipient country to enable it to exercise its right to individual or collective self-defence in accordance with Article 51 of the Charter of the United Nations;

(vi) whether the transfers would contribute to an appropriate and proportionate response by the recipient country to the military and security threats confronting it;

(vii) the legitimate domestic security needs of the recipient country;

(viii) the requirements of the recipient country to enable it to participate in peace-keeping or other measures in accordance with decisions of the United Nations or the Conference on Security and Co-operation in Europe.

(*b*) Each participating State will avoid transfers which would be likely to:

(i) be used for the violation or suppression of human rights and fundamental freedoms;

(ii) threaten the national security of other States and of territories whose external relations are the internationally acknowledged responsibility of another State;

(iii) contravene its international commitments, in particular in relation to sanctions adopted by the Security Council of the United Nations, or to decisions taken by the CSCE Council, or agreements on non- proliferation, or other arms control and disarmament agreements;

(iv) prolong or aggravate an existing armed conflict, taking into account the legitimate requirement for self-defence;

(v) endanger peace, introduce destabilising military capabilities into a region, or otherwise contribute to regional instability;

(vi) be diverted within the recipient country or reexported for purposes contrary to the aims of this document;

(vii) be used for the purpose of repression;

(viii) support or encourage terrorism;

(ix) be used other than for the legitimate defence and security needs of the recipient country.

### III.

(5) Further, each participating State will:

(*a*) reflect, as necessary, the principles in Section II in its national policy documents governing the transfer of conventional arms and related technology;

(*b*) consider mutual assistance in the establishment of effective national mechanisms for controlling the transfer of conventional arms and related technology;

(*c*) exchange information, in the context of security co-operation within the Forum for Security Co-operation, about national legislation and practices in the field of transfers of conventional arms and related technology and on mechanisms to control these transfers.

*Source: FSC Journal,* no. 49 (24 Nov. 1993).

# Appendix 3. Russia's conventional arms export regulations

*Translations of documents 1–23 prepared by Gennadiy Gornostaev. Documents 24–29 translated by SIPRI*

1. Decree of the President of the Russian Federation on military–technical cooperation of the Russian Federation with foreign countries (basic provisions), no. 1008, 5 Oct. 1995
2. Regulations on military–technical cooperation of the Russian Federation with foreign countries, approved by decree no. 1008, 5 Oct. 1995
3. Decree of the President of the Russian Federation on the Interdepartmental Coordinating Council for Military–Technical Policy of the Russian Federation, no. 590, 14 June 1995
4. Regulations on the Interdepartmental Coordinating Council for Military–Technical Policy of the Russian Federation, approved by decree no. 590, 14 June 1995
5. Decree of the President of the Russian Federation approving the Regulations on the status of the Representative of the President of the Russian Federation in the State Company Rosvooruzhenie, no. 450, 4 Mar. 1994
6. Regulations on the Representative of the President of the Russian Federation in the State Company Rosvooruzhenie, approved by decree no. 450, 4 Mar. 1994
7. Regulations on the procedure for imposing embargo on deliveries of armaments and military equipment, the provision of services of a military–technical nature, and on deliveries of raw and other materials and equipment and the transfer of military and dual-purpose technologies to foreign states, including the CIS member states, approved by decree no. 235, 18 Feb. 1993
8. Law of the Russian Federation on the Conversion of the Defence Industry, law no. 2551-1, 20 Mar. 1992, section 4, articles 9, 10
9. Law of the Russian Federation on State Regulation of Foreign Trade Activity, passed by the State Duma on 7 July 1995, articles 6, 12
10. Decision of the Government of the Russian Federation on granting enterprises of the Russian Federation the right to participate in military–technical cooperation with foreign countries, no. 479, 6 May 1994
11. Regulations on the certification and registering of enterprises for the right to export armaments, military equipment and military-purpose work and services, approved by decision no. 479, 6 May 1994
12. Decision of the Government of the Russian Federation on measures for improving the system of control over the export and import of military-purpose products, work and services in the Russian Federation, no. 879, 4 Sep. 1995

13. Regulations on the procedure for licensing export and import of military-purpose products, work and services in the Russian Federation, approved by decision no. 879, 4 Sep. 1995

14. Inventory of military-purpose products, work and services the export and import of which are subject to control and performed under licences issued by the Ministry of Foreign Economic Relations of the Russian Federation, approved by decision no. 879, 4 Sep. 1995

15. Instruction of the Government of the Russian Federation, no. 1683-i, 24 Oct. 1994

16. Instruction of the Government of the Russian Federation, no. 202-i, 19 Feb. 1996

17. Instruction of the Government of the Russian Federation, no. 203-i, 19 Feb. 1996

18. Instruction of the Government of the Russian Federation, no. 204-i, 19 Feb. 1996

19. Instruction of the Government of the Russian Federation, no. 205-i, 19 Feb. 1996

20. Instruction of the Government of the Russian Federation, no. 206-i, 19 Feb. 1996

21. Instruction of the Government of the Russian Federation, no. 207-i, 19 Feb. 1996

22. Instruction of the Government of the Russian Federation, no. 208-i, 19 Feb. 1996

23. Decision of the Government of the Russian Federation affirming the Statute on the procedure for the making available of information by the Russian Federation on deliveries of conventional arms in accordance with the Wassenaar Arrangement, no. 923, 3 Aug. 1996

24. Decree of the Government of the Russian Federation approving the Statute on the Ministry of Foreign Economic Relations and Trade of the Russian Federation, no. 402, 7 Apr. 1997

25. Decree of the President of the Russian Federation on measures to improve the system of management of military–technical cooperation with foreign states, no. 792, 28 July 1997

26. Decree of the President of the Russian Federation on measures to strengthen state control of foreign trade activity in the field of military–technical cooperation of the Russian Federation with foreign states, no. 907, 20 Aug. 1997

27. Decree of the President of the Russian Federation on the Federal State Unitary Enterprise Promexport, no. 908, 20 Aug. 1997

28. Decree of the President of the Russian Federation on the Federal State Unitary Enterprise the State Company Rosvooruzhenie, no. 910, 20 Aug. 1997

29. Statute of the Interdepartmental Coordinating Council for Military–Technical Cooperation between the Russian Federation and Foreign States, 20 Aug. 1997

## 1. DECREE OF THE PRESIDENT OF THE RUSSIAN FEDERATION ON MILITARY–TECHNICAL COOPERATION OF THE RUSSIAN FEDERATION WITH FOREIGN COUNTRIES (BASIC PROVISIONS)

In order to further develop the military–technical cooperation of the Russian Federation with foreign states and strengthen state control in this area, I decree as follows:

1. To introduce alterations to the Regulations on military–technical cooperation of the Russian Federation with foreign states approved by the Decree of the President of the Russian Federation of 12 May 1992, no. 507, 'On military–technical cooperation of the Russian Federation with foreign states', and approve it in the new wording.

2. To establish that export (import) of weapons, military equipment and work (services) of a military designation shall be carried out solely by the State Company for Trade in Armaments and Military–Technical Cooperation Rosvooruzhenie and enterprises which develop and manufacture weapons and military equipment which have such right in accordance with the manner determined by the President of the Russian Federation.

The Ministry of Defence of the Russian Federation shall be authorized to provide services to foreign states in training their national military personnel and technical staff.

3. To consider null and void the direction of the President of the Russian Federation of 24 December 1992, no. 818/ip.

President of the Russian Federation
B. Yeltsin
Moscow, Kremlin
5 Oct. 1995
No. 1008

---

Source: *Sobranie zakonodatelstva Rossiyskoy Federatsii* [Collection of legislative acts of the Russian Federation], no. 41 (1995), pp. 7203–204 (article 3876).

---

## 2. REGULATIONS ON MILITARY–TECHNICAL COOPERATION OF THE RUSSIAN FEDERATION WITH FOREIGN COUNTRIES, APPROVED BY THE DECREE OF THE PRESIDENT OF THE RUSSIAN FEDERATION OF 5 OCTOBER 1995, NO. 1008

(with alterations introduced by the decrees of the President of the Russian Federation of 8 May 1996, no. 686; of 14 August 1996, no. 1177; and of 6 September 1996, no. 1326)

### I. GENERAL PROVISIONS

1. The present Regulations shall specify the terms of reference of the federal authorities in the area of military–technical cooperation of the Russian Federation with foreign countries (hereinafter referred to as military–technical cooperation) and the procedure for its implementation.

2. State regulation in the area of military–technical cooperation shall comprise a complex of organizational, legal, technical and other measures, conducted by the government and aimed to protect the national interests of the Russian Federation, compliance with the established procedure for military–technical cooperation, and coordination of the activities of the federal executive authorities which are competent to resolve questions pertaining to military–technical cooperation (hereinafter referred to as participants of military–technical cooperation) and juridical persons of the Russian Federation who have obtained the right to participate in military–technical cooperation (hereinafter referred to as subjects of military–technical cooperation).

3. Military–technical cooperation with foreign countries shall be established, suspended, discontinued and renewed on the basis of decisions of the President of the Russian Federation taken, as a rule, on the recommendation of the Government of the Russian Federation. In fulfilment of decisions of the President of the Russian Federation on the establishment or renewal of military–technical cooperation with foreign countries, the Government of the Russian Federation, as a rule, signs international agreements with foreign countries, intergovernmental agreements or other treaties and legal documents.

Military–technical cooperation with foreign countries shall also be conducted on the basis of international agreements of the former Soviet Union unless otherwise determined by special decisions of the President of the Russian Federation.

4. International agreements of the Russian Federation on military–technical cooperation shall be concluded in compliance with the Federal Law of the Russian Federation.

Proposals for the conclusion of international agreements on military–technical cooperation shall be submitted to the President or Government of the Russian Federation through the Ministry of Foreign Economic Relations in coordination with the Ministry of Foreign Affairs.

On international agreements:

5. The inventories of armaments, military equipment and research and development work carried out for military purposes offered for the first time for export from the Russian Federation shall be approved by the President of the Russian Federation upon recommendation of the Government of the Russian Federation. Decisions on the export or temporary export of military-purpose products or the results of military-purpose research and development work not included in the aforesaid inventories shall also be taken by the President of the Russian Federation. The export of these products shall be carried out under licences issued by the Ministry of Foreign Economic Relations.

6. Decisions on the export or temporary export of military-purpose products and work (services) to countries with which military–technical cooperation has not been established or is suspended or discontinued, as well as decisions on technical assistance in constructing (fitting out) special projects on the territory of such states shall be taken solely by the President of the Russian Federation.

7. The inventory of military-purpose products and work (services) the export and import of which are subject to control and carried out under licences issued by the Ministry of Foreign Economic Relations shall be approved by the Government of the Russian Federation.

8. Decisions on the export and import or temporary export or import of military-purpose products and work (services) included in the inventory specified in para. 7 of these Regulations shall be taken by the Government of the Russian Federation on the recommendation of the Ministry of Foreign Economic Relations.

The export of spare parts, complementary items, training and auxiliary stores for military-purpose equipment previously delivered to foreign countries or manufactured under Russian licences, as well as the import of such military-purpose products and their technical maintenance and repair, may be carried out under licences of the Ministry of Foreign Economic Relations without special decisions of the Government of the Russian Federation.

The export of spare parts, complementary items, training and auxiliary stores for military-purpose products previously delivered to foreign countries in order to establish temporary stocks for these countries on their own territories may be carried out under licences of the Ministry of Foreign Economic Relations without special decisions of the Government of the Russian Federation.

9. Proposals on matters of military–technical cooperation elaborated and coordinated by participants in military–technical cooperation shall be submitted to the Government of the Russian Federation through the Ministry of Foreign Economic Relations.

10. Requests and applications on behalf of the governments of foreign countries for the delivery of products or the execution of work and services of a military–technical nature shall be accepted by the Government of the Russian Federation, ambassadors of the Russian Federation and the Ministry of Foreign Economic Relations.

Participants and subjects of military–technical cooperation authorized by the Government of the Russian Federation to conduct negotiations with foreign partners, after receiving requests (applications) from foreign partners who are vested by their governments with the relevant powers, shall submit these requests (applications) to the Ministry of Foreign Economic Relations.

11. Activities in the area of military–technical cooperation shall be conducted in compliance with the security regime established by the legislation of the Russian Federation.

## II. PROCEDURE FOR DECISIONS ON QUESTIONS OF MILITARY–TECHNICAL COOPERATION

12. Decisions on questions of military–technical cooperation, depending on their importance, shall be taken by the President of the Russian Federation, the Government of the Russian Federation or federal executive authorities.

13. The President of the Russian Federation, in accordance with the constitution and upon recommendation of the Government of the Russian Federation, shall take decisions on:

– approval of conceptual approaches to military–technical cooperation;
– the conclusion of interstate agreements on military–technical cooperation;
– the determination of the list of foreign countries with which military–technical cooperation is prohibited or restricted;
– the establishment of military–technical cooperation with states with which it has not been conducted before;
– the suspension, termination and resumption of military–technical cooperation;
– approval of the inventory of military–purpose products and work (services) permitted for export;
– demonstration and delivery of armaments and military equipment which are not included in the inventory of armaments and military equipment permitted for export;
– the transfer to foreign countries of licences for the manufacture of military-purpose products;
– cooperation with foreign countries in the area of developing armaments, military equipment and other military-purpose products; and
– the provision of military–technical assistance to foreign countries.

14. The Government of the Russian Federation shall take decisions on:

– the establishment of bilateral and multilateral intergovernmental commissions on military–technical cooperation by agreement with interested foreign states;
– the conclusion of intergovernmental agreements on military–technical cooperation with foreign countries in fulfilment of decisions taken by the President of the Russian Federation;
– the authorization of federal executive bodies with the powers of state customer in the area of military–technical cooperation;
– the determination of the terms, volumes and dates of export–import operations with military-purpose products and work (services) as well as research and development work on military-purpose products performed in the foreign customers' interests in line with the decisions of the President of the Russian Federation;

– the leasing of military-purpose products permitted for export and the transfer of military-purpose products to foreign states for testing on their territory;
– the authorization of juridical persons of the Russian Federation which are the developers and/or manufacturers of military-purpose products to participate in military–technical cooperation, and depriving them of this right in compliance with the procedure established by the President of the Russian Federation;
– the establishment of state control over the export (import) of military-purpose products and work (services);
– the establishment of procedure for settling state debts incurred in export (import) operations with military-purpose products and work (services) or performing other types of military–technical cooperation;
– the organizing of exhibitions and demonstrations of armaments and military equipment both abroad and on the territory of the Russian Federation;
– the confirmation of normative documents regulating the procedure for and organization of military–technical cooperation;
– the delivery or transfer to third countries by foreign countries of samples of armaments and military equipment manufactured under Russian licences; and
– the establishment of procedure for licensing different types of foreign economic activity in the field of military–technical cooperation and the export (import) of military-purpose products and work (services), and reimbursement of the subjects of military–technical cooperation of the Russian Federation for losses caused by the termination (suspension) of the export (import) of military-purpose products or work (services) associated with the discontinuation or suspension of military–technical cooperation between the Russian Federation and foreign countries.

15. The Interdepartmental Council coordinating the military–technical policy of the Russian Federation shall carry out its functions in conformity with its statute, approved by the President of the Russian Federation.

16. The Ministry of Foreign Economic Relations of the Russian Federation shall:

– submit proposals on matters concerning military–technical cooperation to the President and Government of the Russian Federation;

– elaborate drafts of legislative acts and other legal documents concerning issues of military–technical cooperation;

– coordinate the activities of the participants and subjects of military–technical cooperation;

– carry out the licensing of different types of foreign economic activity in the area of military–technical cooperation as well as the export (import) of military-purpose products and work (services);

– submit proposals on setting up bilateral and multilateral intergovernmental commissions on military–technical cooperation and organize their activities;

– exercise control over the pricing of and the level of prices for exported (imported) samples of military-purpose products;

– prepare draft intergovernmental agreements on military–technical cooperation, conduct negotiations and sign these agreements on the basis of decisions of and on behalf of the Government of the Russian Federation;

– organize the fulfilment of undertakings of the Russian Federation arising from interstate treaties and intergovernmental agreements on military–technical cooperation;

– ensure control over the foreign economic activity of the subjects of military–technical cooperation and their fulfilment of the contractual obligations of the Russian Federation under intergovernmental agreements and treaties; and

– carry out other functions in the area of military–technical cooperation as stipulated by the legislation of the Russian Federation.

17. The Ministry of Foreign Affairs shall:

– control the observance by the federal executive authorities within whose terms of reference matters relating to military–technical cooperation fall of the international obligations of the Russian Federation, and provide information concerning military–technical cooperation in the appropriate form to the United Nations Organization and other international organizations;

– participate in elaborating proposals on questions relevant to military–technical cooperation;

– participate in the work of intergovernmental commissions on military–technical cooperation;

– ensure control over the protection of the political interests of the Russian Federation in the area of military–technical cooperation; and

– elaborate proposals on the list of foreign states for which it would be appropriate or necessary to ban or restrict military–technical cooperation, and in agreement with the Ministry of Foreign Economic Relations submit such proposals to the Government of the Russian Federation.

18. The Ministry of the Economy shall elaborate proposals on the range and volumes of armaments and military equipment for export (import) within the state export–import defence order in collaboration with the Ministry of Foreign Economic Relations, the Ministry of Defence Industry, the Ministry of Defence and the Ministry of Finance.

19. The Ministry of Defence Industry shall:

– carry out licensing of all types of activity in the area of development and production of armaments, military equipment and ammunition;

– consider applications from enterprises and organizations which develop and manufacture armaments and military equipment for the right to participate in military–technical cooperation, and participate in decisions on extending such rights in conformity with the procedure established by the President of the Russian Federation;

– coordinate the activities of enterprises and organizations dealing with research on and development and production of armaments and military equipment for export;

– participate in working out proposals on the range and volumes of armaments and military equipment for export (import) within the framework of the export–import state defence order;

– participate, upon the instructions of the Government of the Russian Federation, in the elaboration of draft interstate and intergovernmental agreements on cooperation ties in the area of design and manufacture of armaments and military equipment, and ensure their implementation; and

– take measures jointly with the Ministry of Defence to ensure the patent protection of military-purpose products developed and manufactured by enterprises of the defence branches of industry and being their intellectual property.

20. The Ministry of Defence shall:

– participate in elaborating proposals on matters of military–technical cooperation;

– provide information and data regarding the military–technical cooperation of the Russian Federation with foreign countries;

– undertake practical activities relating to the provision of assistance in the operation and military use of the armaments and military equipment supplied to foreign countries;

– repair at its enterprises the armaments and military equipment previously supplied to foreign countries, lease transport facilities to foreign partners, send military experts to foreign countries, train national military specialists and technical personnel, arrange demonstrations of armaments and military equipment and conduct at its shooting-ranges the military field exercises and firing trials of foreign countries' army detachments;

– exercise military–technical control over the design and manufacture of the armaments and military equipment supplied, as well as research and development work carried out, within the framework of agreements and contracts signed with foreign customers; and

– prepare proposals on the inventories of armaments, military equipment and research and development work on the list for export from the Russian Federation.

21. The Federal External Intelligence Service shall:

– promote military–technical cooperation by carrying out political, international, legal and economic analyses of the aspects of this cooperation; and

– collect and process information on questions of military–technical cooperation and help in checking the reliability of foreign partners.

22. The Federal Security Service of the Russian Federation shall put into effect a complex of necessary measures for protecting the interests of the Russian Federation in the area of military–technical cooperation.

23. Decision making on military–technical assistance to foreign countries and direct provision of this assistance shall be carried out in accordance with the procedure prescribed by the present Regulations. Official requests from the proper authorities of foreign countries concerning the provision of military–technical assistance and the purchase of military-purpose products and services in the Russian Federation shall be passed through the trade and diplomatic channels to the Ministry of Foreign Economic Relations, which, with the participation of the interested federal executive authorities and organizations, shall carry out expert examination of such requests and on the basis of that examination submit relevant proposals to the Government of the Russian Federation.

24. Subjects of military–technical cooperation shall carry out foreign economic activity in the area of military–technical cooperation within the limits of their authority, as well as within the framework of decisions taken by the President of the Russian Federation and the Government of the Russian Federation.

25. The right to employ on a contractual basis foreign juridical and physical persons to provide consulting and intermediary services shall be granted solely to the subjects of military–technical cooperation of the Russian Federation with foreign countries.

## III. CONTROL OVER ACTIVITIES IN THE AREA OF MILITARY–TECHNICAL COOPERATION

26. Control and coordination of the activities of the subjects of military–technical cooperation of the Russian Federation shall be carried out with a view to ruling out unfair competition and averting the possibility of inflicting political, economic and military damage on the Russian Federation. Direct control and coordination of the activities of the participants and subjects of military–technical cooperation shall be exercised by the Ministry of Foreign Economic Relations in collaboration with other federal executive authorities within their terms of reference.

27. The foreign economic activity of the subjects of military–technical cooperation shall be liable to control and coordination at the following stages:

– negotiations with foreign customers;

– preparation and signing of contractual documents; and

– fulfilment of contractual obligations.

28. The stage of conducting negotiations with foreign customers shall include the search for foreign partners, the conduct of promotional, exhibition and marketing activities, including demonstrations of armaments and military equipment for export, the transfer in the course of negotiations of the tactical and technical characteristics and specifications of armaments and military equipment or the basic parameters of research and development work to create (update) samples of armaments and military equip-

ment, the establishment of export prices for military-purpose products or work (services), and the signing of protocols of intention.

29. The stage of preparation and signing of contractual documents includes the preparation, expert examination and coordination of draft contracts with foreign customers and the coordination of agreements with Russian suppliers (enterprises and organizations which manufacture and develop armaments and military equipment), and the signing of these contracts and agreements.

30. The stage of fulfilment of obligations specified in contractual documents includes the implementation of agreements with Russian suppliers on the manufacture of military-purpose products, the execution of work and the provision of services of a military–technical nature specified in a contract concluded with foreign customers, military–technical control over the contract, the carrying out of the export and import of military-purpose products and work (services) licensed, and the arrangement of mutual settlements with foreign customers and Russian suppliers.

31. All subjects of military–technical cooperation can conduct foreign economic activity in the area of military–technical cooperation at any of these stages only by special permission of and in conformity with the instructions of the Ministry of Foreign Economic Relations.

32. Violations of the rules and methods of military–technical cooperation established by the present regulations shall be the responsibility of the directors of the subjects of military–technical cooperation and shall be a statutory ground for depriving the subjects of military–technical cooperation of the right to conduct the export and import of military-purpose products and work (services).

33. The subjects of military–technical cooperation shall submit reports on the results of work executed at each stage to the Ministry of Foreign Economic Relations.

*Source: Sobranie zakonodatelstva Rossiyskoy Federatsii* [Collection of legislative acts of the Russian Federation], no. 41 (1995), pp. 7204–11 (article 3876).

## 3. DECREE OF THE PRESIDENT OF THE RUSSIAN FEDERATION ON THE INTERDEPARTMENTAL COORDINATING COUNCIL FOR MILITARY–TECHNICAL POLICY OF THE RUSSIAN FEDERATION

In conformity with the Decree of the President of the Russian Federation of 3 March 1995, no. 236, 'On the introduction of alterations and amendments to the Decree of the President of the Russian Federation of 30 December 1994, no. 2251, On the State Committee of the Russian Federation on Military–Technical Policy and in the Regulations approved by the Decree' (Collection of legislative acts of the Russian Federation, no. 10 (1995), article 865), I decree as follows:

1. To approve the enclosed Regulations on the Interdepartmental Coordinating Council for Military–Technical Policy of the Russian Federation and its composition.

2. This Decree shall be effective from the date of signature.

President of the Russian Federation
B. Yeltsin
Moscow, Kremlin
14 June 1995
No. 590

*Source: Sobranie zakonodatelstva Rossiyskoy Federatsii* [Collection of legislative acts of the Russian Federation], no. 25 (1995), p. 4519 (article 2379).

## 4. REGULATIONS ON THE INTERDEPARTMENTAL COORDINATING COUNCIL FOR MILITARY–TECHNICAL POLICY OF THE RUSSIAN FEDERATION

Approved by the Decree of the President of the Russian Federation, 14 June 1995, no. 590 (with alterations introduced by the decrees of the President of the Russian Federation of 31 January 1996, no. 131; of 8 May 1996, no. 686; of 14 August 1996, no. 1177; and of 6 September 1996, no. 1326)

1. The Interdepartmental Coordinating Council for Military–Technical Policy of the Russian Federation (hereinafter referred to as the ICC) has been formed with a view to elaborating coordinated proposals corresponding to the political, military and economic interests of the Russian Federation on the following matters:

– the state military–technical policy of the Russian Federation, including that of military–technical cooperation with foreign countries;
– control over activities of the federal executive bodies participating in the implementation of state military–technical policy;
– the settlement of problems in the area of the design, manufacture, export and import of armaments and military equipment; and
– state support of the defence scientific and industrial potential of the Russian Federation.

2. The ICC in its activities shall be guided by the Constitution of the Russian Federation as well as the international commitments of the Russian Federation, federal laws, decrees and instructions of President of the Russian Federation and the present Regulations.

3. The main task of the ICC shall be to elaborate proposals to be submitted to the President of the Russian Federation and the Government of the Russian Federation:

– the determination of political, military and economic priorities in the state military–technical policy and the resolution and settlement of problems in the area of the development, manufacture, export and import of armaments and military equipment;
– ensuring the coordination of the activities of the federal executive bodies in the area of the military–technical cooperation of the Russian Federation with foreign countries and the fulfilment by them of the international commitments of the Russian Federation in this area, as well as obligations in the field of the reduction and liquidation of armaments; and
– working out measures for the execution of armament programmes, the implementation of programmes for the conversion of defence enterprises and the industrial utilization of armaments and military equipment.

4. The ICC in order to carry out the tasks assigned shall:

– consider relevant proposals of the State Committee of the Russian Federation on Military–Technical Policy and of other federal executive bodies on matters of state military–technical policy, on the certification of Russian juridical persons for the right to conduct foreign economic activity in the area of the military–technical cooperation of the Russian Federation with foreign countries, on settling disagreements between Russian participants of military–technical cooperation, and on other matters falling within the competence of the ICC; and
– analyse, sum up and process the information essential for elaborating proposals on the questions within its competence.

5. The ICC shall be entitled to:

– demand from federal executive bodies, executive authorities of the Russian Federation and enterprises, institutions and organizations, irrespective of their form of property and departmental subordination, the information, documents and materials necessary for fulfilling the tasks assigned to it;
– hear reports from the heads of relevant federal executive bodies on the execution of state military–technical policy and the fulfilment of the international obligations of the Russian Federation in the area of military–technical cooperation with foreign countries and in the field of reduction and liquidation of armaments;
– set up, if necessary, working groups to elaborate the questions within its competence; and
– submit proposals on state military–technical policy and the coordination of the activities of the federal executive authorities to the Government of the Russian Federation.

6. The First Deputy Chairman of the Government of the Russian Federation shall be appointed Chairman of the ICC, and shall bear personal responsibility for the fulfilment of the tasks imposed on the ICC.

The Deputy Chairman of the Government of the Russian Federation shall be appointed the First Deputy Chairman of the ICC.

The Chairman of the State Committee of the Russian Federation on Military–Technical Policy shall be appointed Deputy Chairman of the ICC.[*]

7. The ICC shall also include the chief executives of the following: the Ministry of

---

[*] In accordance with Decree no. 1177 of the President of the Russian Federation of 14 Aug. 1996 the State Committee on Military–Technical Policy was abolished and its functions transferred to the Ministry of Foreign Economic Relations.

Foreign Affairs, Ministry of Defence, Ministry of Finance, Ministry of the Economy, Ministry of Atomic Energy, Ministry of Defence Industry, State Customs Committee, Russian Space Agency, Federal External Intelligence Service, Federal Security Service, Security Service of the President of the Russian Federation, Federal Agency for Governmental Communication and Information under the President of the Russian Federation, Department of the Defence Branches of Industry under the Government Administration of the Russian Federation, the State Company for Trade in Armaments and Military–Technical Cooperation Rosvooruzhenie and the Deputy Secretary of the Security Council.

Members of the ICC shall participate in Council meetings without the right of substitution.

The composition of the ICC shall be approved by the President of the Russian Federation.

8. The work of the ICC shall be carried out through meetings called as and when needed but not less than once a month.

Officials of federal executive bodies, enterprises, institutions and organizations, irrespective of their forms of property and departmental subordination, can be invited upon the instructions of the ICC Chairman to its meetings to participate in discussions of particular items on the agenda, with the right of deliberative vote.

9. Information on the agenda and materials included in it for consideration at the ICC meetings shall be forwarded to all participants of the meeting by the ICC Secretary not later than two weeks before the date of the meeting.

10. Decisions on each question on the agenda of the meeting shall be made by the ICC members present by simple majority vote. Decisions shall be made only when not less than half of the total number of ICC members participate in the meeting.

The ICC members shall have equal rights in making decisions. ICC decisions aimed at the elaboration of proposals on matters within its terms of reference shall be mandatory for federal executive bodies.

In the event of fundamental differences arising between the members, the Chairman of the ICC shall be entitled to postpone consideration of the matter for further clarification and submit it for consideration again.

The results of questions considered at the ICC meeting (with an indication of voting results on each item) shall be recorded in the appropriate protocols or drawn up as separate decisions of the ICC.

Protocols and decisions shall be signed by the Chairman of the ICC and in his absence by the Deputy Chairman.

11. The functions of the working body (secretariat of the ICC) shall be placed in a subdivision of the State Committee of the Russian Federation on Military–Technical Policy.*

One of the Deputy Chairmen of the State Committee of the Russian Federation on Military–Technical Policy* shall be responsible for the work of the ICC and function as the Secretary of the ICC.

12. Informational, organizational and technical support for the ICC's activities shall be provided by the ICC secretariat, which shall be entrusted with the following tasks: summing up data supplied pertaining to matters within the ICC's competence, developing proposals for planning the ICC's work, and drawing up the agenda of ICC meetings.

## Composition of the Interdepartmental Coordinating Council for Military–Technical Policy of the Russian Federation

Approved by the Decree of President of the Russian Federation of 14 June 1995, no. 590

– The First Deputy Chairman of the Government of the Russian Federation (Chairman of the Council);
– the Deputy Chairman of the Government of the Russian Federation (the First Deputy Chairman of the Council);
– the Deputy Chairman of the ICC;
– the Minister of Defence Industry;
– the First Deputy Minister of Defence;
– the Director General of the Russian Space Agency;
– the Director General of the State Company Rosvooruzhenie;
– the Chairman of the State Customs Committee;
– the Deputy Minister of Foreign Affairs;
– the Deputy Secretary of the Security Council;

---

* In accordance with Decree no. 1177 of the President of the Russian Federation of 14 Aug. 1996 the State Committee on Military–Technical Policy was abolished and its functions transferred to the Ministry of Foreign Economic Relations.

– the Head of the Department of Defence Branches of Industry in the Government Administration of the Russian Federation;

– the Minister of Atomic Energy;

– the Minister of Finance;

– the First Deputy Chief of the Security Service of the President of the Russian Federation;

– the Deputy Director of the Federal Security Service;

– the General Director of the Federal Agency for Governmental Communication and Information under the President of the Russian Federation;

– the First Deputy Minister of the Economy; and

– the Deputy Director of the Federal External Intelligence Service.

———

*Source: Sobranie zakonodatelstva Rossiyskoy Federatsii* [Collection of legislative acts of the Russian Federation], no. 25 (1995), pp. 4519–23 (article 2379).

tion in the State Company Rosvooruzhenie, comprising seven staff members.

3. The Head of the Administration of the President of the Russian Federation shall, upon recommendation of the Representative of the President of the Russian Federation in the State Company Rosvooruzhenie, approve the structure and staff of the Office of the Representative of the President of the Russian Federation in the State Company Rosvooruzhenie.

President of the Russian Federation
B. Yeltsin
Moscow, Kremlin
4 Mar. 1994
No. 450

———

*Source: Sobranie aktov prezidenta i pravitelstva Rossiykoy Federatsii* [Collection of legislative acts of the President and Government of the Russian Federation], no. 10 (1994), p. 880 (article 778).

## 5. DECREE OF THE PRESIDENT OF THE RUSSIAN FEDERATION APPROVING THE REGULATIONS ON THE STATUS OF THE REPRESENTATIVE OF THE PRESIDENT OF THE RUSSIAN FEDERATION IN THE STATE COMPANY ROSVOORUZHENIE

In order to assure the state interests in organizing and conducting military–technical cooperation, as well as the activity of the State Company Rosvooruzhenie, I hereby decree:

1. To approve the enclosed Regulations on the status of the Representative of the President of the Russian Federation in the State Company Rosvooruzhenie.

2. To set up in the Administration of the President of the Russian Federation the working office of the Representative of the President of the Russian Federation in the State Company Rosvooruzhenie, which shall operate as a department of the Administration of the President of the Russian Federation.

To establish the Office of the Representative of the President of the Russian Federa-

## 6. REGULATIONS ON THE REPRESENTATIVE OF THE PRESIDENT OF THE RUSSIAN FEDERATION IN THE STATE COMPANY ROSVOORUZHENIE

Approved by the Decree of the President of the Russian Federation of 4 March 1994, no. 450

1. The Representative of the President of the Russian Federation in the State Company Rosvooruzhenie in accordance with the present Regulations shall represent the interests of the state in this company.

2. The Representative of the President of the Russian Federation in the State Company Rosvooruzhenie shall be appointed to the post and dismissed from it by the President of the Russian Federation and shall be subordinate to him.

3. The Representative of the President of the Russian Federation in the State Company Rosvooruzhenie shall act in conformity with the Constitution of the Russian Federation, the laws of the Russian Federation, the decrees and directions of President of the Russian Federation and the present regulations.

4. The Representative of the President of the Russian Federation in the State Company Rosvooruzhenie shall:

– supervise the execution by the State Company Rosvooruzhenie of decrees and directions of the President of the Russian Federation and decisions and directions of the Government of the Russian Federation regulating relations in the area of military–technical cooperation with foreign states;

– ensure interaction with federal executive bodies and officials of the Russian Federation in elaborating and carrying out measures aimed at implicit observance of state interests in the activity of the State Company Rosvooruzhenie; and

– prepare and submit to the President of the Russian Federation proposals on matters of the suspension, termination and resumption of military–technical cooperation with foreign states and on other problems requiring the decision of the President of the Russian Federation.

5. The Representative of the President of the Russian Federation in the State Company Rosvooruzhenie, in performing the duties imposed on him, shall be entitled to:

– familiarize himself with any documents pertaining to the activity of the State Company Rosvooruzhenie;

– upon the instructions of the President of the Russian Federation, organize and verify the State Company Rosvooruzhenie's execution of the decrees and directions of the President of the Russian Federation and of the directions and decisions of the Government of the Russian Federation regulating relations in the area of military–technical cooperation with foreign states, and submit the reports of inspections directly to the President of the Russian Federation;

– elaborate proposals to perfect conceptual approaches to military–technical cooperation and submit them to the President of the Russian Federation;

– participate in the development of military–technical cooperation and its establishment with foreign states with which it has not existed before;

– participate in organizing and perfecting cooperation with foreign states in the area of joint developments of armaments and military equipment;

– submit reports to the directorate of the State Company Rosvooruzhenie and the Government of the Russian Federation in cases of non-observance of the state interests in the activity of the State Company Rosvooruzhenie or of the non-execution or improper execution by the company of decrees and directions of the President of the Russian Federation and directions and regulations of the Government of the Russian Federation connected with the activity of this company;

– inform the President of the Russian Federation of these facts, as well as of the causes of violations revealed;

– ask federal executive bodies, institutions and organizations of the Russian Federation for necessary information and documents on the activity of the State Company Rosvooruzhenie and receive replies in the established manner;

– participate in the work of collegiate bodies of the State Company Rosvooruzhenie in negotiations of representatives of the company with official foreign delegations or representatives of foreign firms when concluding transactions on the export and import of armaments and military equipment;

– in accordance with the established manner, use the services of specialists of the Administration of the President of the Russian Federation, ministries and departments of the Russian Federation;

– in accordance with the established manner, make use of data banks of the Administration of the President of the Russian Federation;

– participate in the work of federal executive bodies within whose competence decision making in the area of military–technical cooperation with foreign states falls; and

– submit proposals on the organization and implementation of military–technical cooperation to the federal executive bodies.

6. The Representative of the President of the Russian Federation in the State Company Rosvooruzhenie shall be vested with other powers by separate directions of the President of the Russian Federation.

7. The Representative of the President of the Russian Federation in the State Company Rosvooruzhenie shall not be entitled to interfere directly in matters pertaining to the administrative management and organizational structure of the financial and economic activities of the company.

8. In order to assure the activities of the Representative of the President of the Russian Federation in the State Company Rosvooruzhenie, the working staff of the Representative of the President in Rosvooruzhenie (hereinafter referred to as the staff), comprising seven persons, shall be formed as a division in the Administration of the President of the Russian Federation.

9. The chief and members of the working staff shall be appointed by the Chief of the Administration of the President of the Russian Federation upon recommendation of the Representative of the President of the Russian Federation in the State Company Rosvooruzhenie.

The Representative of the President of the Russian Federation in the State Company Rosvooruzhenie shall:

– nominate candidates for appointment to posts on the working staff, as well as submit suggestions on the dismissal of members of the working staff;

– define the official duties of the members of the working staff;

– in accordance with established procedure, send experts from the working staff on business trips, including trips abroad; and

– in accordance with established procedure, submit proposals on encouraging employees on the working staff.

10. The Representative of the President of the Russian Federation in the State Company Rosvooruzhenie shall be equated on the scale of ranks with a federal minister so far as the material, technical and informational support and the interaction with federal executive bodies are concerned.

11. The material, technical and social support of the Representative of President of the Russian Federation in the State Company Rosvooruzhenie and his working staff shall be provided by the appropriate subdivisions and divisions of the Administration of the President of the Russian Federation.

*Source: Sobranie aktov prezidenta i pravitelstva Rossiyskoy Federatsii* [Collection of legislative acts of the President and Government of the Russian Federation], no. 10 (1994), pp. 881–83 (article 778).

## 7. REGULATIONS ON THE PROCEDURE FOR IMPOSING EMBARGO ON DELIVERIES OF ARMAMENTS AND MILITARY EQUIPMENT, THE PROVISION OF SERVICES OF A MILITARY–TECHNICAL NATURE, AND ON DELIVERIES OF RAW AND OTHER MATERIALS AND EQUIPMENT AND THE TRANSFER OF MILITARY AND DUAL-USE TECHNOLOGIES TO FOREIGN STATES, INCLUDING THE CIS MEMBER STATES

Approved by the Decree of the President of the Russian Federation of 18 February 1993, no. 235 (with alterations introduced by decrees of the President of the Russian Federation of 30 December 1994, no. 2251; of 3 March 1995, no. 236; of 8 May 1996, no. 680; of 14 August 1996, no. 1177; and of 6 September 1996, no. 1326)

The present regulations shall specify the procedure for the imposition of embargoes by the Russian Federation on deliveries of armaments and military equipment, on the provision of services of a military–technical nature, including business trips of Russian military experts and training of foreign specialists, and on deliveries of raw and other materials and equipment and the transfer of military and dual-purpose technologies to foreign states, including the CIS member states.

1. The position to which the Russian Federation adheres when voting in the UN Organization on the declaration of embargoes shall be previously coordinated with the interested ministries and departments of the Russian Federation and, where necessary, submitted for discussion by the Security Council of the Russian Federation.

The President of the Russian Federation, upon recommendation of the Government of the Russian Federation prepared by the Ministry of Foreign Affairs, in a case where a resolution has been passed by the UN Security Council, will take a decision on placing an embargo on deliveries of armaments and military equipment, on providing services of a military–technical nature, and on the delivery of raw and other materials and equipment and the transfer of military and dual-purpose technologies to foreign states, including the CIS member states.

2. The President of the Russian Federation, upon recommendation of the Government of the Russian Federation, shall consider proposals prepared by the Interdepartmental Coordinating Council for Military–Technical Policy and the Export Control Committee under the Government of the Russian Federation and take decisions on imposing embargoes on military–technical cooperation with foreign states, including the CIS member states, as well as on delivering to these countries raw and other materials and equipment and technologies of either military or dual purpose, proceeding from the national interests of the Russian Federation.

3. The practical implementation of decisions of the President of the Russian Federation shall be the responsibility of the Ministry of Foreign Economic Relations, the Federal Security Service, the State Customs Committee, the Ministry of Defence, the Ministry of the Economy, the Ministry of Finance, the Ministry of Defence Industry, the Ministry of Foreign Affairs and the Federal External Intelligence Service, which shall put into effect the necessary measures immediately after the President of the Russian Federation takes the decision on the imposition of an embargo.

The measures shall include the termination and prevention of deliveries from the Russian Federation of armaments and military equipment to states with which military–technical cooperation is under embargo, the discontinuance of services of a military–technical nature, the termination of deliveries of raw and other materials and equipment and of the transfer of military and dual-purpose technologies, the refusal of licences to participants of foreign economic activity in this area, and the suspension of the relevant intergovernmental agreements and contracts.

Verification of the observance of sanctions imposed shall be the responsibility of the Interdepartmental Coordinating Council for Military–Technical Policy.

4. The Ministry of Foreign Economic Relations and the Ministry of Defence Industry, by agreement with the Ministry of Foreign Affairs and the Ministry of Defence, shall submit to the Government of the Russian Federation proposals for embargoes in the field of military–technical cooperation. These proposals shall stipulate compensation to enterprises and organizations of the Russian Federation for losses caused by the suspension of relevant intergovernmental agree-ments, as well as contracts concluded in order to implement these agreements, with the states subject to the embargo imposed.

The above proposals shall also specify the possibilities for sale to third countries of military-purpose products manufactured for but not delivered to the countries subject to the embargo imposed.

---

*Source: Sobranie aktov prezidenta i pravitelstva Rossiyskoy Federatsii* [Collection of legislative acts of the President and Government of the Russian Federation], no. 8 (1993), pp. 799–800 (article 658).

## 8. LAW OF THE RUSSIAN FEDERATION ON THE CONVERSION OF THE DEFENCE INDUSTRY

(Section 4)

### Article 9. Types of foreign economic activities

1. Converted enterprises have the right to engage independently in foreign economic activities in accordance with the legislation of the Russian Federation. This right applies to:

– the export of raw and other materials and equipment released in the course of conversion if it is impossible to use them for the manufacture of civil products, taking into account the requirements of article 10 of the present Law;

– the import of up-to-date machinery, equipment and new technologies and complementary articles for the manufacture of civilian goods;

– the transfer (exchange and sale) in the established manner of technologies, licences, know-how and scientific and technical information which before conversion were used in the production of armaments and military hardware; and participation at conferences, symposia, exhibitions and fairs with demonstrations of new materials, equipment, instruments and advertising material of technologies previously used for the production of armaments and military equipment;

– the design, production and sale of armaments and military equipment under licences in the order established by the legislation of the Russian Federation; and

– participation, in cooperation with foreign firms, in the design, production and sale of military-purpose products in accordance with the legislative acts of the Russian Federation which protect the military and technological interests of the Russian Federation.

2. The activities of enterprises with foreign investments shall be regulated by the RSFSR law on foreign investments in the RSFSR and other legislative acts of the Russian Federation.

### Article 10. Protection of the military–economic and scientific–technological potential of the Russian Federation

1. In order to avoid damage to the military–economic and scientific–technological potential of the Russian Federation through the pursuit of foreign economic activities by converted enterprises and to ensure the non-proliferation of weapons of mass destruction, these enterprises must strictly follow the restrictions imposed on the export (transfer or exchange) of civil-purpose products and technologies which can be used to build weapons of mass destruction. Restrictions on the export (transfer or sale) of these types of products and technologies are established by the Supreme Soviet of the Russian Federation and the Government of the Russian Federation.

2. In their foreign economic activities converted enterprises shall be guided by the following provisions:

– the export of strategic raw materials, other materials and equipment is carried out under licences issued in each particular case in accordance with the legislation of the Russian Federation;

– the transfer of technologies, licences, know-how and scientific and technical information for the manufacture of civil-purpose products and/or their use in commercial, scientific and technological cooperation activities with foreign firms are allowed only if the protection of the military–economic interests of the Russian Federation is assured; and

– the sale of armaments and military hardware, special systems, complexes, functional blocks and assemblies which are part of armaments and military hardware, as well as technologies for their production, is carried

out in the manner established by the Government of the Russian Federation.

President of the Russian Federation
B. Yeltsin
Moscow,
20 Mar. 1992
No. 2551-1

---

*Source: Vedemosti Syezda Narodnykh Deputatov Rossiyskoy Federatsii i Verkhovnogo Soveta Rossiyskoy Federatsii* [Gazette of the Congress of People's Deputies and Supreme Soviet of the Russian Federation], no. 18 (1992), pp. 1319–20 (article 964).

---

## 9. LAW OF THE RUSSIAN FEDERATION ON STATE REGULATION OF FOREIGN TRADE ACTIVITY

Passed by the State Duma, 7 July 1995

### Article 6. Terms of reference of the Russian Federation in the field of foreign trade activity

In accordance with its terms of reference the Russian Federation shall:

1) elaborate the concept and strategy for the development of foreign trade relations and the basic principles of the foreign trade policy of the Russian Federation;

2) make provisions to safeguard the economic security, economic sovereignty and economic interests of the Russian Federation and the economic interests of the subjects of the Russian Federation and of Russian citizens;

3) ensure state regulation of foreign trade including financial, currency, credit and customs (tariff and non-tariff) regulation and the performance of export control; formulate policy on the certification of exported and imported goods;

4) establish standards and criteria for the safety and/or hazardous nature of imported goods for human use, these standards and criteria to be mandatory on the entire territory of the Russian Federation, and rules for their enforcement;

5) determine the procedure for import and export of armaments, military equipment and

stores; provide technical assistance in the building of military facilities abroad, transferring technical documentation and organizing licensed production and modernization and repair of military equipment; and provide other services in the field of military–technical cooperation and cooperation with foreign states in the field of rocket and space engineering;

6) establish procedure for the export and import of fissionable materials, toxic, explosive, poisonous and psychotropic substances, strong drugs, biologically active materials (blood, internal organs and other materials), genetically active materials (cultures of fungi, bacteria, viruses, animal and human semen and other materials), animals and plants of endangered species, parts and derivatives, as well as the procedure for the use thereof;

7) establish procedure for the import and export of toxic wastes and for the use thereof;

8) establish procedure for the export of certain kinds of primary goods, materials, equipment, technologies and scientific and technical information and for the provision of services which are used or can be used for the creation of armaments and military equipment or which are intended for peaceful purposes but can be used for the creation of nuclear, chemical and other types of weapons of mass destruction or missile systems for their delivery;

9) establish procedure for the export of certain strategically important raw materials under the international obligations of the Russian Federation, for the import of raw materials to be processed on the customs territory of the Russian Federation and for the export of products obtained by processing these materials;

10) establish procedure for the import and export of precious metals, precious stones and articles made therefrom, precious metals and precious stones scrap, waste from their processing and chemical compounds containing precious metals;

11) establish indicators for statistical reports relating to foreign trade activity, to be mandatory on the entire territory of the Russian Federation;

12) grant state credits and other kinds of economic assistance to foreign states, their juridical persons and international organizations, conclude international agreements for external borrowing by the Russian Federation and the granting of state credits to the Russian Federation by foreign states, and establish the maximum amount of state credits of the Russian Federation and external borrowing of the Russian Federation;

13) form and use the official gold and currency reserves of the Russian Federation;

14) draw up the balance of payments of the Russian Federation;

15) attract state, banking and commercial credits under the guarantees of the Government of the Russian Federation and control their use;

16) establish the limit for the external debt of the Russian Federation, service this debt and make arrangements for the repayment by foreign states of their debts to the Russian Federation;

17) conclude international agreements in the field of foreign economic relations;

18) participate in the activity of international economic and scientific–technical organizations and in the implementation of resolutions adopted by these organizations;

19) establish and control the operation of trade representations of the Russian Federation abroad and the representations of the Russian Federation at the international economic and scientific–technical organizations; and

20) own, use and manage the federal state property of the Russian Federation abroad.

### Article 12. Federal executive bodies responsible for state regulation of foreign trade activity

The state foreign trade policy shall be carried out through the application of economic and administrative methods of regulation of foreign trade activity under this Law, other federal laws and other normative legal acts of the Russian Federation.

In accordance with the Constitution of the Russian Federation and federal laws, the President of the Russian Federation shall:

1) direct the foreign trade policy of the Russian Federation;

2) include a section on the state foreign trade policy in the annual messages to the Federal Assembly of the Russian Federation on the situation in the country and basic directions of the internal and external policy of the state;

3) regulate cooperation in the military–technical field;

4) establish procedure for the export of precious metals, precious stones and fissionable materials;

5) have the right to impose economic sanctions recognized by international law for the purpose of safeguarding the national security of the Russian Federation;

6) when considering it necessary, under part 1 of article 85 of the Constitution of the Russian Federation, use conciliation procedures to settle differences between bodies of state power of the Russian Federation and those of the subjects of the Russian Federation on matters concerning state foreign trade policy and, if no agreement is reached, have the right to submit the dispute for settlement to an appropriate court; and

7) when considering it necessary, under part 2 of article 85 of the Constitution of the Russian Federation, suspend acts passed by the executive bodies of the Russian Federation subjects on matters concerning state foreign trade policy pending the settlement of the matter by an appropriate court.

The Government of the Russian Federation shall:

1) ensure the pursuit in the Russian Federation of a common foreign trade policy, take measures to carry out this policy, adopt appropriate decisions and enforce them;

2) draft a federal programme for the development of foreign trade activity and submit this programme for approval by the Federal Assembly of the Russian Federation;

3) take provisional measures to protect the internal market of the Russian Federation;

4) establish customs tariff rates within the limits laid down by federal laws;

5) impose quantitative restrictions on export and import in accordance with federal laws;

6) within its competence take decisions on holding negotiations and concluding international treaties;

7) manage the federal property of the Russian Federation abroad; and

8) in accordance with para. (g), part 1 of article 114 of the Constitution of the Russian Federation, exercise other powers vested in it by the Constitution, federal laws and decrees of the President of the Russian Federation in the field of state management of foreign trade activity.

Proposals concerning the foreign trade policy of the Russian Federation, regulation of the foreign trade activity of its participants and the conclusion of international treaties in the field of foreign trade relations shall be elaborated by a federal executive body directly charged by the Government of the Russian Federation with the coordination and regulation of foreign trade activity, together with other federal executive bodies within the limits of their competence. Wherever the interests of the subjects of the Russian Federation are involved, the said proposals shall be elaborated with the participation of the appropriate executive bodies of the subjects of the Russian Federation.

The federal executive body indicated in the fourth part of this article shall be responsible for the direct implementation of such objectives of the state foreign trade policy as the protection of the economic interests of the Russian Federation and of the subjects of the Russian Federation and Russian citizens, and the elaboration and implementation of measures connected with the regulation of foreign trade activity.

The federal executive body indicated in the fourth part of this article shall be the only body of state power to issue licences for export and import operations in respect of which quantitative restrictions are set or authorization is needed under the provisions of this federal law.

_____

Source: Sobranie zakonodatelstva Rossiyskoy Federatsii [Collection of legislative acts of the Russian Federation], no. 42 (1995), pp. 7409–10, 7413 (article 3923).

## 10. DECISION OF THE GOVERNMENT OF THE RUSSIAN FEDERATION ON GRANTING ENTERPRISES OF THE RUSSIAN FEDERATION THE RIGHT TO PARTICIPATE IN MILITARY–TECHNICAL COOPERATION WITH FOREIGN COUNTRIES

(with alterations introduced by Decision of the Government of the Russian Federation of 4 September 1995, no. 879)

In order to increase the effectiveness of the military–technical cooperation of the Russian

Federation with foreign countries, the Government of the Russian Federation hereby decides as follows:

1. To approve the enclosed regulations on the certification and registering of enterprises for the right to export armaments, military equipment and military-purpose work and services, and to put it into effect from 1 May 1994.

2. To grant enterprises which develop and manufacture armaments and military equipment and are certified and registered as participants in foreign economic activity in the area of military–technical cooperation the right to:

– search for potential foreign customers in the countries with which military–technical cooperation is not prohibited;

– arrange demonstrations and provide during negotiations tactical and technical characteristics and specifications of armaments and military equipment permitted for export;

– quote approximate prices mutually agreed in the established manner;

– carry out promotional and other marketing activities;

– sign contracts and, under licences obtained in the established manner, independently export the armaments and military equipment manufactured by them above the volume of the state defence order, as well as military-purpose work and services; and

– select intermediary agents from the organizations which are permitted in the established manner to carry out foreign economic activity in the area of military–technical cooperation.

Chairman of the Government of the Russian Federation V. Chernomyrdin
Moscow
6 May 1994
No. 479

*Source: Sobranie zakonodatelstva Rossiyskoy Federatsii* [Collection of legislative acts of the Russian Federation], no. 4 (1994), p. 557 (article 364).

## 11. REGULATIONS ON THE CERTIFICATION AND REGISTERING OF ENTERPRISES FOR THE RIGHT TO EXPORT ARMAMENTS, MILITARY EQUIPMENT AND MILITARY-PURPOSE WORK AND SERVICES

Approved by the Decision of the Government of the Russian Federation of 6 May 1994, no. 479 (with alterations and amendments introduced by the decrees of the President of the Russian Federation of 30 December 1994, no. 2251; of 3 March 1995, no. 236; of 5 October 1995, no. 1008; of 8 May 1996, no. 686; of 14 August 1996, no. 1177; and of 6 September 1996, no. 1320)

1. The present regulations, elaborated in order to ensure implementation of the Decree of the President of the Russian Federation of 12 May 1992, no. 507, 'On the military–technical cooperation of the Russian Federation with foreign states', shall determine the procedure for certifying and registering enterprises which are developers and manufacturers of armaments and military equipment (hereinafter referred to as enterprises) for the right to export armaments, military equipment and military-purpose work and services (hereinafter referred to as military–technical cooperation).

2. The certification and registration shall be done with the aim of:

– assessing the potential foreign economic activities of enterprises in the area of military–technical cooperation;

– protecting the state interests of the Russian Federation in the course of the activities of enterprises in the area of military–technical cooperation; and

– creating the necessary conditions for the coordination and supervision of the activities of enterprises in the area of military–technical cooperation.

3. The certification of enterprises for the right to engage in military–technical cooperation shall be carried out by the Interdepartmental Coordinating Council for Military–Technical Policy upon recommendation of the Ministry of Defence Industry.

4. Enterprises shall be certified for the right to engage in military–technical cooperation only within the range of armaments and military equipment being developed and manufactured by them.

Export deliveries of products of a military–technical nature produced by other enterprises shall be permitted if these products are, in conformity with standard technical documentation, the constituent parts of a system or a complex of armaments manufactured by the exporter enterprise.

5. In order to be certified, enterprises shall submit to the Ministry of Defence Industry an application and a set of documents in accordance with the appendix [not reproduced here].

The Ministry of Defence Industry shall be entitled with the help of experts to carry out inspections of enterprises as well as to demand additional information from these enterprises.

6. The Ministry of Defence Industry shall send copies of applications with the relevant documents attached to the Ministry of Foreign Economic Relations, the Ministry of Defence, the Federal Security Service, the Ministry of the Economy, the State Customs Committee and the State Company Rosvooruzhenie and appoint a date for consideration of the applications, which is to take place within three weeks from the date of receipt. The date of the examination shall be brought to the notice of the applicant a week before it takes place.

Applications from enterprises shall be considered by the Certifying Commission under the Ministry of Defence Industry (hereinafter referred to as the Commission), made up of representatives of the Ministry of Defence Industry, the Ministry of Foreign Economic Relations, the Ministry of Defence, the Federal Security Service, the Ministry of the Economy and the State Company Rosvooruzhenie. The staff composition of the Commission shall be approved by the Ministry of Defence Industry.

7. As a result of the meeting of the Commission, an appropriate protocol shall be drawn up and signed by members of the Commission or by the persons authorized to do so by the relevant organization.

The Ministry of Defence Industry within a week from the date of the meeting shall submit to the Interdepartmental Coordinating Council for Military–Technical Policy a proposal to grant a particular enterprise the right to engage in military–technical cooperation with the protocol of the meeting and the draft decision of the Government of the Russian Federation enclosed.

8. In the event of refusal of the right to engage in military–technical cooperation, the reasons for this shall be recorded in the protocol of the meeting of the Commission as well as, as and when needed, the expert opinions of members of the Commission or persons authorized by relevant organizations. A written reply stating reasons shall be sent to the enterprise within a week from the moment of the decision being taken.

9. On the basis of the decision of the Government of the Russian Federation to grant an enterprise the right to engage in military–technical cooperation, the Ministry of Foreign Economic Relations within a period of one month shall register the enterprise as a participant in foreign economic activities in the area of military–technical cooperation. Registered enterprises shall be given a certificate of registration, signed by the Deputy Minister of Foreign Economic Relations and stamped with its seal.

The Ministry of Foreign Economic Relations shall keep the register of enterprises entitled to engage in foreign economic activities in the area of military–technical cooperation and send notification of the inclusion of enterprises in the register to the Ministry of Defence Industry, the State Customs Committee, the Federal Security Service, the Federal External Intelligence Service, the Ministry of Finance, the Central Bank, the Ministry of the Economy, the Ministry of Defence, the Ministry of Foreign Affairs, the State Committee for State Property Management, the State Company Rosvooruzhenie and the trade representations of the Russian Federation abroad.

10. The enterprises registered as participants in foreign economic activities in the area of military–technical cooperation shall submit to the Ministry of Defence Industry, the Ministry of Foreign Economic Relations, the Ministry of Defence, the Ministry of Finance and the State Company Rosvooruzhenie the following documents:

– certified copies of contracts and supplementary agreements within 10 days after the conclusion of the transaction;

– documented data on the progress of contracts (stages of their execution) and on their completion; and

– information on the entry of currency capital to the accounts of enterprises as payment for work executed under contracts (not to be submitted to the Ministry of Defence).

11. The Interdepartmental Coordinating Council for Military–Technical Policy, upon recommendation of the Ministry of Foreign Economic Relations, the State Customs Committee, the Federal Security Service, the Federal External Intelligence Service, the Ministry of Defence Industry, the Ministry of Finance, the Central Bank, the Ministry of the Economy, the Ministry of Defence, the Ministry of Foreign Affairs and the State Company Rosvooruzhenie, shall have the right to suspend the activity of enterprises engaged in military–technical cooperation for periods of up to three months in the event of:

– violation by an enterprise of the legislation of the Russian Federation, decrees of the President of the Russian Federation or decisions of the Government of the Russian Federation, of the instructions of the Ministry of Foreign Economic Relations, the Ministry of Finance, the Central Bank or the State Customs Committee concerning military–technical cooperation or export control, or of monetary or tariff regulations in the area of foreign economic activity;
– violation of the legislation of foreign countries, deliberately or by negligence, which inflicts or may inflict damage on the economic, military or political interests of the Russian Federation;
– non-fulfilment of obligations under the state defence order, effecting unlawful transactions;
– violation of the secrecy regime;
– non-fulfilment of financial obligations in settling accounts with suppliers of component items, connected with the execution of export deliveries; or
– unfair competition among Russian exporters on foreign markets.

The Interdepartmental Coordinating Council for Military–Technical Policy shall send notification of its decision to the Ministry of Foreign Economic Relations, the State Customs Committee, the Federal Security Service, the Federal External Intelligence Service, the Ministry of Defence Industry, the Ministry of Finance, the Central Bank, the Ministry of the Economy, the Ministry of Defence, the Ministry of Foreign Affairs, the State Committee for State Property Management and the State Company Rosvooruzhenie, and instruct the relevant ministries and departments to prepare draft decisions for the Government of the Russian Federation on

depriving the enterprises of the right to engage in military–technical cooperation.

On the basis of a decision by the Government of the Russian Federation to deprive an enterprise of the right to carry on military–technical cooperation, the Ministry of Foreign Economic Relations shall strike it off the register of enterprises having the right to engage in foreign trade in the area of military–technical cooperation and notify the Ministry of Defence Industry, the State Customs Committee, the Federal Security Service, the Federal External Intelligence Service, the Ministry of Finance, the Central Bank, the Ministry of the Economy, the Ministry of Defence, the Ministry of Foreign Affairs, the State Committee for State Property Management, the State Company Rosvooruzhenie and trade representatives of the Russian Federation abroad of this.

_____

*Source: Sobranie zakonodatelstva Rossiyskoy Federatsii* [Collection of legislative acts of the Russian Federation], no. 4 (1994), pp. 558–63 (article 364).

## 12. DECISION OF THE GOVERNMENT OF THE RUSSIAN FEDERATION ON MEASURES FOR IMPROVING THE SYSTEM OF CONTROL OVER THE EXPORT AND IMPORT OF MILITARY-PURPOSE PRODUCTS, WORK AND SERVICES IN THE RUSSIAN FEDERATION (BASIC PROVISIONS)

In order to improve the system of control over the export and import of military-purpose products, work and services and in connection with the establishment of the State Committee of the Russian Federation on Military–Technical Policy* the Government of the Russian Federation hereby decides:

1. To approve the:
Regulations on the procedure for licensing the export and import of military-purpose products, work and services in the Russian

_____

* In accordance with Decree no. 1177 of the President of the Russian Federation of 14 Aug. 1996 the State Committee on Military–Technical Policy was abolished and its functions transferred to the Ministry of Foreign Economic Relations.

Federation. These regulations shall be effective from 1 September 1995;

Inventory of military-purpose products, work and services, the export and import of which are subject to control and carried out under licences issued by the State Committee of the Russian Federation on Military–Technical Policy.

2. Licences for export and import of military-purpose products, work and services issued earlier by the Ministry of Foreign Economic Relations shall remain valid up to the expiration of their term.

3. To consider null and void the Decision of the Government of the Russian Federation of 24 July 1992, no. 517, 'On the procedure for licensing in the Russian Federation of deliveries of special component articles for the manufacture of armaments and military equipment within the territories of the CIS member states' (Collection of legislative acts of the President and Government of the Russian Federation, no. 5 (1992), article 247); the Resolution of the Council of Ministers of the Government of the Russian Federation of 28 January 1993, no. 80, 'On the procedure for licensing export and import of military-purpose products and work (services) on the territory of the Russian Federation' (Collection of legislative acts of the President and Government of the Russian Federation, no. 6 (1993), article 484); and item 3 of the Resolution of the Government of the Russian Federation of 6 May 1994, no. 479, 'On granting the enterprises of the Russian Federation the right to participate in military–technical cooperation with foreign countries' (Collection of legislative acts of the President and Government of the Russian Federation, no. 4 (1994), article 364).

Chairman of the Government of the Russian Federation V. Chernomyrdin

Moscow
4 Sep. 1995
No. 879

*Source: Sobranie zakonodatelstva Rossiyskoy Federatsii* [Collection of legislative acts of the Russian Federation], no. 37 (1995), pp. 6788–89 (article 3626).

## 13. REGULATIONS ON THE PROCEDURE FOR LICENSING EXPORT AND IMPORT OF MILITARY-PURPOSE PRODUCTS, WORK AND SERVICES IN THE RUSSIAN FEDERATION

Approved by the Decision of the Government of the Russian Federation of 4 September 1995, no. 879 (with the additions introduced by the Decision of the Government of the Russian Federation of 8 May 1996, no. 686; of 11 June 1996, no. 697; of 14 August 1996, no. 1177; and of 6 September 1996, no. 1326)

1. The present Regulations establish the procedure for licensing in the Russian Federation of the export and import of military-purpose equipment, work and services, and shall be applied to all juridical and physical persons of the Russian Federation.

2. The export and import of military equipment, work and services shall be carried on in conformity with decisions of the Government of the Russian Federation concerning licences issued by the Ministry of Foreign Economic Relations.

The transit of military-purpose equipment, as well as its transport across the customs border of the Russian Federation, shall be done without licensing; military pass permits, issued in accordance with the established manner, shall be used.

A licence for the export or import of military-purpose products, work and services shall be issued for each foreign trade transaction. In specific cases the export and import of military-purpose products, work and services shall be carried out without special decisions of the Government of the Russian Federation, exclusively on the basis of licences issued by the Ministry of Foreign Economic Relations. The following cases fall into this category:

– the transport for repair purposes of Russian samples of armaments, military equipment and training and auxiliary equipment of a military–technical nature, including component parts, across the customs border of the Russian Federation;

– the export and import of special component items to armaments and military equipment in order to ensure production of and repairs to military-purpose products by Russian industrial enterprises on the basis of inter-factory cooperation agreements with enterprises of foreign countries;

– the export of special component items to ensure production of military-purpose products in foreign countries under Russian licences;

– the export of spare parts, training and auxiliary stores to the armaments and military equipment formerly delivered to foreign countries or in service in the armed forces of the members of the Commonwealth of Independent States, as well as work on their technical maintenance and repairs, including repairs with the use of mass-produced component items replacing parts withdrawn from production; and

– the import of spare parts to ensure the operation of and repairs to armaments, military equipment, military-purpose training and auxiliary stores used by the Russian Federation Army.

The import of items of armaments and military equipment from the members of the Commonwealth of Independent States for the needs of the Ministry of Defence and Ministry of Internal Affairs of the Russian Federation, the Federal Agency for Governmental Communication and Information under the President of the Russian Federation and the Federal Border Guard Service of the Russian Federation shall be conducted by these ministries and departments within the framework of the state defence order and approved by the Government of the Russian Federation under licences issued by the Ministry of Foreign Economic Relations.

3. Licences shall be granted exclusively to juridical persons of the Russian Federation who have obtained in the established manner the right to conduct foreign economic activity in the area of military–technical cooperation.

Official registration of these juridical persons shall be effected by the State Committee on Military–Technical Policy.

Enterprises, ministries and departments mentioned in item 2 of the present Regulations importing military-purpose products to the Russian Federation from the CIS member states, as well as industrial enterprises listed in the register of the Ministry of Defence Industry as developers or manufacturers of armaments, military equipment and ammunition and which are exporting and importing special component items on the basis of inter-factory cooperation and export of spare parts, technical maintenance and repair of arma-

ments and military equipment agreements within the territories of the CIS members, shall not be subjects of obligatory registration at the Ministry of Foreign Economic Relations as participants in foreign economic activity in the area of military–technical cooperation.

Mutual deliveries of special component items for the manufacture of armaments and military equipment within the CIS framework shall be done by enterprises of the Russian Federation according to the order established in intergovernmental agreements on scientific and technical cooperation between enterprises of the defence branches of industry.

To fulfil contractual obligations with a foreign customer an applicant is entitled to conclude agreements to manufacture and deliver products, conduct work and provide services of a military–technical nature directly only with Russian developers or manufacturers of armaments, military equipment and ammunition listed in the register of the Ministry of Defence Industry as well as with analogous developers and manufacturers from the Commonwealth of Independent States vested with corresponding powers.

4. Applications for licences and licences for export and import of military-purpose products, work and services shall be drawn up in accordance with the procedures established by the Ministry of Foreign Economic Relations of the Russian Federation.

Trade classification codes are not indicated in applications for licences or in the licences for export and import of military-purpose products, work and services. While exporting or importing armaments and military equipment these codes shall be indicated solely in the customs declarations in accordance with the customs legislation of the Russian Federation.

5. Applications for licences for the export of military-purpose products, work and services shall be coordinated with the Ministry of Defence.

Applications for licences for the import of military-purpose products, work and services shall be coordinated with the Ministry of Defence, as well as with ministries or departments interested in the purchases (the Ministry of the Interior, the Federal Agency for Governmental Communication and Information under the President of the Russian Federation and the Federal Border Guard Service).

In licensing the export and import of military-purpose products, work and services, control and responsibility shall be distributed between ministries and departments of the Russian Federation as follows.

The Ministry of Foreign Economic Relations shall be responsible for the legality of licences issued for the export and import of military-purpose products, work and services in conformity with the present Regulations.

The Ministry of Defence shall:

– if necessary, confirm to the federal executive authorities and applicants the classification of the exported or imported products, work and services according to the category of military-purpose products, work and services;

– evaluate the expediency of exporting the military-purpose products in the volumes requested and bear responsibility for the conformity of products specified in the export licence application with the types of armaments and military equipment permitted for export in accordance with the established manner;

– confirm that an applicant in his application for an export licence has correctly classified the military-purpose production to be exported either as standard armaments or military equipment or as spare parts for them, and that the products declared are not subject to licensing in the order stipulated by the documents on export control; and

– confirm that the military-purpose products, work and services listed in the licence application are actually the subjects of activities of the juridical persons participating in the foreign trade transaction.

The Ministry of Defence Industry shall provide the Ministry of Foreign Economic Relations with information on juridical persons registered as developers or manufacturers of armaments, military equipment and ammunition.

Ministries and departments of the Russian Federation which have participated in working on an application for a licence to import military-purpose products, work and services shall give the applicant the original end-user certificate when making each foreign trade deal and be responsible for the use of these products, work and services for the purposes declared.

6. The grounds for issuing a licence to export military-purpose products, work and services shall be:

– a decision of the Government of the Russian Federation;

– an application, drawn up and coordinated in the established order;

– a signed or initialled contract;

– a permit (licence) of the authorized state body of the country on whose territory the foreign firm (which has concluded a contract with an applicant for the foreign economic operation with military-purpose products, work and services) is registered;

– the original of the end-user's international or national import certificate, issued by the authorized state body and containing the obligations of the recipient country to use the military-purpose products, work and services imported from the Russian Federation only for needs of that country, as well as to prevent their re-export or transfer to third countries without the consent of the Russian side; and

– signed or initialled agreements of the applicant with the developers or manufacturers of military-purpose products, work and services registered by the Ministry of Defence Industry.

7. The grounds for issuing a licence for the import of military-purpose products, work and services shall be:

– a decision of the Government of the Russian Federation;

– an application, drawn up and coordinated in the established manner;

– a signed or initialled contract;

– the original of the end-user certificate issued by the ministry or department of the Russian Federation in whose interests the import is to be carried out; and

– a signed or initialled agreement of an applicant with a ministry or department of the Russian Federation in whose interests the import is to be carried out.

When importing military-purpose products, work and services to the Russian Federation for subsequent re-export to third countries, an original Russian end-user's certificate shall not be presented.

8. In the event of improper registration of documents required for a licence to be issued, the Ministry of Foreign Economic Relations shall be entitled to request the applicant to submit additional documents and information necessary for making a decision on issuing a licence for the export or import of military-purpose products, work and services.

Responsibility for the authenticity of information presented to the Ministry of Foreign Economic Relations in order to obtain a licence lies with the juridical person who applied for a licence.

9. Consideration of applications and the drawing up and issuing of licences shall be done on a payment basis. The amount of fees and the procedure for using the receipts shall be established by the Ministry of Foreign Economic Relations in coordination with the Ministry of Finance.

Fees shall not be charged for the examination of applications and the drawing up and issuing of licences to export or import military-purpose products, work and services in conformity with agreements between the Government of the Russian Federation and governments of members of the Commonwealth of Independent States.

Licences for the export and import of military-purpose products, work and services issued by the Ministry of Foreign Economic Relations shall be printed on blank forms made of special paper protected against counterfeiting. The blank forms shall be considered strictly accountable documents.

The transfer of licences issued to other juridical or physical persons shall be prohibited.

Copies of licences for the export of military-purpose products, work and services shall be passed by the Ministry of Foreign Economic Relations to the Ministry of Defence Industry.

10. A licence for the export or import of military-purpose products, work and services shall be issued for a period of up to 12 months.

The validity of the licence shall end on the date indicated therein.

The period of validity of the licence can be extended at the request, stating reasons, of an applicant according by order of the Ministry of Foreign Economic Relations. The extension of a licence and any other alteration to the licence shall be effected by the Ministry of Foreign Economic Relations in written form and coordinated with the State Customs Committee.

11. A licence for the export and import of military-purpose products, work and services or a notification of a refusal of a licence, with reasons given, shall be sent to the applicant within 25 days of the date of receipt of the application by the Ministry of Foreign Economic Relations. In the event of a request being made by the Ministry of Foreign Economic Relations for additional documents or information necessary for making the decision on the issuance of a licence, the period indicated shall start from the date of their receipt by the Ministry of Foreign Economic Relations and shall not exceed 15 days.

12. The Ministry of Foreign Economic Relations shall be entitled to cancel the licence or suspend it in the event of violation by the applicant of the rules and procedures established by the legislation of the Russian Federation for military–technical policy, and its decision shall be final.

13. The Ministry of Foreign Economic Relations shall exercise control over the level of export prices and establish the procedure and time-limits for presentation by the [licensee] of the necessary information on their use for the purposes of statistical accounting and reports on the work done in the area of military–technical cooperation of the Russian Federation with foreign countries.

14. Control over the shipment of military–purpose export and import products across the customs border of the Russian Federation shall be effected by the State Customs Committee of the Russian Federation.

15. Violations of the provisions of the present regulation shall be punished by law.

_____

*Source: Sobranie zakonodatelstva Rossiyskoy Federatsii* [Collection of legislative acts of the Russian Federation], no. 37 (1995), pp. 6789–97 (article 3626).

---

## 14. INVENTORY OF MILITARY-PURPOSE PRODUCTS, WORK AND SERVICES THE EXPORT AND IMPORT OF WHICH ARE SUBJECT TO CONTROL AND PERFORMED UNDER LICENCES ISSUED BY THE MINISTRY OF FOREIGN ECONOMIC RELATIONS OF THE RUSSIAN FEDERATION

Approved by the Decision of the Government of the Russian Federation of 4 September 1995, no. 879

1. Tanks and other self-propelled armoured vehicles with or without weapons
1.1. Group and complete repair sets of spare parts to commodities in category 1

1.2. Special auxiliary and support equipment, spare parts and component items to commodities in category 1

1.3. Technical documentation (normative–technical, design, technological and programming) to commodities in categories 1, 1.1 and 1.2

1.4. Military-purpose work and services carried out for a customer and connected with the commodities in categories 1, 1.1, 1.2 and 1.3

2. Motor cars and other self-propelled military-purpose vehicles (wheeled or tracked)

2.1. Group and complete repair sets of spare parts to commodities in category 2

2.2. Special auxiliary and rear equipment, spare parts and component items to commodities in category 2

2.3. Technical documentation (normative–technical, design, technological and programming) to commodities in categories 2, 2.1 and 2.2

2.4. Military-purpose work and services carried out for a customer connected with the commodities in categories 2, 2.1, 2.2 and 2.3

3. Military-purpose means for fitting out troops with engineering facilities (bridge-building machinery, construction engineering machines, anti-mine detachments, vehicles, repair shops, pontoons, etc.)

3.1. Group and complete repair sets of spare parts to commodities in category 3

3.2. Special auxiliary and rear equipment, spare parts and component items to commodities in category 3

3.3. Technical documentation (normative–technical, design, technological and programming) to commodities in categories 3, 3.1 and 3.2

3.4. Military-purpose work and services carried out for a customer connected with the commodities in categories 3, 3.1, 3.2 and 3.3

4. Aircraft, helicopters and other military-purpose flying vehicles

4.1. Group and complete repair sets of spare parts to commodities in category 4

4.2. Special auxiliary and support equipment, spare parts and component items to commodities in category 4

4.3. Technical documentation (normative–technical, design, technological and programming) to commodities in categories 4, 4.1 and 4.2

4.4. Military-purpose work and services carried out for a customer connected with the commodities in categories 4, 4.1, 4.2 and 4.3

5. Warships and submarines, auxiliary military ships and submarines

5.1. Group and complete repair sets of spare parts to commodities in category 5

5.2. Special auxiliary and rear equipment, spare parts and component items to commodities in category 5

5.3. Technical documentation (normative–technical, design, technological and programming) to commodities in categories 5, 5.1 and 5.2

5.4. Military-purpose work and services carried out for a customer connected with the commodities in categories 5, 5.1, 5.2 and 5.3

6. Combat weapons (artillery units, missile and bomb launchers, torpedo tubes, howitzers, grenade discharges, mortars and similar weapons for conducting combat actions)

6.1. Group and complete repair sets of spare parts to commodities in category 6

6.2. Special auxiliary and support equipment, spare parts and component items to commodities in category 6

6.3. Technical documentation (normative–technical, design, technological and programming) to commodities in categories 6, 6.1 and 6.2

6.4. Military-purpose work and services carried out for a customer connected with the commodities in categories 6, 6.1, 6.2 and 6.3

7. Military-purpose small arms (except for commodity items 9303 and 9304 of the CN FEA) of 14.5-mm calibre and less

7.1. Group and complete repair sets of spare parts to commodities in category 7

7.2. Special auxiliary and support equipment, spare parts and component items to commodities in category 7

7.3. Technical documentation (normative–technical, design, technological and programming) to commodities in categories 7, 7.1 and 7.2

7.4. Military-purpose work and services carried out for a customer connected with the commodities in categories 7, 7.1, 7.2 and 7.3

8. Bombs, grenades, torpedoes, mines, missiles and similar weapons for conducting combat actions

8.1. Group and complete repair sets of spare parts to commodities in category 8

8.2. Special auxiliary and support equipment, spare parts and component items to commodities in category 8

8.3. Technical documentation (normative–technical, design, technological and programming) to commodities in categories 8, 8.1 and 8.2

8.4. Military-purpose work and services carried out for a customer connected with the commodities in categories 8, 8.1, 8.2 and 8.3

9. Gunpowder

9.1. Technical documentation (normative–technical, design, technological and programming) to commodities in category 9

9.2. Military-purpose work and services carried out for a customer connected with the commodities in categories 9 and 9.1

10. Military-purpose finished explosives (except gunpowder)

10.1. Technical documentation (normative–technical, design, technological and programming) to commodities in category 10

10.2. Military-purpose work and services carried out for a customer connected with the commodities in categories 10 and 10.1

11. Military-purpose explosive and pyrotechnic means (Bickford and detonating cords, percussion and detonating caps, fuses, electric detonators, fireworks, signal rockets and similar explosive and pyrotechnic means)

11.1. Technical documentation (normative–technical, design, technological, programming) to commodities in category 11

11.2. Military-purpose work and services carried out for a customer connected with the commodities in categories 11 and 11.1

12. Military-purpose telescopic and laser gun sights, periscopes, optical tubes, lasers

12.1. Group and complete repair sets of spare parts to commodities in category 12

12.2. Special auxiliary and support equipment, spare parts and component items to commodities in category 12

12.3. Technical documentation (normative–technical, design, technological and programming) to commodities in categories 12, 12.1 and 12.2

12.4. Military-purpose work and services carried out for a customer connected with the commodities in categories 12, 12.1, 12.2 and 12.3

13. Military-purpose navigational devices

13.1. Group and complete repair sets of spare parts to commodities in category 13

13.2. Special auxiliary and support equipment, spare parts and complementary articles to commodities in category 13

13.3. Technical documentation (normative–technical, design, technological, programming) to commodities in categories 13, 13.1 and 13.2

13.4. Military-purpose work and services carried out for a customer connected with the commodities in categories 13, 13.1, 13.2 and 13.3

14. Military-purpose hydroacoustic, radio-locating, radio-navigational and range-guide radio devices

14.1. Group and complete repair sets of spare parts to commodities in category 14

14.2. Special auxiliary and support equipment, spare parts and complementary articles to commodities in category 14

14.3. Technical documentation (normative–technical, design, technological and programming) to commodities in categories 14, 14.1 and 14.2

15. Military-purpose parachutes (including dirigible parachutes) and rotational parachutes

15.1. Group and complete repair sets of spare parts to commodities in category 15

15.2. Special auxiliary and support equipment, spare parts and complementary articles to commodities in category 15

15.3. Technical documentation (normative–technical, design, technological and programming) to commodities in categories 15, 15.1 and 15.2

15.4. Military-purpose work and services carried out for a customer connected with the commodities in categories 15, 15.1, 15.2 and 15.3

16. Military-purpose transmitting devices for radio-telephone and radio-telegraph communication, radio or TV broadcasting, whether or not including the receiving, sound recording or sound reproducing equipment, and television cameras

16.1. Group and complete repair sets of spare parts to commodities in category 16

16.2. Special auxiliary and support equipment, spare parts and complementary articles to commodities in category 16

16.3. Technical documentation (normative–technical, design, technological and programming) to commodities in categories 16, 16.1 and 16.2

16.4. Military-purpose work and services carried out for a customer connected with the commodities in categories 16, 16.1, 16.2 and 16.3

17. Protective means against war gases

17.1. Group and complete repair sets of spare parts to commodities in category 17

17.2. Special auxiliary and support equipment, spare parts and complementary articles to commodities in category 17

17.3. Technical documentation (normative–technical, design, technological and programming) to commodities in categories 17, 17.1 and 17.2

17.4. Military-purpose work and services carried out for a customer connected with the commodities in categories 17, 17.1, 17.2 and 17.3

18. Specially developed equipment, devices and facilities for the manufacture and repair of ammunition, armaments and military machinery

18.1. Group and complete repair sets of spare parts to commodities in category 18

18.2. Special auxiliary and support equipment, spare parts and complementary articles to commodities in category 18

18.3. Technical documentation (normative–technical, design, technological and programming) to commodities in categories 18, 18.1 and 18.2

19. Military uniforms and attributes

19.1. Group and complete repair sets of spare parts to commodities in category 19.

---

*Source: Sobranie zakonodatelstva Rossiyskoy Federatsii* [Collection of legislative acts of the Russian Federation], no. 26 (1994), pp. 6794–97 (article 3626).

---

## 15. INSTRUCTION OF THE GOVERNMENT OF THE RUSSIAN FEDERATION

Pursuant to the Decision of the Government of the Russian Federation of 5 May 1994, no. 479, on granting enterprises the right to participate in military–technical cooperation of the Russian Federation with foreign countries (Collection of legislative acts of the President and Government of the Russian Federation, no. 4 (1994), article 364), approves the proposal of the Interdepartmental Commission on Military–Technical Cooperation of the Russian Federation with Foreign Countries* on granting to the Moscow Aviation Production Association the right to participate in the established manner in military–technical cooperation with foreign countries within the schedule of armaments

---

* The Interdepartmental Commission on Military and Technical Cooperation with Foreign Countries was abolished and its functions were transferred to the Interdepartmental Coordinating Council for Military Technical Policy.

and military equipment, types of work and services specified in the appendix.

Chairman of the Government of the Russian Federation V. Chernomyrdin
24 Oct. 1994
No. 1683-i

## Appendix to the Instruction of the Government of the Russian Federation of 24 October 1994, no. 1683-i

Schedule of armaments and military equipment, types of work and services within which the Moscow Aircraft Production Association shall be granted the right to participate in military–technical cooperation with foreign countries

*Schedule of armaments and military equipment*
MiG-29 plane and its modifications
*Types of work and services*
Complex export deliveries; delivery for export of complementary items and spare parts, training and auxiliary equipment; technical assistance in setting up production; assistance in operation, repairs and modernization; execution of research and development work; training of national technical engineering personnel; sending on mission trips (receiving) consultants and specialists

*Schedule of armaments and military equipment*
MiG-21 plane and its modifications
*Types of work and services*
Export deliveries of component items and spare parts, training and auxiliary equipment; technical assistance in setting up production; assistance in operation, repairs and modernization; execution of research and development work; training of national technical engineering personnel; sending on mission trips (receiving) consultants and specialists.

*Schedule of armaments and military equipment*
MiG-23 plane and modifications; Il-38 plane
*Types of work and services*
Export deliveries of component items and spare parts.

---

*Source: Sobranie zakonodatelstva Rossiyskoy Federatsii* [Collection of legislative acts of the Russian Federation], no. 26 (1994), pp. 3902–3903 (article 2821).

## 16. INSTRUCTION OF THE GOVERNMENT OF THE RUSSIAN FEDERATION

To accept the proposal of the State Committee of the Russian Federation on Defence Industries,[*] approved by the Interdepartmental Coordinating Council for Military–Technical Policy of the Russian Federation, on granting the Metrovagonmash closed-type joint-stock company (Mytishchi, Moscow Region) the right to participate in the established manner in military–technical cooperation with foreign countries for a five-year period, within the schedule of armaments and military equipment, types of work and services specified in the attached appendix.

Chairman of the Government of the Russian Federation V. Chernomyrdin
Moscow
19 Feb. 1996
No. 202-i

### Appendix to the Instruction of the Government of the Russian Federation of 19 February 1996, no. 202-i

Schedule of armaments and military equipment, types of work and services within which the Metrovagonmash closed-type joint-stock company shall be granted the right to participate in military–technical cooperation with foreign countries

*Schedule of armaments and military equipment*
Shilka air defence self-propelled system and its modifications; Kvadrat air defence missile complex and its modifications
*Types of work and services*
Delivery of spare parts, services dealing with repair and maintenance problems; repair and updating of articles, training of operational personnel in repairing and servicing the articles; delivery of chassis, units and assemblies, spare parts, repair and maintenance services; repair and updating of articles, training of maintenance and operational personnel.

*Schedule of armaments and military equipment*
Chassis of the Tunguska air defence missile complex and its modifications; chassis of the

Tor air defence missile complex and its modifications; chassis of the Buk air defence missile complex and its modifications.

———

*Source: Sobranie zakonodatelstva Rossiyskoy Federatsii* [Collection of legislative acts of the Russian Federation], no. 10 (1996), pp. 2482–83 (article 996).

## 17. INSTRUCTION OF THE GOVERNMENT OF THE RUSSIAN FEDERATION

To accept the proposal of the State Committee of the Russian Federation on Defence Industries,[*] approved by the Interdepartmental Coordinating Council for Military–Technical Policy of the Russian Federation, on granting the Izhmash open-type joint-stock company (Izhevsk) the right to participate in the established manner in military–technical cooperation with foreign countries for a five-year period, within the schedule of armaments and military equipment, types of work and services specified in the attached appendix.

Chairman of the Government of the Russian Federation V. Chernomyrdin
Moscow
19 Feb. 1996
No. 203-i

### Appendix to the Instruction of the Government of the Russian Federation of 19 February 1996, no. 203-i

Schedule of armaments and military equipment, types of work and services within which the Izhmash open-type joint-stock company shall be granted the right to participate in military–technical cooperation with foreign countries

*Schedule of armaments and military equipment*
Kalashnikov sub-machine-gun and its modifications; SVD rifle and its modifications
*Types of work and services*
Export delivery; export deliveries of spare parts and auxiliary equipment; handing over licences and technical documentation for pro-

———

[*] By Decree no. 686 of the President of the Russian Federation of 8 May 1996, this Committee became the Ministry of Defence Industry.

[*] By Decree no. 686 of the President of the Russian Federation of 8 May 1996, this Committee became the Ministry of Defence Industry.

duction and technical assistance in setting up production; assistance in operating, training for use, repairs and modernization; furnishing technical documentation including specifications, operating instructions and repair manuals.

*Schedule of armaments and military equipment*
Control testing machines (CИMO2-1, 9B869, 9B871-2, B94 and 9B921)
*Types of work and services*
Export delivery; assistance in operating, training for use, repairs, modernization.

*Schedule of armaments and military equipment*
MPM-M2K mechanized repair shop
*Types of work and services*
Supplying technical documentation including specifications, operating instructions and repair manuals.

---

*Source: Sobranie zakonodatelstva Rossiyskoy Federatsii* [Collection of legislative acts of the Russian Federation], no. 10 (1996), pp. 2483–84 (article 997).

---

## 18. INSTRUCTION OF THE GOVERNMENT OF THE RUSSIAN FEDERATION

To accept the proposal of the State Committee of the Russian Federation on Defence Industries,[*] approved by the Interdepartmental Coordinating Council for Military–Technical Policy of the Russian Federation, on granting the Instrument Making Design Bureau (Tula) the right to participate in the established manner in military–technical cooperation with foreign countries for a five-year period, within the schedule of armaments and military equipment, types of work and services specified in the attached appendix.

Chairman of the Government of the
Russian Federation V. Chernomyrdin
Moscow
19 Feb. 1996
No. 204-i

---

[*] By Decree no. 686 of the President of the Russian Federation of 8 May 1996, this Committee became the Ministry of Defence Industry.

**Appendix to the Instruction of the Government of the Russian Federation of 19 February 1996, no. 204-i**

Schedule of armaments and military equipment, types of work and services within which the Instrument Making Design Bureau shall be granted the right to participate in military–technical cooperation with foreign countries

*Schedule of armaments and military equipment*
Kornet-E anti-tank missile complex and its modifications
*Types of work and services*
Export delivery; handing over licences and technical documentation for manufacturing and technical assistance in setting up production; execution of work on construction of military depots, intended for location, combat application, operation, production and repair, and ensuring their functioning; training of national technical personnel; delivery of technical documentation including specifications on production, repair and operation.

*Schedule of armaments and military equipment*
Metis light anti-tank missile and its modifications; Konkurs-M anti-tank missile complex; Kastet guided armament complex; Bastion tank complex of guided armament and modifications, including Sheksna; Krasnopol complex of guided artillery armament and modifications; complexes of guided armament for the Kitolov-2 120-mm and Kitolov-2M 122-mm calibre artillery systems; Kashtan ship-borne air defence missile artillery complex; Tunguska air defence gun-missile complex and modifications
*Types of work and services*
Handing over licences and technical documentation for manufacture and technical assistance in setting up production; execution of work on the construction and completion of military depots for location, application in combat, operation, production and repair and ensuring their functioning; training of national technical engineering personnel; delivery of technical documentation, including specifications, operating instructions and repair manuals.

*Schedule of armaments and military equipment*
Shmel light infantry flame thrower; Vikhr (9A-4172) anti-tank guided missile; PP-90

and PP-93 9-mm pistol sub-machine-guns and their modifications; Udar, Udar-1 and Udar-T compact 12.3-mm revolvers; 9A-91 9-mm sub-machine-gun

*Types of work and services*
Export delivery; handing over licences and technical documentation for manufacture and technical assistance in setting up production; training of national technical engineering personnel; delivery of technical documentation including specifications, operating instructions and repair manuals.

*Schedule of armaments and military equipment*
Complex of guided armament for the Gran 120-mm calibre mortar; extra-short range missile artillery complex based on the ZRAK Kashtan missile artillery unit; Germes self-propelled multi-purpose complex; modernization of the armament complex of the BMP-3 infantry combat vehicle; single-seat combat module for light-weight category objects; the Yastreb self-propelled air defence unit; the Lezvie self-propelled air defence unit; complex of guided armament for equipping foreign-made tanks; modernization of T-55, T-62 and T-72 tanks exported earlier by equipping them with the Kitolov-2M guided armament complex for the 122-mm artillery system; mounting of the Kornet-E missile complex on various chassis

*Types of work and services*
Execution in the established manner of the research and development ordered in coordination with the Ministry of Defence; manufacture and delivery of pilot models and their units for testing; carrying out tests; setting up and organizing full-scale production; sending on mission trips (receiving) of specialists and consultants; handing over design documentation.

*Note:* By modification of armament complexes is meant the modification, export of which is permitted in the established manner.

*Source: Sobranie zakonodatelstva Rossiyskoy Federatsii* [Collection of legislative acts of the Russian Federation], no. 10 (1996), pp. 2484–86 (article 998).

## 19. INSTRUCTION OF THE GOVERNMENT OF THE RUSSIAN FEDERATION

To accept the proposal of the State Committee of the Russian Federation on Defence Industries,* approved by the Interdepartmental Coordinating Council for Military–Technical Policy of the Russian Federation, on granting the Rostvertol open-type joint-stock company (Rostov-on-Don) the right to participate in the established manner in military–technical cooperation with foreign countries for a five-year period, within the schedule of armaments and military equipment, types of work and services specified in the attached appendix.

Chairman of the Government of the Russian Federation V. Chernomyrdin
Moscow
19 Feb. 1996
No. 205-i

**Appendix to the instruction of the Russian Government of the Russian Federation of 19 February 1996, no. 205-i**

Schedule of armaments and military equipment, types of work and services within which the Rostvertol open-type joint-stock company shall be granted the right to participate in military–technical cooperation with foreign countries

*Schedule of armaments and military equipment*
Mi-35, Mi-26 and Mi-28 helicopters and their modifications

*Types of work and services*
Complex delivery for export; export of spare parts and component items, training and auxiliary equipment; repair and modernization of helicopters and component items; technical assistance in setting up production; assistance in operation, operational training, repairs and modernization; leasing; delivery of technical documentation for operation and repairs; training of national technical engineering and flight personnel; sending on mission trips (receiving) consultants and specialists.

* By Decree no. 686 of the President of the Russian Federation of 8 May 1996, this Committee became the Ministry of Defence Industry.

*Schedule of armaments and military
equipment*
Mi-24 and Mi-25 helicopters and their
modifications
*Types of work and services*
Export of spare parts and component items,
training and auxiliary equipment; repair and
modernization of helicopters and component
items; technical assistance in setting up
production; assistance in operation, operational training, repairs and modernization;
leasing; delivery of technical documentation
for operation and repairs; training of national
technical engineering and flight personnel;
sending on mission trips (receiving) consultants and specialists.

---

*Source: Sobranie zakonodatelstva Rossiyskoy
Federatsii* [Collection of legislative acts of
the Russian Federation], no. 10 (1996),
pp. 2486–87 (article 999).

---

## 20. INSTRUCTION OF THE GOVERNMENT OF THE RUSSIAN FEDERATION

To accept the proposal of the State Committee of the Russian Federation on Defence
Industries,* approved by the Interdepartmental Coordinating Council for Military–
Technical Policy of the Russian Federation,
on granting the Ufa Motor Building Production Association (Ufa) open-type joint-stock
company the right to participate in the
established manner in military–technical
cooperation with foreign countries for a five-
year period, within the schedule of armaments
and military equipment, types of work and
services specified in the attached appendix.

Chairman of the Government of the
Russian Federation V. Chernomyrdin
Moscow
19 Feb. 1996
No. 206-i

**Appendix to the instruction of the
Government of the Russian Federation of
19 February 1996, no. 206-i**

Schedule of armaments and military equipment, types of work and services for which
the Ufa Motor Building Production Associa-

tion open-type joint-stock company shall be
granted the right to participate in military–
technical cooperation with foreign countries

*Schedule of armaments and military
equipment*
Aircraft engines: P25-300, P29BC-300,
P29B-300, P95SH, P195, P13-300, AL-31F
and their modifications
*Types of work and services*
Complex export delivery of aircraft engines
of its own production; export of component
items and spare parts, training and auxiliary
equipment; technical assistance in setting up
production; assistance in operation, repairs
and updating; setting up maintenance and
servicing centres and repair bases; delivery of
technical documentation including specifications, operating instructions and repair
manuals; training of national technical
engineering personnel; sending on mission
trips (receiving) consultants and specialists.

---

*Source: Sobranie zakonodatelstva Rossiyskoy
Federatsii* [Collection of legislative acts of
the Russian Federation], no. 10 (1996),
p. 2488 (article 1000).

---

## 21. INSTRUCTION OF THE GOVERNMENT OF THE RUSSIAN FEDERATION

To accept the proposal of the State Committee of the Russian Federation on Defence
Industries,* approved by the Interdepartmental Coordinating Council for Military–
Technical Policy of the Russian Federation,
on granting the Gidromash open-type joint-
stock company (Nizhniy Novgorod) the right
to participate in the established manner in
military–technical cooperation with foreign
countries for a five-year period, within the
schedule of armaments and military
equipment, types of work and services
specified in the attached appendix.

Chairman of the Government of the
Russian Federation V. Chernomyrdin
Moscow
19 Feb. 1996
No. 207-i

---

* By Decree no. 686 of the President of the
Russian Federation of 8 May 1996, this Committee
became the Ministry of Defence Industry.

**Appendix to the instruction of the Russian Government of the Russian Federation of 19 February 1996, no. 207-i**

Schedule of armaments and military equipment, types of work and services within which the Gidromash open-type joint-stock company shall be granted the right to participate in military–technical cooperation with foreign countries

*Schedule of armaments and military equipment*
Mechanisms of landing gear and other hydraulic systems for the MiG-21, MiG-23, MiG-25, MiG-27, MiG-29, MiG-31, Su-24, Su-25 and Su-27 aircraft and their modifications

*Types of work and services*
Export of mechanisms and units of own production; export of component items and spare parts, training and auxiliary equipment.

*Schedule of armaments and military equipment*
Mechanisms of landing gear and other hydraulic systems for Mi-24, Ka-27 and Ka-28 helicopters and their modifications

*Types of work and services*
Technical maintenance and finishing operations, handing over technological, repair and operational documentation; sending on mission trips (receiving) consultants and specialists.

---

*Source: Sobranie zakonodatelstva Rossiyskoy Federatsii* [Collection of legislative acts of the Russian Federation], no. 10 (1996), p. 2489 (article 1001).

---

## 22. INSTRUCTION OF THE GOVERNMENT OF THE RUSSIAN FEDERATION

To accept the proposal of the State Committee of the Russian Federation on Defence Industries,[*] approved by the Inter-departmental Coordinating Council for Military–Technical Policy of the Russian Federation, on granting the Antey open-type joint-stock company (Moscow) the right to

[*] By Decree no. 686 of the President of the Russian Federation of 8 May 1996, this Committee became the Ministry of Defence Industry.

participate in the established manner in military–technical cooperation with foreign countries for a five-year period, within the schedule of armaments and military equipment, types of work and services specified in the attached appendix.

Chairman of the Government of the Russian Federation V. Chernomyrdin
Moscow
19 Feb. 1996
No. 208-i

**Appendix to the instruction of the Government of the Russian Federation of 19 February 1996, no. 208-i**

Schedule of armaments and military equipment, types of work and services within which the Antey open-type joint-stock company shall be granted the right to participate in military–technical cooperation with foreign countries

*Schedule of armaments and military equipment*
S-300B air defence missile system and its modifications; Tor air defence missile system and its modifications; Senezh-M1E automated air defence control system; automated control system of the fighter plane regiment Rubezh-ME

*Types of work and services*
Complex export delivery; export of component items and spare parts, training and auxiliary equipment; handing over manufacturing licences for items of its own development, technical assistance in the organization of production; assistance in operation, repairs and modernization; conduct in the established manner of research and development work in coordination with the Ministry of Defence; training of national technical engineering personnel; sending on mission trips (receiving) consultants and specialists.

*Schedule of armaments and military equipment*
Military staff vehicle for controlling the fire of self-propelled artillery battalion 1B16M; Ulybka (1B44) meteorological radar complex for atmospheric sounding; Goloturia radar complex for spotting ground targets; Fara-U (1RL-136) radar station for short reconnaissance of ground targets; 9C80 mobile reconnaissance and control post; 1L3 radar control complex of air defence artillery; complex and servicing for Zoopark 1 recon-

naissance radar serving the artillery battalion; 1RL-232 radar station

*Types of work and services*

Complex export delivery; export of component items and spare parts, training and auxiliary equipment; handing over manufacturing licences for items of its own development, technical assistance in the organization of production; assistance in operation, repairs and modernization; conduct in the established manner of research and development work in coordination with the Ministry of Defence; training of national technical engineering personnel; sending on mission trips (receiving) consultants and specialists.

*Schedule of armaments and military equipment*

Osa-AKM air defence missile complex (9A33BM2, 9A33BM3)

*Types of work and services*

Handing over manufacturing licences for items it has developed, technical assistance in the organization of production; assistance in operation, repairs and modernization; conduct in the established manner of research and development work in coordination with the Ministry of Defence; training of national technical engineering personnel; and sending on mission trips (receiving) consultants and specialists.

---

*Source: Sobranie zakonodatelstva Rossiyskoy Federatsii* [Collection of legislative acts of the Russian Federation], no. 10 (1996), pp. 2490–91 (article 1002).

---

**23. DECISION OF THE GOVERNMENT OF THE RUSSIAN FEDERATION AFFIRMING THE STATUTE ON THE PROCEDURE FOR THE MAKING AVAILABLE OF INFORMATION BY THE RUSSIAN FEDERATION ON DELIVERIES OF CONVENTIONAL ARMS IN ACCORDANCE WITH THE WASSENAAR ARRANGEMENT**

In order to implement the Wassenaar Arrangement on Export Controls for Conventional Arms and Dual-Use Goods and Technologies in the part involving the procedure for the exchange of information on deliveries of conventional arms to foreign countries, the Government of the Russian Federation decides to affirm the attached Statute on the Procedure for the Making Available of Information by the Russian Federation on Deliveries of Conventional Arms in Accordance with the Wassenaar Arrangement.

Chairman of the Government of the Russian Federation V. Chernomyrdin
Moscow
3 Aug. 1996
No. 923

**Statute on the procedure for the making available of information by the Russian Federation on deliveries of conventional arms in accordance with the Wassenaar Arrangement**

1. The present statute defines the procedure for the making available of information by the Russian Federation to the states participating in the Wassenaar Arrangement on Export Controls for Conventional Arms and Dual-Use Goods and Technologies (hereafter called the Wassenaar Arrangement) on deliveries of the conventional arms stipulated by the UN Register of Conventional Arms to states that are not participants in the Wassenaar Arrangement (hereafter information on deliveries). The present statute was elaborated for the purpose of guaranteeing the performance of the international obligations of the Russian Federation ensuing from its participation in the Wassenaar Arrangement and extends to all participants and subjects of military–technical cooperation of the Russian Federation with foreign countries.

2. In accordance with the Wassenaar Arrangement, the participating states exchange information on deliveries every six months. In the initial stage of the development of the Wassenaar Arrangement, this information includes the name of the importing state, data on the quantity of conventional arms delivered to the indicated state in the reporting period by categories in accordance with appendix 1, and data on the models and types of these arms (other than the models and types of missiles and missile launchers).

3. Twice a year and no later than 15 January and 15 July the subjects of military–technical cooperation of the Russian Federation with foreign countries will make available to the State Committee on Military–Technical Policy and the Ministry of Defence information on deliveries of conventional

arms in the preceding half-year in conformity with appendix 2 [not reproduced here].

4. The State Committee on Military–Technical Policy correlates the information received and twice a year, no later than 10 February and 10 August, in coordination with the Ministry of Defence, will send to the Ministry of Foreign Affairs correlated information in the form indicated in appendix 2 to the present statute.

5. In the period stipulated by the Wassenaar Arrangement, the Ministry of Foreign Affairs will convey to the states through diplomatic channels in accordance with appendix 3 [not reproduced here] information on deliveries over the past half-year taking into consideration the established requirements in the Russian Federation for the conveyance of such information and also with the mandatory consent of importers to the provision of such information.

6. Information on deliveries will be confidential in all stages of its collection, processing and conveyance. Within the boundaries of the Russian Federation, information on deliveries must be registered under the classification 'Secret'.

The principle of confidentiality will extend to any use of information on deliveries, including in discussion with states participating in the Wassenaar Arrangement, and correspondence on these matters will have diplomatic status and the corresponding immunities and privileges.

7. Control of the implementation of this statute will be carried out by the State Committee of the Russian Federation on Military–Technical Policy in cooperation with the Ministry of Foreign Affairs.

8. The State Committee on Military–Technical Policy in coordination with the Ministry of Foreign Affairs will inform the participants and subjects of the military–technical cooperation of the Russian Federation with foreign countries of changes in the body of states participating in the Wassenaar Arrangement.

## Appendix 1

Categories of Conventional Arms stipulated by the UN Register of Conventional Arms for which information on deliveries is exchanged

### I. Battle tanks
Tracked or wheeled self-propelled armoured vehicles possessing high mobility in rough terrain and a high level of protection, having a dry weight of no less than 16.5 tonnes, and armed with a gun of a calibre of not less than 75 mm with a high initial velocity of the projectile for direct fire.

### II. Armoured fighting vehicles
Tracked, semi-tracked or wheeled self-propelled vehicles possessing armoured protection and cross-country capability in rough terrain, designed and equipped for the transport of an infantry squad of four or more persons and/or armed with a built-in or regularly mounted gun with a calibre of not less than 12.5 mm or a missile launcher.

### III. Large-calibre artillery systems
Cannons, howitzers and artillery pieces combining the qualities of cannons and howitzers, mortars, and reactive systems for salvo fire capable of destroying ground targets primarily from covered gun positions and having a calibre of 100 mm or more.

### IV. Combat aircraft
Aircraft with unchangeable or changeable wing geometry designed, equipped or modified for the destruction of targets through the use of guided missiles, unguided missiles, bombs, machine-guns, guns or other means of destruction, including variants of such aircraft that perform special functions of radioelectronic warfare, suppression of air defence or reconnaissance.

The term 'combat aircraft' does not include trainer aircraft for basic flight training with the exception of those that are designed, equipped, or modified as indicated above.

### V. Attack helicopters
Rotary-wing aircraft designed, equipped or modified for the destruction of targets through the use of guided or unguided anti-tank weapons and weapons of the 'air-to-ground', 'air-to-submarine' or 'air-to-air' classes and equipped with a complex system of fire control and aiming for this weapon, including variants of aircraft that perform special functions of reconnaissance or radioelectronic warfare.

### VI. Warships
Surface ships or submarines armed and equipped for use for military purposes, having a standard displacement of 750 tonnes or more or having a standard displacement of less than 750 tonnes equipped for the launching of missiles with a range of not less than 25 000 metres or torpedoes of the same range.

*VII. Missiles and missile launchers*
Guided or unguided missiles and ballistic or cruise missiles capable of delivering a warhead or means of destruction a distance of not less than 25 000 metres and systems designed or modified specifically for the launch of such guided or unguided missiles, if they are not covered by categories I–VI.

This category:

– also includes remotely piloted aircraft with the characteristics of missiles specified above;

– does not include missiles of the 'ground-to-air' class.

---

*Source: Sobranie zakonodatelstva Rossiyskoy Federatsii* [Collection of legislative acts of the Russian Federation], no. 33 (1996), pp. 8026–30 (article 3997).

---

### 24. DECREE OF THE GOVERNMENT OF THE RUSSIAN FEDERATION APPROVING THE STATUTE ON THE MINISTRY OF FOREIGN ECONOMIC RELATIONS AND TRADE OF THE RUSSIAN FEDERATION

The Government of the Russian Federation decrees:

1. The appended Statute on the Ministry of Foreign Economic Relations and Trade of the Russian Federation shall be approved.

2. The Decree of the Council of Ministers, Government of the Russian Federation, of 26 April 1993, no. 85, 'On approval of the Statute on the Ministry of Foreign Economic Relations and Trade of the Russian Federation', shall be deemed invalid.

Chairman of the Government of the Russian Federation V. Chernomyrdin
Moscow
7 Apr. 1997
No. 402

### Statute on the Ministry of Foreign Economic Relations and Trade of the Russian Federation

1. The Ministry of Foreign Economic Relations and Trade shall be a federal agency of executive power which shall pursue a unified state policy and exercise control in the area of foreign economic relations with respect to foreign trade activity, military–technical cooperation with foreign countries, and within the limits of its authority other types of foreign economic activity, and also in the area of foreign trade, and shall coordinate in this area the activity of other federal agencies of executive power in compliance with the legislation of the Russian Federation.

The ministry . . . shall be directly responsible for coordination and regulation of foreign trade activity.

The ministry shall work in coordination with other federal agencies of executive power, the corresponding agencies of executive power of components of the Russian Federation, and organizations.

2. The ministry . . . shall be guided in its work by the Constitution of the Russian Federation, the federal law On State Regulation of Foreign Trade Activity and other federal laws, edicts, and directives of the President of the Russian Federation, decrees and directives of the Government of the Russian Federation, the present Statute, the generally recognized principles and norms of international law and the international agreements of the Russian Federation.

3. The system of the Ministry of Foreign Economic Relations and Trade shall include authorized ministries in components and individual regions of the Russian Federation . . . , the State Inspectorate on Trade, Product Quality and Consumer Protection (Gostorginspektsiya) operating on the basis of the statute approved by the Government of the Russian Federation, and organizations indicated in Appendices nos. 1, 2, and 3.

[The appendices were not published in the original source and are not translated here.]

The ministry shall provide leadership of the work of the Russian Federation's representatives for trade and economic issues in foreign states and personnel support.

In order to provide for effective participation of the Russian Federation in international economic organizations and the performance of tasks in the area of developing foreign economic relations, the Ministry of Foreign Economic Relations and Trade, in coordination with the Ministry of Foreign Affairs, shall send its representatives to permanent missions of the Russian Federation in international organizations. Operational leadership of the activity of these representatives shall be provided by the Ministry of Foreign Economic Relations and Trade.

4. The main tasks of the Ministry of Foreign Economic Relations and Trade shall be:

1) the development, in conjunction with other federal agencies of executive power within the limits of their authority, of proposals to implement a unified state foreign economic policy and provide for its implementation as a constituent part of the foreign policy of the Russian Federation;

2) the development of proposals for the forming of a state policy in the area of military–technical cooperation between the Russian Federation and foreign countries and provision for its implementation;

3) the development and implementation of state policy in the area of domestic trade and public catering;

4) the development and implementation of measures for state regulation of foreign trade activity and control over the performance of this activity;

5) the regulation of relations in the area of military–technical cooperation;

6) the performance of functions of the state client in export and import deliveries of products and goods for state needs in the area of military–technical cooperation between the Russian Federation and foreign countries;

7) participation in providing for export control in the Russian Federation;

8) the development of proposals in the main areas of export policy and the activity of the mechanism for state support for industrial exports, including within the framework of regional programmes;

9) the protection of the economic interests of the Russian Federation, components of the Russian Federation, and Russian participants in foreign trade activity on the foreign market as well as the interests of domestic commodity producers and consumers from unfair foreign competition through the implementation of measures envisioned by the legislation of the Russian Federation;

10) the development, in compliance with established procedure, of proposals for international agreements of the Russian Federation concerning questions of foreign economic relations, including military–technical cooperation, provision for the fulfilment of the commitments of the Russian side in these agreements, and the exercise of the rights of the Russian side ensuing from them, and also monitoring of the fulfilment of their obligations by other parties to the agreements;

11) coordination in conjunction with the Ministry of Foreign Affairs of the actions of federal agencies of executive power in the area of international negotiations on questions of trade in goods and services in order to implement a unified foreign trade policy;

12) participation in the development and implementation of measures to provide for the effective integration of the economy of the Russian Federation into the world economy;

13) in compliance with established procedure, coordination of the foreign trade activity of components of the Russian Federation in questions of joint jurisdiction of the Russian Federation and components of the Russian Federation;

14) coordination of the actions of participants and subjects of military–technical cooperation;

15) control over the implementation of foreign trade activity by subjects of military–technical cooperation;

16) participation in the development and performance of the mechanism for regulating foreign exchange–credit relations with foreign states and attracting foreign investments;

17) coordination of work on questions of protecting the consumer market from poor-quality imported goods and preparing the appropriate normative documents;

18) interaction with the agencies of executive power of components of the Russian Federation concerning questions of improving the organization of trade service and monitoring compliance by trade organizations with the requirements of the legislation of the Russian Federation, and the dissemination of progressive technologies and advanced domestic and foreign expertise in this area;

19) the organization of information support in the area of foreign trade activity and domestic trade; and

20) the organization of training, retraining and improvement of qualifications of personnel in educational institutions included in the system of the ministry.

5. The Ministry of Foreign Economic Relations and Trade, in compliance with the tasks assigned to it, shall:

1) prepare proposals for the formation and provide for the implementation of a unified foreign trade policy, including with respect to individual foreign countries and groups of countries;

2) develop proposals for forecasting the development of the foreign trade of the

Russian Federation and provide for the study of market conditions and forecasts of tendencies in the development of the world market for goods and services;

3) with the participation of other federal agencies of executive power, develop proposals for state support for industrial exports, including within the framework of regional programmes, and participate in the development of plans for export credits;

4) take measures to prevent discrimination in the markets of foreign countries with respect to Russian participants in foreign trade activity and protect their interests;

5) contribute to the creation of favourable conditions for the access of Russian goods and services to foreign markets, envisioning the appropriate provisions in bilateral and multilateral agreements with foreign countries;

6) participate in the organization of international and foreign trade/industrial exhibitions in the Russian Federation and Russian exhibitions on the territories of foreign states;

7) with the participation of other federal agencies of executive power and organizations, develop drafts of legislative and other normative acts in the area of regulation and coordination of foreign economic activity, military–technical cooperation, and foreign trade;

8) in conjunction with other federal agencies of executive power and in compliance with established procedure, develop and submit drafts of annual federal programmes for the development of foreign trade activity and military–technical cooperation between the Russian Federation and foreign countries, and coordinate the activity of participants and subjects of military–technical cooperation in implementing them;

9) participate in the development of conceptual approaches to problems of military–technical cooperation, and analyse and generalize the results of foreign activity of subjects of military–technical cooperation;

10) participate in the coordination of planning scientific research and experimental design to maintain and develop the export potential of the defence industry in order to create the latest export models of arms and modernized weapons and military equipment that were previously delivered abroad;

11) participate in the preparation of drafts of programmes for cooperation with foreign countries in the development and production of weapons and military equipment;

12) develop proposals for establishing the state defence order with respect to the delivery of weapons and military equipment for export, and organize its fulfilment;

13) in compliance with established procedure, take measures for non-tariff regulation of foreign trade activity, including issuing licences to conduct export and import operations and licences for the export and/or import of goods in cases envisioned by the legislation of the Russian Federation; and provide for licensing in the area of military–technical cooperation;

14) provide control over the export and import of goods to which non-tariff regulations apply;

15) participate in the preparation of proposals for improving the customs legislation of the Russian Federation;

16) in compliance with established procedure, submit suggestions for improving the customs tariff of the Russian Federation and procedures for changing and introducing rates of customs duties;

17) provide organizational–technical support for the activity of the Commission of the Government of the Russian Federation on Protective Foreign Trade Measures, the Commission of the Government of the Russian Federation on Customs and Tariff Issues, and the Interdepartmental Coordinating Council for Military–Technical Policy;

18) in conjunction with other federal agencies of executive power and in compliance with established procedure, develop and submit proposals for changing and supplementing the commodity classification for foreign economic activity and adapting it to the requirements of state regulation of foreign economic activity;

19) contribute to improving the state system of product certification and standardization and the development of international and domestic norms, rules and standards;

20) develop proposals for improving the policy for the export and import of goods and services, including for federal state needs;

21) provide accounting for export and import contracts and make recommendations on legal questions of documenting foreign trade transactions;

22) participate in the regulation and control of Russian investments abroad;

23) develop and submit, in compliance with established procedure, proposals for balancing settlements with foreign states;

24) participate in providing for control over the granting to foreign states, the receipt from them, and the repayment of state loans, including special ones;

25) participate in the process of regulating the foreign indebtedness of the Russian Federation and the debts of foreign states to the Russian Federation;

26) provide for control of the conditions of commercial and foreign exchange finance and the level of foreign trade prices under contracts concluded at the expense of foreign borrowing by the Russian Federation, or against credit granted to foreign states, their legal entities and international organizations, and with respect to individual types of goods to whose export and/or import state monopoly or quantitative restrictions have been applied;

27) exercise control over the level of foreign trade prices for the basic types of military-purpose products and services;

28) participate in the development of proposals concerning the mechanisms and procedures for making budget allocations to finance the export part of the state defence order;

29) participate in the organization of a system of insurance and guarantees of export credits and credits for the production of export products;

30) in conjunction with other federal agencies of executive power, participate in the implementation of state policy in the area of non-proliferation of weapons of mass destruction and other weapons of the most dangerous kinds;

31) in order to prepare proposals concerning the introduction of protective measures with respect to the import of goods, conduct research, including consultation with the corresponding agencies of foreign states, and in accordance with the results of the research submit proposals in compliance with established procedure for the introduction of protective measures;

32) in compliance with established procedure, prepare proposals for introducing retaliatory measures in the area of foreign trade activity with respect to foreign states which violate the economic interests of the Russian Federation, components of the Russian Federation, municipal formations or Russian participants in foreign trade activity, or the political interests of the Russian Federation, and in cases where these states fail to meet their commitments to the Russian Federation adopted under international agreements;

33) participate in consideration of questions pertaining to the participation of the Russian Federation in international economic sanctions against one or a number of states;

34) interact with other federal agencies of executive power for purposes of introducing temporary technical measures to regulate the export and import of goods, work and services (standards, systems of quality compliance, safety standards, rules for packaging and marking of goods, forms of documents accompanying products and information they must contain, requirements for pre-dispatch inspection, expanded customs formalities, ecological, veterinary, phytosanitary and sanitary standards, measures for providing for national security, and methods of providing for the compliance of goods, work and services with the aforementioned standards (certification));

35) develop proposals for the introduction of quantitative restrictions on exports and imports, the establishment of state monopolies on exports and/or imports of individual goods, and bans and restrictions on exports and/or imports on the basis of the national interests of the Russian Federation;

36) in conjunction with the Ministry of Foreign Affairs or in coordination with it, in compliance with established procedure, submit proposals for international agreements of the Russian Federation concerning questions of trade–economic and military–technical cooperation, including questions of payment and credit relations, the regulation of foreign indebtedness, the repayment of debts of foreign states, and questions of cooperation in the construction and operation of facilities abroad and on the territory of the Russian Federation with the participation of foreign firms and organizations;

37) in compliance with established procedure, conduct negotiations on international agreements of the Russian Federation on questions within the ministry's jurisdiction;

38) prepare and submit in conjunction with the Ministry of Foreign Affairs or in coordination with it, in compliance with established procedure, proposals concerning measures to provide for the fulfilment of the international agreements of the Russian Federation in the area of foreign economic cooperation;

39) monitor the fulfilment by other parties of international agreements of the Russian Federation and, in the event of their violation, submit in conjunction with the Ministry of Foreign Affairs, in compliance with estab-

lished procedure, proposals for the necessary measures;

40) participate in the development, coordination and fulfilment of international agreements of the Russian Federation in the area of transport, including the transit of goods, and also legal, tariff, and other measures for regulating foreign trade shipments;

41) participate in the organization of the development of technical and economic justifications for the implementation of cooperative projects in foreign countries carried out on the basis of international technical assistance agreements of the Russian Federation;

42) analyse the long-term international agreements of the Russian Federation that envision the export of arms and military equipment, taking into account mutual indebtedness, the possibilities of commodity exchange, and other forms of settlement, and develop proposals for fulfilling the international agreements of the Russian Federation in the area of military–technical cooperation;

43) prepare and submit, in compliance with established procedure and in conjunction with the Ministry of Foreign Affairs or in coordination with it, proposals to establish, terminate, curtail or resume military–technical cooperation with individual foreign states;

44) participate in the development of proposals to deliver arms and military equipment, including that which has not been delivered abroad previously, and to transfer licences for their production to foreign countries;

45) analyse the potential needs of foreign states for weapons and military equipment and the ability of the defence industry of the Russian Federation to satisfy demand on the arms market;

46) develop recommendations for participation in financial–industrial groups of specialized export–import companies;

47) coordinate work to verify the reliability of foreign partners in the area of military–technical cooperation;

48) coordinate marketing, tender and advertising activity in the area of military–technical cooperation;

49) participate, if necessary, in the negotiations of subjects of military–technical cooperation with foreign partners;

50) control the work of subjects of military–technical cooperation in sending delegations abroad to manage this cooperation;

51) submit, in compliance with established procedure, proposals for the creation of intergovernmental commissions on trade–economic, scientific–technical and military–technical cooperation with foreign countries, and organize and support the work of the secretariats of their Russian units (with the exception of those Russian units whose work is organized and supported by the Ministry for Cooperation with Countries of the Commonwealth of Independent States);

52) receive, in compliance with established procedure, foreign delegations to Russia and send delegations to foreign countries in order to resolve issues within the ministry's jurisdiction;

53) prepare and submit, in compliance with established procedure, proposals for the establishment and maintenance of relations with international economic organizations; participate in the work of these organizations and their agencies and in interaction with the aforementioned agencies, in conjunction with federal agencies of executive power, including on questions of their rendering technical assistance to the Russian Federation; and participate in the work of international organizations on questions of military–technical cooperation;

54) organize and coordinate the work of federal agencies of executive power in conducting relations with the World Trade Organization and other international economic organizations in relation to which the ministry has been determined to be the leading organization;

55) provide, in compliance with established procedure, for the coordination of the foreign economic activity of components of the Russian Federation, participate in the implementation of regional and inter-regional programmes for the development of foreign trade activity, and promote the organization of expert evaluations of projects submitted by the regions within the framework of the programmes being developed;

56) contribute to the participation of representatives of components of the Russian Federation in the work of intergovernmental commissions on trade–economic and scientific–technical cooperation and enlist them for participation in conferences, seminars, exhibitions and other measures taken in Russia and in foreign states to realize the export potential of components of the Russian Federation;

57) in conjunction with the Ministry for Cooperation with Countries of the Commonwealth of Independent States and the Ministry of Foreign Affairs, work to coordinate foreign trade policy with countries of the Commonwealth of Independent States and submit the corresponding proposals to the Government of the Russian Federation in coordination with federal agencies of executive power and organizations;

58) draw up proposals for the development of military–technical cooperation with countries of the Commonwealth of Independent States;

59) implement a scientific–technical and investment policy aimed at the modernization and technical retooling of trade and public catering organizations;

60) determine the technological specifications for retail trade services and production of public catering products for the domestic consumer market;

61) organize and coordinate work for certifying trade and public catering services;

62) analyse the condition of domestic trade and public catering, develop forecasts of present and future developments and submit reports on these questions to the Government of the Russian Federation;

63) coordinate the work of federal agencies of executive power on questions of protecting the consumer market from poor-quality goods, including those that are imported, and prepare the appropriate normative documents;

64) prepare proposals concerning the volumes of production of domestic goods, the improvement of their competitiveness, and the shipment and import of consumer goods, and forward these proposals to federal agencies of executive power and organizations;

65) as a state client, work to prepare and implement target programmes concerning problems of domestic trade and through the system of wholesale trade organize the competitive placement of orders for consumer goods, including those purchased through imports for state needs;

66) develop and approve instructions and other departmental acts concerning preparations for the supply of the country's population with food and necessities under exceptional circumstances, and interact with agencies of executive power of components of the Russian Federation concerning questions of mobilization and provision for the defence needs of the state and the stability of the operation of trade and public catering organizations during states of emergency;

67) organize the conduct of federal and interstate (with the participation of countries of the Commonwealth of Independent States) wholesale trade fairs for consumer goods, festivals and competitions in professional skill, and render assistance in the work of regional and inter-regional trade fairs, festivals and competitions;

68) in order to protect the rights and interests of consumers, through the State Trade Inspectorate, provide for state control of compliance with the norms and rules of trade and public catering, price policy and discipline, and the quality and safety of consumer goods in industrial, trade and public catering organizations;

69) interacting with federal agencies of executive power, agencies of executive power of components of the Russian Federation, and organizations, take measures to satisfy the demand for goods on the domestic consumer market and participate in the development of federal and regional target programmes, including programmes to improve the system of shipment of consumer goods into regions of the Far North and localities on an equal footing, and handle the consequences of emergency situations by providing goods;

70) participate in work to implement programmes for the deepening of economic reforms in the area of domestic trade and public catering, demonopolization, the creation of a competitive environment and privatization of enterprises, and prepare proposals to encourage the formation of market relations and the development of entrepreneurship and to improve the system of price setting, taxation and bookkeeping;

71) coordinate the activity of federal agencies of executive power in developing and providing for the functioning of a system of foreign trade information financed through the federal budget;

72) organize the conduct of scientific research and provide the necessary commercial, scientific–technical, economic and legal information, including from participants in foreign economic activity for payment, and carry out work to create information systems and databases on questions of foreign economic activity, including military–technical cooperation, and on trade;

73) participate in the development and revision of forms for state statistical reporting, the range of indicators of foreign eco-

nomic activity, and the development of the domestic consumer market;

74) analyse statistical information on foreign economic activity and the development of domestic trade officially drawn up by the State Committee on Statistics, the State Customs Committee and other state agencies;

75) participate in international cooperation in the area of foreign and domestic trade statistics;

76) provide for training, retraining and improvement of the qualifications of personnel in foreign economic activity, domestic trade and public catering, including in educational institutions that are part of the ministry system, and participate in international cooperation in this area;

77) determine requirements for the level of training of candidates for the conferment of senior rank in professions in the area of domestic trade and public catering, and confer these ranks;

78) develop and approve the procedure for certification of workers of organizations of the ministry system that are budget-financed and provide for the accreditation of secondary specialized educational institutions that are a part of the ministry system;

79) organize and conduct audits and inspections of the operational–commercial and financial–economic activity of organizations, authorized representatives and state inspection teams included in the ministry system and its foreign staff, and provide for regular control according to plans coordinated with other federal agencies of executive power and the efficient utilization in the ministry system of budgeted funds and of the property provided for its use free of charge, including through representatives of the Russian Federation for trade and economic issues in foreign states;

80) organize the auditing (inspecting) of the foreign economic activity of Russian participants and subjects of military–technical cooperation and their representative offices in foreign states;

81) organize representative offices of the Russian Federation for trade and economic issues in foreign states and, in compliance with established procedure, foreign economic organizations for special communications, and take measures for protecting information in compliance with the legislation of the Russian Federation; and

82) perform other functions envisioned by the legislation of the Russian Federation.

6. The Ministry of Foreign Economic Relations and Trade shall utilize free of charge, in compliance with established procedure, the official premises in the Russian Federation assigned to it and real estate abroad within the limits necessary for the functioning of the missions of the Russian Federation for trade and economic issues in foreign states.

7. The Ministry of Foreign Economic Relations and Trade shall implement measures for the social protection, social development, improvement of working conditions, housing and cultural–domestic conditions, and medical services for workers of the ministry system.

8. The Ministry of Foreign Economic Relations and Trade in order to perform the tasks assigned to it shall have the right to:

1) request and receive from federal agencies of executive power, agencies of executive power of components of the Russian Federation, and organizations, including those of the defence complex, information and materials necessary to solve problems within the ministry's jurisdiction;

2) publish, in compliance with established procedure and within the limits of its jurisdiction, normative acts that are binding for other federal agencies of executive power, agencies of executive power of components of the Russian Federation, organizations and citizens;

3) in compliance with established procedure, enlist experts and consultants and conclude agreements with organizations and citizens for the performance of work in areas included in the ministry's jurisdiction;

4) give opinions on projects for solutions concerning the privatization of trade enterprises that are federally owned;

5) in compliance with established procedure, submit proposals concerning candidacies for representatives of the interests of the state on the management bodies of joint-stock companies that have federally owned stocks;

6) appoint and conduct, within the framework of its jurisdiction and in coordination with the State Committee for the Management of State Property, document and physical audits (inspections and inventories) and schedule audits of enterprises that have economic jurisdiction over or operational management of state and federally owned property;

7) prepare and submit, in compliance with the procedure established by the legislation of the Russian Federation, proposals regarding

the creation, reorganization and abolition of organizations within the ministry system; and decide on the creation, reorganization and abolition of state institutions within the ministry system if these decisions are implemented within the framework of the budget allocations and the personnel allotted to it;

8) in compliance with established procedure, issue permits to open representative offices in the Russian Federation to foreign organizations and firms;

9) maintain an independent central encryption agency and a departmental network for special communications;

10) interact within the limits of its authority with state agencies and organizations, including associations and unions, officials and private individuals both within the Russian Federation and abroad; and

11) utilize, in compliance with established procedure, off-budget funds received for the issuing of licences and certificates and from the provision of services, including those rendered by foreign institutions, to finance the development of the ministry's material and technical base, for social needs and for material incentives for its workers.

9. The Ministry of Foreign Economic Relations and Trade shall be headed by a minister appointed and dismissed by the President of the Russian Federation at the suggestion of the Chairman of the Government of the Russian Federation.

The minister shall bear personal responsibility for performance of the tasks and functions assigned to him.

The minister shall have deputies appointed to the position and discharged from it by the Government of the Russian Federation. The minister shall distribute duties among the deputy ministers.

10. The Minister of Foreign Economic Relations and Trade shall:

1) publish, within the limits of his authority and in compliance with the legislation of the Russian Federation, orders, directives and instructions that shall be mandatory for workers of the ministry and organizations included in its system;

2) submit, in compliance with established procedure, proposals to appoint or change the personnel of Russian units of intergovernmental commissions for trade–economic, military–technical and scientific–technical cooperation between the Russian Federation and foreign states;

3) submit, in coordination with federal agencies of executive power and in compliance with the legislation of the Russian Federation, proposals for temporary measures for the protection of the domestic market;

4) submit, in coordination with federal agencies of executive power and in compliance with established procedure, proposals for changes and additions to the list of goods, work and services, exports and imports that are provided under licences or under a special policy, and also the policy for conducting the corresponding export–import operations;

5) determine the list and extent of information on the condition of foreign economic activity of the Russian Federation and also the deadlines for the submission of this information to the Ministry of Foreign Economic Relations and Trade by federal agencies of executive power and agencies of executive power of components of the Russian Federation;

6) establish the duties and determine the responsibility of leaders of structural subdivisions of the ministry;

7) submit to the Government of the Russian Federation proposals for the appointment and discharge of leaders of trade and economic missions of the Russian Federation in foreign states;

8) appoint and discharge management workers of the ministry's central staff, deputy managers of trade and economic missions of the Russian Federation to foreign countries, other management workers of the foreign staff, and officials;

9) approve the provisions on the structural subdivisions of the ministry and, in compliance with established procedure, the regulations of the enterprises, institutions and other organizations included in the ministry system; and, in compliance with the legislation of the Russian Federation, conclude contracts with the managers of these enterprises;

10) approve the structure and distribution of the ministry's central staff, its foreign staff, the territorial agencies of the State Trade Inspection, and authorized representatives within the limits of the numbers established by the Government of the Russian Federation and the wage budget for workers, and within the limits of the budget approved for the relevant period;

11) in compliance with established procedure, submit the names of exceptional workers of the ministry's system for the con-

ferment of honorary titles and state awards of the Russian Federation; and

12) exercise other rights in compliance with the legislation of the Russian Federation.

11. The Ministry of Foreign Economic Relations and Trade shall form a board including the minister (chairman of the board), deputy ministers (according to their positions) and managers of the main subdivisions of the ministry. The board may also include representatives of other federal agencies of executive power, organizations, scholars and specialists.

Members of the board, except for individuals included on it by virtue of their position, shall be approved by the Government of the Russian Federation at the suggestion of the minister.

At its meetings the board shall consider the most important issues relating to the ministry's work and adopt the appropriate decisions concerning it. Board decisions shall be adopted by a majority of votes of its members, documented with protocols, and implemented, as a rule, by orders from the minister.

In the event of disagreements between the minister and the board, the minister shall carry out his own decision, reporting disagreements to the Government of the Russian Federation. Members of the board who have a special opinion regarding a decision adopted may also report it to the Government of the Russian Federation.

12. The Ministry of Foreign Economic Relations and Trade may form coordination, scientific–consultative and expert councils for problems of foreign economic relations and domestic trade and participate in their work.

The members of the coordination, scientific–consultative and expert councils of the ministry and the provisions concerning them shall be approved by the minister.

Organizational–technical support for the activity of the aforementioned councils shall be provided by the ministry's central staff.

13. The costs of maintaining the central staff of the Ministry of Foreign Economic Relations and Trade, its agents, the state inspections and the foreign staff shall be financed with funds from the federal budget earmarked for state management and from other sources within the framework of the legislation of the Russian Federation.

14. The Ministry of Foreign Economic Relations and Trade shall be a legal entity and have a budget and other accounts in the Central Bank of the Russian Federation, accounts in other banks and credit organizations, including in foreign currency, and a stamp with a depiction of the state seal of the Russian Federation and its own name.

------

*Source: Rossiyskaya Gazeta*, 22 Apr. 1997, p. 5 (in Russian).

## 25. DECREE OF THE PRESIDENT OF THE RUSSIAN FEDERATION ON MEASURES TO IMPROVE THE SYSTEM OF MANAGEMENT OF MILITARY–TECHNICAL COOPERATION WITH FOREIGN STATES

In order to increase the efficiency of management of military–technical cooperation with foreign states and ensure the state's monopoly on the export and import of armaments and military equipment, I decree:

1. To entrust the Chairman of the Government of the Russian Federation with direct coordination of the activities of the State Company for the Export and Import of Armament and Military Equipment Rosvooruzhenie and monitoring of implementation of military–technical cooperation with foreign states.

2. To establish that the State Company . . . Rosvooruzhenie is under the jurisdiction of the Government of the Russian Federation; the General Director of the State Company . . . Rosvooruzhenie shall be appointed to the position and removed from the position by the President of the Russian Federation on the recommendation of the Chairman of the Government of the Russian Federation.

3. The Government of the Russian Federation shall bring the charter of the State Company . . . Rosvooruzhenie in line with the civilian legislation of the Russian Federation and this edict.

4. The main state–legal administration of the President of the Russian Federation shall within a two-week period submit proposals on making changes and additions stemming from this edict to edicts and directives of the President of the Russian Federation.

5. This edict shall enter into force as of the day of its official publication.

President of the Russian Federation
B. Yeltsin
Moscow, Kremlin
28 July 1997
No. 792

———

*Source: Rossiyskaya Gazeta*, 2 Aug. 1997, p. 6 (in Russian).

## 26. DECREE OF THE PRESIDENT OF THE RUSSIAN FEDERATION ON MEASURES TO STRENGTHEN STATE CONTROL OF FOREIGN TRADE ACTIVITY IN THE FIELD OF MILITARY–TECHNICAL COOPERATION OF THE RUSSIAN FEDERATION WITH FOREIGN STATES

In order to further develop military–technical cooperation between the Russian Federation and foreign states and to strengthen state control over the foreign trade activities of organizations in the Russian Federation whose products have military applications, I decree as follows:

1. To establish that the following organizations shall carry out the export (import) of weapons, military technology, work and services with military applications, data and results of intellectual endeavour in the military–technical field, as well as licensing for weapon production and military items and corresponding technologies (hereafter referred to as 'production with military applications'):

– enterprises which are the developers and manufacturers of weapons and military items which have duly received the right as established by the President of the Russian Federation; and

– state intermediaries—federal state unitary enterprises established in accordance with presidential decrees of the Russian Federation and which have the right to conduct business.

2. To confirm the attached regulations on the proper implementation of foreign trade activities of organizations in the Russian Federation whose products have military applications and the regulations on the proper licensing of organizations in the Russian Federation to conduct foreign trade in products with military applications.

3. To establish that the Government of the Russian Federation shall coordinate the activities of federal bodies of executive power in order to implement policy regarding military–technical cooperation between the Russian Federation and foreign states and control the foreign trade activities of Russian Federation organizations whose products have military applications.

4. To rename the Interdepartmental Coordinating Council for Military–Technical Policy of the Russian Federation, established in accordance with the Decree of the President of the Russian Federation of 3 March 1995, no. 236, 'On the introduction of alterations and amendments to the Decree of the President of the Russian Federation of 30 December 1994, no. 2251, On the State Committee of the Russian Federation on Military–Technical Policy and in the Regulations approved by the decree', to the Interdepartmental Coordinating Council on Military–Technical Cooperation between the Russian Federation and Foreign States.

5. To confirm the attached regulations on the Interdepartmental Coordinating Council on Military–Technical Cooperation between the Russian Federation and Foreign States and its composition.

6. To rename:

– the State Company Rosvooruzhenie . . . to the Federal State Unitary Enterprise the State Company Rosvooruzhenie; and

– the export trading association Promexport to the Federal State Unitary Enterprise Promexport.

7. To create the Federal State Unitary Enterprise Rossiyskiye Tekhnologii.

8. To establish that the federal state unitary enterprises the State Company Rosvooruzhenie, Promexport and Rossiyskiye Tekhnologii, established having the right to conduct business, are government intermediaries in the export (import) of production with military applications.

9. To preserve the right of the Ministry of Defence, as set out by the Government of the Russian Federation, to provide assistance to national military and technical personnel of foreign governments. The Ministry of Defence shall be permitted to sell weapons

and military items (surplus and related items which have been taken out of service in the Armed Forces of the Russian Federation as a result of reforms) through the government intermediaries specified in para. 8 of this decree.

10. The Government of the Russian Federation shall:

– undertake decisions to implement the requirements specified under paras 1 and 2 of this decree within two months;

– before 1 January 1998 provide the President of the Russian Federation for his approval a single draft list of items with military applications whose transfer to foreign customers is permitted and a list of states to which the transfer of items listed is permitted;

– confirm within two months the composition of the federal state unitary enterprises . . . the State Company Rosvooruzhenie, Promexport and Rossiyskiye Tekhnologii, ensuring that:

(*a*) the activities of each of the above-mentioned unitary federal state enterprises are such that the possibility of competition between them is excluded;

(*b*) the designation and removal from office of the heads of the above-mentioned unitary federal state enterprises is carried out by the President of the Russian Federation on the recommendation of the Chairman of the Government of the Russian Federation;

(*c*) oversight commissions formed by the Government of the Russian Federation with representatives of federal executive bodies shall exercise direct control over the activities of the above-mentioned unitary federal state enterprises and their financial condition; and

(*d*) the Government of the Russian Federation, on the recommendation of the relevant oversight commission, shall determine the maximum prices for work (services) carried out (provided) by the above-mentioned federal state unitary enterprises, their staff lists and their expenditure, as well as the banks with which the enterprises in question open accounts;

– undertake decisions relevant to this decree within two months.

11. The Ministry of Foreign Economic Relations and Trade shall ensure that a register of organizations in the Russian Federation which have the right to conduct foreign trade in products with military applications is kept, and that this register includes organizations in the Russian Federation which had the right to conduct foreign trade in products with military applications before publication of this decree, and shall provide the organizations with corresponding documentation.

12. To acknowledge that the following decrees are no longer in force:

– the Decree of the President of the Russian Federation of 30 December 1994, no. 2251, 'On the State Committee of the Russian Federation on Military–Technical Policy' (Collection of legislative acts of the Russian Federation, no. 1 (1995), article 45);

– the Decree of the President of the Russian Federation of 3 March 1995, no. 236, 'On the insertion of changes and additions to the Decree of the President of the Russian Federation of 30 December 1994, no. 2251, On the State Committee of the Russian Federation on Military–Technical Policy, and to the attached regulations, confirmed by this decree' (Collection of legislative acts of the Russian Federation, no. 10 (1995), article 865);

– the Decree of the President of the Russian Federation of 14 June 1995, no. 590, 'On the Interdepartmental Coordinating Council for Military–Technical Policy of the Russian Federation' (Collection of legislative acts of the Russian Federation, no. 25 (1995), article 2379);

– the Decree of the President of the Russian Federation of 5 October 1995, no. 1008, 'On military–technical cooperation of the Russian Federation with foreign countries (basic provisions)' (Collection of legislative acts of the Russian Federation, no. 41 (1995), article 3876);

– the Decree of the President of the Russian Federation of 5 October 1995, no. 1009, 'On the centralized social and material–technical fund of the State Committee of the Russian Federation on Military–Technical Policy' (Collection of legislative acts of the Russian Federation, no. 41 (1995), article 3877); and

– the Decree of the President of the Russian Federation of 31 January 1996, no. 131, 'On an insertion into Presidential decree of 14 June 1995, no. 590, On the Interdepartmental Coordinating Council for Military–Technical Policy of the Russian Federation' (Collection of legislative acts of the Russian Federation, no. 6 (1996), article 535).

13. The Ministry of Foreign Affairs of the Russian Federation shall inform foreign governments of decisions taken.

14. This decree shall enter into force on the day of signature.

All decisions on the export of items with military applications from the territory of the Russian Federation, including their re-export or transfer to third countries, are taken solely by the President of the Russian Federation on the advice of the Government of the Russian Federation. These decisions shall precede confirmation of a single list of products with military applications which may be transferred to foreign purchasers and a list of governments which may receive and transfer such items.

President of the Russian Federation
B. Yeltsin
Moscow, Kremlin
20 Aug. 1997
No. 907

Source: Diplomaticheskiy Vestnik, no. 9 (Sep. 1997), pp. 5–7 (in Russian).

## 27. DECREE OF THE PRESIDENT OF THE RUSSIAN FEDERATION ON THE FEDERAL STATE UNITARY ENTERPRISE PROMEXPORT

For the purpose of selling armaments and military hardware abroad which are released from operation in the Armed Forces of the Russian Federation, and the use of the receipts therefrom for the implementation of military reform, I hereby decree:

1. To establish that the sale abroad of armaments and military hardware (as well as the spare parts for them and ammunition) released from operation in the Armed Forces of the Russian Federation in connection with the implementation of military reform shall be carried out mainly by the state intermediary the Federal State Unitary Enterprise Promexport.

2. To establish that the decision to release armaments and military hardware, spare parts for them and ammunition from operation in the Armed Forces of the Russian Federation as a result of the implementation of measures related to military reform shall be made by the Ministry of Defence of the Russian Federation.

3. To establish that the receipts from the sale of the armaments and military hardware released from operation in the Armed Forces of the Russian Federation, spare parts for them and ammunition shall be entered in full (except for commission fees and transportation and insurance expenses) into a special account of the Ministry of Defence of the Russian Federation to finance measures connected with the implementation of the military reform, including measures to ensure social protection of servicemen.

4. To appoint Vyacheslav Yevgeniyevich Filimonov Director General of the Federal State Unitary Enterprise Promexport upon presentation by the Chairman of the Government of the Russian Federation.

V. Y. Filimonov shall coordinate with the Ministry of Defence of the Russian Federation a procedure for organizing work for the sale abroad of the armaments and military hardware released from operation in the Armed Forces of the Russian Federation, spare parts for them and ammunition, and also coordinate with the above ministry the personnel to be made responsible for export operations.

5. The Government of the Russian Federation shall take the decisions needed to implement this Decree within one month.

6. This Decree shall enter into force as of the day of its official publication.

President of the Russian Federation
B. Yeltsin
20 Aug. 1997
No. 908

Source: Rossiyskaya Gazeta, 22 Aug. 1997 (in Russian).

## 28. DECREE OF THE PRESIDENT OF THE RUSSIAN FEDERATION ON THE FEDERAL STATE UNITARY ENTERPRISE THE STATE COMPANY ROSVOORUZHENIE

With the aim of implementing measures to strengthen state control over foreign trade activity in the field of military–technical cooperation of the Russian Federation with foreign states, I hereby decree:

1. To establish that the Federal State Unitary Enterprise the State Company Rosvooruzhenie is the legal successor of the State Company for Arms and Military Hardware Exports and Imports Rosvooruzhenie.

2. The Government of the Russian Federation shall set up, within two months, a supervisory commission to control the activity of the Federal State Unitary Enterprise the State Company Rosvooruzhenie out of representatives of federal executive bodies.

3. To appoint, at the presentation of the Chairman of the Government of the Russian Federation, Yevgeniy Ananyev Director General of the federal state unitary enterprise the State Company Rosvooruzhenie.

Ye. Ananyev shall coordinate with the Ministry of Defence of the Russian Federation within two months and submit to the supervisory commission, to be set up in accordance with point 2 of this decree, the draft list of the staff of this commission and proposals concerning its leadership.

4. This decree shall enter into force as of the day of its official publication.

President of the Russian Federation
B. Yeltsin
20 Aug. 1997
Decree no. 910

*Source: Rossiyskaya Gazeta*, 22 Aug. 1997 (in Russian).

---

# 29. STATUTE OF THE INTERDEPARTMENTAL COORDINATING COUNCIL FOR MILITARY–TECHNICAL COOPERATION BETWEEN THE RUSSIAN FEDERATION AND FOREIGN STATES

1. The Interdepartmental Coordinating Council for Military–Technical Cooperation between the Russian Federation and Foreign States [here referred to by the Russian acronym, KMS] has been formed in order to formulate agreed proposals in the political, military and economic interests of the Russian Federation in:

– state policy on military–technical cooperation with foreign states;
– supervision of the work of the federal executive authorities in the field of military–technical cooperation between the Russian Federation and foreign states and implementation by them of the Russian Federation's international commitments in this field;

– settlement of problems of an interdepartmental nature in the field of military–technical cooperation between the Russian Federation and foreign states;
– the granting to Russian organizations of the right to engage in foreign trade activity in the field of military–technical cooperation, and stripping them of this right; and
– state support for the export of products (works and services) of a military nature.

2. The KMS in its work shall be guided by the Constitution of the Russian Federation, federal laws, decrees and instructions of the President of the Russian Federation, decisions and instructions of the Government of the Russian Federation, the international obligations of the Russian Federation, and this statute.

3. The principal objective of the KMS shall be to prepare for submission to the President of the Russian Federation and the Government of the Russian Federation proposals on:

– defining priority directions in military–technical cooperation between the Russian Federation and foreign states and resolving problems arising in this field;
– concluding international treaties of the Russian Federation on matters of military–technical cooperation;
– monitoring the implementation of military–technical cooperation with foreign states, including supervision over the transfer abroad of information and results of intellectual activity in the military–technical field;
– determining which military-purpose products can be permitted to be transferred to foreign customers;
– drawing up a list of states to be allowed to take delivery of military-purpose products indicated in the unified list of military-purpose products whose transfer to foreign customers is permitted;
– granting Russian organizations the right to engage in foreign trade in the field of military–technical cooperation and stripping them of this right;
– coordinating the activities of the federal executive authorities whose terms of reference cover matters relating to military–technical cooperation between the Russian Federation and foreign states and of Russian organizations that have been granted the right to engage in foreign economic activity in the field of military–technical cooperation between the Russian Federation and foreign states;

– organizing and holding exhibitions and demonstrations of arms and *matériel* in the Russian Federation and abroad; and

– deciding the composition and procedure of supervisory commissions to monitor the activities and financial status of state intermediaries—federal state unitary enterprises, having the right to conduct business, engaged in foreign trade activity in the field of military–technical cooperation between the Russian Federation and foreign states.

4. The KMS, in order to carry out its principal function, shall examine:

– proposals from federal executive authorities on matters of military–technical cooperation between the Russian Federation and foreign states, and on the settlement of differences between Russian participants and subjects of military–technical cooperation;

– drafts of a unified list of military-purpose products whose transfer to foreign customers is permitted, and a list of countries which may take delivery of military products as indicated in the unified list of military-purpose products whose transfer to foreign customers is permitted;

– proposals from federal executive authorities on the granting to Russian organizations of the right to engage in foreign economic activity in the field of military–technical cooperation; and

– reports from federal executive authorities on abuses by subjects of military–technical cooperation between the Russian Federation and foreign states of the established procedures for cooperation, as well as monetary, tax and other legislation of the Russian Federation, and take appropriate decisions on these reports, including stripping the organizations concerned of the right to take part in military–technical cooperation.

5. The KMS shall have the right to:

– ask for and receive information, documents and materials necessary for the fulfilment of its tasks from federal executive authorities, executive authorities of subjects of the Russian Federation, and enterprises, institutions and organizations regardless of their form of ownership or departmental affiliation;

– hear heads of appropriate federal executive authorities on matters of military–technical cooperation between the Russian Federation and foreign states;

– set up working groups from among representatives of federal executive authorities, executive authorities of subjects of the Russian Federation, enterprises, institutions and organizations regardless of their form of ownership or department affiliation, to study questions necessary for the performance of the tasks entrusted to the KMS; and

– submit to the Government of the Russian Federation proposals on matters of military–technical cooperation between the Russian Federation and foreign states.

6. The Chairman of the Government of the Russian Federation shall be Chairman of the KMS and shall bear personal responsibility for the fulfilment of the tasks entrusted to it.

The Deputy Chairman of the Government of the Russian Federation responsible for questions of military–technical cooperation between the Russian Federation and foreign states shall be Deputy Chairman of the KMS.

The Deputy Minister for Foreign Economic Relations and Trade of the Russian Federation responsible for questions of military–technical cooperation between the Russian Federation and foreign states shall be Secretary of the KMS.

7. The KMS shall be made up of heads of federal executive authorities responsible for questions of military–technical cooperation between the Russian Federation and foreign states and senior executives of the Administration of the President of the Russian Federation.

The composition of the KMS shall be approved by the President of the Russian Federation upon recommendation of the Chairman of the Government of the Russian Federation.

Members shall attend KMS sessions without the right of proxy.

8. The organizational form of KMS activity shall be sessions convened as the need arises but not less than once a month.

Officials of federal executive authorities not represented on the KMS, of executive authorities of the subjects of the Russian Federation, enterprises, institutions and organizations, regardless of the form of their ownership and departmental jurisdiction, may be invited, upon the instruction of the KMS Chairman, to take part in the discussion of particular items on the agenda in an advisory capacity.

9. Information on the agenda and materials on matters to be discussed at a KMS session shall be sent out by the KMS Secretary to all participants not later than two weeks before the session.

10. Decisions on each item on the agenda of a session shall be passed by a simple majority of votes of members present. Two-thirds of the total number of its members shall constitute a quorum.

Members shall have an equal voice in decision making. KMS decisions for the elaboration of proposals within its competence and backed by relevant regulatory acts shall be mandatory for execution by the federal executive authorities.

If there are fundamental differences of principle between members of the KMS, the Chairman shall have the right to postpone the item concerned for further work and re-examination.

The results of consideration of items on the agenda of a KMS session with voting results shown against each shall be entered into appropriate minutes or recorded in separate KMS decisions.

The protocols and decisions shall be signed by the KMS Chairman and, in his absence, by the Deputy Chairman.

11. The working body of the KMS shall be an appropriate division in the Ministry of Foreign Economic Relations and Trade.

12. Information, organization and material and technical support for the KMS activities shall be provided by the KMS working body in agreement with the Ministry of Defence.

13. The KMS's working body shall:

– sum up information reaching the KMS on matters within its terms of reference;

– frame proposals on KMS work planning and draw up the agendas for KMS sessions;

– prepare the necessary materials and draft decisions on matters put before the KMS;

– ensure that decisions are carried out;

– carry out the instructions of the KMS Chairman;

– study, together with the federal executive authorities concerned, conclusions when decisions are being drafted on military–technical cooperation between the Russian Federation and foreign states, bearing in mind the country's political, military and economic security; and

– in timely fashion, bring KMS protocols and decisions to the notice of the federal executive authorities concerned.

Approved by Decree No. 907 of the President of the Russian Federation, 20 Aug. 1997

**Composition of the Interdepartmental Coordinating Council for Military–Technical Cooperation between the Russian Federation and Foreign States**

– V. S. Chernomyrdin, Chairman of the Government of the Russian Federation (Chairman);

– Ya. M. Urinson, Deputy Chairman of the Government of the Russian Federation and Minister of the Economy (Deputy Chairman);

– I. S. Ivanov, State Secretary, First Deputy Minister of Foreign Affairs;

– N. D. Kovalev, Director of the Federal Security Service of Russia;

– Yu. N. Koptev, Director General of the Russian Space Agency;

– V. V. Korabelnikov, Deputy Chief of the General Staff of the Armed Forces;

– A. L. Kudrin, First Deputy Minister of Finance;

– V. N. Mikhailov, Minister of Nuclear Energy;

– A. V. Ogarev, Deputy Head of the Administration of the President of the Russian Federation;

– R. G. Orekhov, Deputy Head of the Administration of the President of the Russian Federation and Head of the Main State Legal Department of the President of the Russian Federation;

– V. A. Pakhomov, Deputy Minister of Foreign Economic Relations and Trade (Executive Secretary of the Council);

– I. O. Rybkin, Secretary of the Security Council;

– I. D. Sergeyev, Minister of Defence;

– V. I. Trubnikov, Director of the Federal External Intelligence Service; and

– M. E. Fradkov, Minister of Foreign Economic Relations and Trade.

*Source: Rossiyiskaya Gazeta, 26 Aug. 1997 (in Russian).*

# About the authors

**Dr Ian Anthony** (UK) is Leader of the SIPRI Arms Transfers Project. He is editor of the SIPRI volumes *Arms Export Regulations* (1991) and *The Future of Defence Industries in Central and Eastern Europe* (1994) and author of *The Naval Arms Trade* (SIPRI, 1990) and *The Arms Trade and Medium Powers: Case Studies of India and Pakistan 1947–90* (1991). He has written or co-authored chapters for the *SIPRI Yearbook* since 1988.

**Dr Elena Denezhkina** (Russia) was educated at the Kiev Institute of National Economy and Leningrad State University and has worked on the Soviet/Russian defence industry in a number of capacities. She is at present a Research Fellow at the Centre for Russian and East European Studies at the University of Birmingham. Her recent publications include 'Problems of conversion and privatization in the military–industrial complex of St Petersburg', eds P. Opitz and W. Pfaffenberger, *Adjustment Processes in Russian Defence Enterprises within the Framework of Conversion and Transition* (1994) and *Some Problems of Conversion of Russian Defence Industry: its Economic Restructuring and Management* (1996). *The Transformation of Russian Shipbuilding* (for the Bonn International Centre for Conversion) and *Lost Illusions: the Restructuring of the Russian Military Complex* are forthcoming.

**Dr Gennadiy Gornostaev** (Russia) has worked in the Department of Military Economy and Politics in the Institute of World Economy and International Relations (IMEMO) of the Academy of Sciences in Moscow, and is at present Head of Department in the All-Russian Institute for the Study of External Economic Relations in the Russian Ministry of the Economy, as well as being director of a private company, Marketing Consultancy. He has published numerous articles on the Western arms market and Russia and on arms control. His *Russian Military–Industrial Companies: Problems and Prospects of International Cooperation* and *Corporate Structures in the Field of the Russian Defence Industry* (both in Russian) are forthcoming.

**Academician General Yuriy Kirshin** (Russia) has been Professor of the Academy of the Soviet General Staff, Director of the Strategy Section of the Soviet General Staff, and Editor of the *Soviet Military Encyclopaedia*. He is President of the independent Association for Military–Political and Military–Historical Studies, of which he was the founder in 1992, and Vice-President of the Academy of Military Sciences. His recent publications include 'Ensuring military security in a democratic society', *NATO Review* (1997) and 'The Soviet Armed Forces on the eve of the Great Patriotic War', ed. B. Wegner, *From Peace to War: Germany, Soviet Russia and the World, 1939–41* (1997).

**Dr Irina Kobrinskaya** (Russia) is a Senior Research Fellow at the Institute of US and Canada Studies of the Russian Academy of Sciences in Moscow. She has published numerous books and articles, including most recently 'After Madrid', *Nezavisimaya Gazeta*, 5 July 1997 and *Russia and Central Eastern Europe after the Cold War*

(1997); 'Internal factors of Russian foreign policy' is forthcoming as a contribution to *Transforming Russia*, to be published by the Carnegie Endowment for International Peace (1998).

**Dr Sergey Kortunov** (Russia) is Deputy Head of Staff and Head of the Strategic Assessments and International Policy Department of the Defence Council of Russia (part of the Staff of the President of the Russian Federation). His responsibilities include international security matters, arms control and disarmament, strategic foreign policy planning, development of the new Russian military doctrine and the national security concept. He served previously in the International Organizations Department of the Ministry of Foreign Affairs, working on multilateral and bilateral arms control and disarmament and serving on several Soviet/Russian delegations to the UN General Assembly and negotiations on the INF, START I, CFE and CWC treaties. He was responsible for implementation of the INF Treaty and was involved in preparing the START-II Treaty.

**Dr Peter Litavrin** (Russia) has worked at senior level in the Russian Ministry of Foreign Affairs since 1987, in which capacity he has been a member of the Panel of Experts on the UN Register of Conventional Arms and of the UN Panel of Experts on Small Arms. From 1974 to 1986 he was a Research Fellow at the Institute of US and Canada Studies of the Russian Academy of Sciences, and he has published 'Russia and the Register', *Disarmament* (1994) and 'Russian arms exports: new aspects of an old business', *International Affairs* (Moscow, 1994).

**Prof. Alexander A. Sergounin** (Russia) is Professor of Political Science at the University of Nizhniy Novgorod. He has published generally in the fields of political science and international relations, and, most recently, 'Conversion in Russia: regional implications', eds B. Møller and L. Voronkov, *Defence Doctrines and Conversion* (1996) (in English). *Russia's Regionalization: the International Dimension* is forthcoming (1998, in English) from the Copenhagen Peace Research Institute.

**Dr Sergey V. Subbotin** (Russia) is Assistant Professor of Political Science at the University of Nizhniy Novgorod, working on Russian and US arms transfers policies. He has published widely—most recently, with Alexander Sergounin, 'In search of a new Russian arms export policy', *International Spectator* (1994); 'Indo-Russian military cooperation: economic and strategic considerations', *Political Economy Journal of India* (1995); and *Sino-Russian Military Cooperation and the Evolving Security System in East Asia* (published by the University of Nizhniy Novgorod, 1996).

# Index